RSC Paperbacks

FOOD
The Chemistry of Its Components

Fourth Edition

T. P. COULTATE
formerly of
South Bank University, London, UK

ROYAL SOCIETY OF CHEMISTRY

ISBN 0-85404-615-1

A catalogue record for this book is available from the British Library

First published 1984
Reprinted 1988

Second Edition 1989
Reprinted 1990, 1992, 1993, 1995

Third Edition 1996
Reprinted 1999, 2001

Fourth Edition © The Royal Society of Chemistry 2002

Published by The Royal Society of Chemistry,
Thomas Graham House, Science Park, Milton Road,
Cambridge CB4 0WF, UK

Registered Charity Number 207890

For further information see our web site at www.rsc.org

Typeset by Keytec Typesetting Ltd.
Printed in Great Britain by TJ International Ltd, Padstow, Cornwall

Acknowledgements

As far as the preparation of the fourth edition is concerned my retirement from full-time teaching has allowed me the best of both worlds. I have been able to devote much more time to the tasks of updating, redrawing and the introduction of new material but at the same time I have continued to enjoy the use of South Bank University's library and other facilities. Even more valuable has been the continued encouragement and support from my erstwhile colleagues. As with previous editions they have provided specialist knowledge and have given the manuscript severe but indispensable scrutiny. Special contributions to this edition have come from South Bank Professors Martin Chaplin, Jill Davies, Sibel Roller and Mike Hill. From the University of Reading Dr Jenny Ames provided valuable information on the Maillard reaction. From 'The Fat Duck' in Bray, near Maidenhead, the chef/proprietor Heston Blumenthal provided invaluable insights into the contribution that science can make to gastronomy, in contrast to the author's more down to earth experience. To all these, many thanks, but of course the opinions eventually expressed here, and the inevitable errors, are mine, not theirs.

The Royal Society of Chemistry must be thanked for their unfailing efforts on my behalf. Janet Freshwater, Alan Cubitt and the rest of the team at Cambridge, and also their copy editor John Rhodes, have, as always, done a splendid job.

My wife Ann continues her enthusiastic support and now that both my sons are 'off the payroll' I can enjoy unhindered access to the word processor.

This is an appropriate point at which to mention my mother, Olive. She passed away last year, at the age of 90, about the time I was starting work on this edition. Food was a major interest in her life, seeking out the best ingredients, cooking them well and sharing the results with her family. Although no scientist she was fascinated by the 'why' and 'how' of food (as long as it was English and not 'mucked about with') and never deviated from her mother's philosophy that plenty of good food never killed anybody. I suspect that my approach to food chemistry has always been subconsciously influenced by her, carried along by my father Sidney's ongoing contribution to any merit there may be in the way I write.

Tom Coultate, March 2002

Contents

Chapter 1

Introduction

For the chemists of the 18th and 19th centuries an understanding of the chemical nature of our food was a major objective. They realised that this knowledge was essential if dietary standards, and with them health and prosperity, were to improve. Inevitably it was the food components present in large amounts, the carbohydrates, fats, and proteins, that were first nutrients to be described in chemical terms. However, it was also widely recognised that much of the food, and drink, on sale to the general public was very likely to have been adulterated. The chemists of the day took the blame for some of this:

> *There is in this city [London] a certain fraternity of chemical operators who work underground in holes, caverns and dark retirements ... They can squeeze Bordeaux from the sloe and draw champagne from an apple.*
>
> The Tattler, 1710

but by the middle of the 19th century the chemists were deeply involved in exposing the malpractices of food suppliers. Chemistry was brought to bear on the detection of dangerous colourings in confectionery, additional water in milk, beer, wines and spirits, and other many other unexpected food ingredients. A major incentive for the development of chemical analysis was financial. The British government's activities were largely funded by excise duties on alcohol and tea and large numbers of chemists were employed to protect this revenue.

As physiologists and physicians began to relate their findings to the chemical knowledge of foodstuffs the need for reliable analytical techniques increased. 20th century laboratory techniques were essential for the study of vitamins and the other components that occur in similarly small amounts, the natural, and artificial colour and flavour compounds.

Until the use of gas chromatography (GC) became widespread in the 1960s the classical techniques of 'wet chemistry' were the rule but since

1

that time increasingly sophisticated instrumental techniques have taken over. The latest methods are often now so sensitive that many food components can now detected and quantified at such low levels (parts per billion, 1 microgram per kilogram, are commonplace) that one can have serious doubts over whether the presence of a particular pesticide residue, environmental toxin or the like at the detection limits of the analysis really has any biological or health significance. It is perhaps some consolation for the older generation of food chemists that some of the methods of proximate analysis* that they, like the author, struggled with in the 1960s, are still in use today, albeit in automated apparatus rather than using extravagant examples of the glassblower's art.

By the time of World War II it appeared that most of the questions being asked of food chemists by nutritionists, agriculturalists, and others had been answered. This was certainly true as far as questions of the 'what is this substance and how much is there?' variety were concerned. However, as reflected in this book, over the past few decades new questions have been asked and many answers are still awaited. Apart from the question of undesirable components, both natural and man-made food chemists nowadays are required to explain the *behaviour* of food components. What happens when food is processed, stored, cooked, chewed, digested and absorbed? Much of the stimulus to this type of enquiry has come from the food-manufacturing industry. For example, the observation that the starch in a dessert product provides a certain amount of energy has been overtaken in importance by the need to know which type of starch will give just the right degree of thickening and what is the molecular basis for the differences between one starch and another. Furthermore that dessert product must have a long shelf-life and look as pretty as the one served in the expensive restaurant.

In recent years these apparently rather superficial aspects of food supply have begun to take on a wider significance. The nutritionists, physiologists and other scientists now recognise what consumers have always known – there is more to the business of feeding people than compiling a list of nutrients in the correct proportions. This is as true if one is engaged in famine relief as it is in a five-star restaurant. To satisfy a nutritional need a foodstuff must be acceptable, and to be acceptable it must first look and then taste 'right'.

In recent years we have become increasingly conscious of two other aspects of our food. As some of us have become more affluent, our food intake is no longer limited by our income and we have begun to suffer from the Western 'disease' of *overnutrition*. Our parents are appalled when

---

* The determination of total fat, protein, carbohydrate, ash (*i.e.* metals) and water.

our children follow sound dietetic advice and discard the calorie-laden fat from around their sliced ham and food scientists are called upon to devise butter substitutes with minimal fat content. Closely associated with this issue is the intense public interest in the 'chemicals' in our food. Sadly, the general public's appreciation of the terminology of chemistry leaves much to be desired. It is unlikely that many people would buy *coleslaw* from the delicatessen if the label actually listed the 'active ingredients':

ethanoic acid,
α-D-glucopyranosyl-(1,2)-β-D-fructofuranose,
p-hydroxybenzyl and indoylmethyl glucosinolates,
S-propenyl and other S-alkyl cysteine sulfoxides*,
β-carotene (and other carotenoids),
phosphatidylcholine.

(The relationship of this list to the recipe for coleslaw will emerge later.)

The issue of 'chemicals' in food is closely linked to the pursuit of 'naturalness' as a guarantee of 'healthiness'. The enormous diversity of the diets consumed by *Homo sapiens* as a colonist of this planet makes it impossible to define the *ideal* diet. Diet related disease, including starvation, is a major cause of death but it appears that while choice of diet can certainly influence the manner of our passing diet has no influence on it's inevitability. As chemists work together with nutritionists, doctors, epidemiologists and other scientists to understand what it is we are eating and what it does to us we will come to understand the essential compromises the human diet entails. After all, our success on this planet is to some extent at least owed to an extraordinary ability to adapt our eating habits to what is available in the immediate environment. Whether that environment is an Arctic waste, a tropical rain forest, or a hamburger-infested inner city, humans actually cope rather well.

This book sets out to introduce the chemistry of our diet. The early chapters (2 to 5), covering food's 'macro-components' are more overtly chemical in character because these are the substances whose chemical properties exert the major influence on the obvious physical characteristics of foodstuffs. If we are to understand the properties of food gels we are going to need a firm grasp of the chemical properties of polysaccharides. Similarly we will not understand the unique properties that cocoa butter gives to chocolate without getting involved in the crystallography of triglycerides.

---

* The replacement of 'ph' with 'f' in the sulphur compounds mentioned in this book is in deference to rules of the International Union of Pure and Applied chemistry (IUPAC) rather than the influence of Webster's Dictionary.

In the next seven chapters we consider substances drawn together by the nature of their contribution to food, as colours, vitamins *etc.* rather than broad chemical classifications. Although this means that less attention can be devoted the chemical behaviour of these substances individually there is still much that the chemist contribute to our understanding of taste, appearance, nutritional value *etc.* Science is rarely as tidy as one might wish and it is inevitable that some food components have found their way into chapters where they do not really belong. For example, some of the flavonoids mentioned in Chapter 6 make no contribution to the colour of food. However, in terms of chemical structure they are closely related to the flavonoid anthocyanin pigments and Chapter 6 could be regarded as good a location for them as any. The final chapter is devoted to water. Apparently the simplest food component, and certainly the most abundant, it remains one of the most poorly understood. It could, as has often been suggested, be placed at the front of this book but it is feared that much of water's chemistry would appear so intimidating as to prevent many readers penetrating to the tastier chapters.

This book does not set out to be a textbook of nutrition, its author is in no way qualified to make it one. Nevertheless chemists cannot ignore nutritional issues and wherever possible the links between the subtleties of chemical nature of food components and nutritional and health issues have been pointed out. Similarly students of nutrition should find a greater insight into the chemical elements of their subject valuable. It is also to be hoped that health pundits who campaign for the reduction of this or that element in our diet will have a better appreciation of what exactly it is they are asking for, and what any knock-on effects might be.

Food chemists should never overlook the fact that the object of their study is not just another, albeit fascinating, aspect of applied science. It is all about what we eat, not just to provide nutrients for the benefit our bodies but also to give pleasure and satisfaction to the senses. Not even the driest old scientist compliments the cook on the vitamin content of the dish, it's the texture and flavour, and the company around the table, that wins every time.

A Brief Note on Concentrations

The concentrations of chemical components are expressed in a number of different styles in this book, depending on the context and the concentrations. Some readers may find the following helpful.

(a) However they are expressed, concentrations always imply the amount contained, rather than added. Thus '5 g of X per 100 g of foodstuff' implies that 100 g of the foodstuff contains 95 g of substance(s) that are *not* X.

(b) The abbreviation 'p.p.m.' means 'parts per million' *i.e.* grams per million grams, or more realistically milligrams per kilogram. One 'p.p.b.', or part per billion, corresponds to one microgram per kilogram.

(c) Amounts contained in 100 g (or 100 cm^3) are often referred to as simple percentages. Where necessary the terms 'w/w', 'v/v' or 'w/v' are added to indicate whether volumes or weights or both are involved. Thus '5% w/v' means that 100 cm^3 of a liquid contains 5 g of a solid, either dissolved or in suspension. Note that the millilitre (ml) and litre (l) are no longer considered acceptable. Although the replacement for the ml, 'the centimetre cubed' ('cm^3'), is widely recognised there is little sign that the cubic decimetre, or 'dm^3' has taken over from the litre, except in the teaching laboratories that must always be seen to 'toe the party line'.

(d) Most often a strictly mathematical style is adopted, with 'per' expressed as the power of minus one. Since mathematically:

$$x^{-1} = 1/x$$

5 μg kg^{-1} becomes a convenient way of writing 5 micrograms per kilogram. This brief but mathematically rigorous style comes into its own when the rates of intake of substances such as toxins have to be related to the size of the animal consuming them, as in '5 milligrams per day per kilogram body weight', which abbreviates to:

$$5 \text{ mg day}^{-1} \text{ kg}^{-1}$$

A quantity, say 10 mg, per cubic centimetre, cm^3, would be written: 10 mg cm^{-3}.

FURTHER READING

Many readers, especially advanced students, will want to learn more about particular topics than can be accommodated in this volume. The 'Further Reading' sections at the end of each chapter will generally meet such needs and provide a gateway into the relevant primary literature, *i.e.* research papers. The review journal '*Trends in Food Science and Technology*' is also an invaluable source of up-to-date information for students, as it was for the author when preparing this edition. Nowadays students have easy access to computerised databases, downloadable journals *etc.* that open up the scientific literature in ways that those of us brought up on the bound volumes of '*Chemical Abstracts*' can only marvel at.

The internet can provide a wealth of information but it must be used with very great caution. It has always been the case that one should not necessarily believe something just because it was printed in a book, but such cynicism is an absolute essential when looking at web sites. For example, an internet search using 'choline' and 'vitamin' as its keywords will locate the official statement that choline is not a vitamin (*see* Chapter 8) together with innumerable commercial sites that claim the opposite in order to enhance sales of choline as a 'supplement'. The temptation to include addresses of relevant and authoritive websites was resisted since, unlike the printed scientific literature, they tend to be rather ephemeral and unlikely to remain available throughout the lifetime of a textbook.

Chapter 2

Sugars

Sugars, such as sucrose and glucose, together with polysaccharides, such as starch and cellulose, are the principal components of the class of substances we call carbohydrates. In this chapter we will be concerned with the sugars and some of their derivatives; the polysaccharides, essentially a special class of their derivatives, will be considered in Chapter 3. Although chemists never seem to have the slightest difficulty in deciding whether or not a particular substance should be classified as a carbohydrate, a concise, formal definition has proved elusive. The empirical formulae of most of the carbohydrates we encounter in food-stuffs approximate to $(CH_2O)_n$, hence the name. More usefully, it is simpler to regard them as aliphatic polyhydroxy compounds which carry a carbonyl group (and, of course, derivatives of such compounds). The special place of sugars in our everyday diet will be apparent from the data presented in Table 2.1.

Sugars are automatically associated in most peoples' minds with sweetness and it is this property that normally reveals to us their presence in a food. In fact there are a great many foods, especially elaborately processed products, where the sugar content may be less obvious. Fortunately it is now common to find information about the total sugar content (as in: 'Total carbohydrates ... of which sugars ...') on the labels of packaged food products but this tells us little about which sugars are present. In fact only a few different sugars are common in foods, as shown in Figure 2.1.

MONOSACCHARIDES

The monosaccharides constitute the simplest group of carbohydrates, and, as we shall see, most of them can be referred to as sugars. The monosaccharides have a backbone of between three and eight carbon atoms, but only those with five or six carbon atoms are common. The

7

Table 2.1 *The total sugar contents of a variety of foods and beverages. These figures are taken from 'McCance and Widdowson' (see Appendix II) and in all cases refer to the edible portion. They should be regarded as typical values for the particular food rather than absolute figures which apply to any samples of the same food. The levels of sugars in eggs, meat, cured or fresh, poultry, game and fish, and fats such as butter and margarine are nutritionally insignificant although even traces can sometimes reveal themselves through browning at high temperatures caused by the Maillard reaction, see page 33.*

Food	Total sugars (%)	Food	Total sugars (%)
White bread	2.6	Cabbage (raw)	4.0
Corn flakes	8.2	Beetroot (raw)	7.0
Sugar coated cornflakes	41.9	Onions (raw)	5.6
Digestive biscuits	13.6	Cooking apples (raw)	8.9
Gingernuts	35.8	Eating apples (raw)	11.8
Rich fruit cake	48.4	Bananas	20.9
Cow's milk (whole)	4.8	Grapes	15.4
Human milk	7.2	Oranges	8.5
Cheese (hard)	0.1	Raisins	69.3
Cheese (processed)	0.9	Peanuts	6.2
Yoghurt (plain)	7.8	Honey	76.4
Yoghurt (fruit)	15.7	Jam	69.0
Ice cream (dairy)	22.1	Plain chocolate	59.5
Lemon sorbet	34.2	Cola	10.5
Cheesecake (frozen)	22.2	Beer (bitter)	2.3
Beef sausages	1.8	Lager	1.5
New potatoes	1.3	Red wine	0.3
Canned baked beans	5.9	White wine (medium)	3.4
Frozen peas	2.7	Port	12.0

names of monosaccharides that have a carbonyl group have the suffix '-ose', and in the absence of any other identification the number of carbon atoms is indicated by terms such as triose, tetrose, and pentose. The chain of carbon atoms is always straight, never branched (2.1):

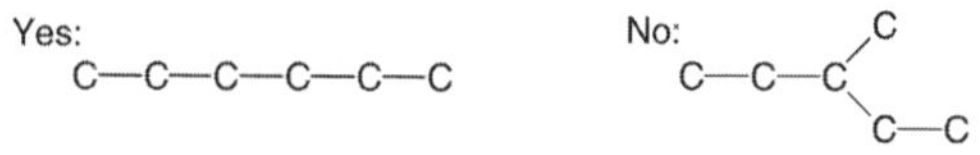

Yes:

$$C—C—C—C—C—C$$

No:

$$C—C—C\begin{array}{c}C\\C—C\end{array} \qquad (2.1)$$

All but one of the carbon atoms carry a hydroxyl group, the exception forming the carbonyl group. It is the presence of carbonyl group that confers reducing properties on monosaccharides and many other sugars and the carbonyl is often referred to as the 'reducing group' in considera-

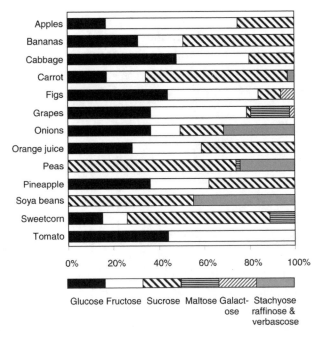

Figure 2.1 *Representative percentage distributions of sugars in the total sugar fraction of various plant foods. There is considerable variation in these proportions depending on variety, season etc. The contribution of sugars to the total weight of plant material also varies widely between different species.*

tions of sugar structure. The prefixes 'aldo-' and 'keto-' are used to show whether the carbonyl carbon is on the first or a subsequent carbon atom, *i.e.* the sugar is an aldehyde or a ketone. Thus we refer to sugars as, for example, aldohexoses or ketopentoses. To complicate matters further, the two triose monosaccharides are almost never named in this way but are referred to as glyceraldehyde (2,3-dihydroxypropanal) (2.2) and dihydroxy-acetone (1,3-dihydroxypropan-2-one) (2.3).

Of greatest concern to us will be the aldo[(2.4) and (2.6)] and keto[(2.5) and (2.7)] pentoses [(2.4) and (2.5)] and hexoses [(2.6) and (2.7)], shown here with the conventional numbering of the carbon atoms:

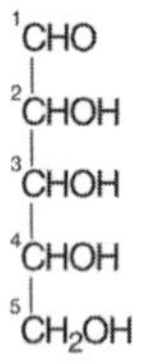

$$^1CHO \qquad ^1CH_2OH \qquad ^1CHO \qquad ^1CH_2OH$$
$$^2CHOH \qquad ^2CO \qquad ^2CHOH \qquad ^2CO$$
$$^3CHOH \qquad ^3CHOH \qquad ^3CHOH \qquad ^3CHOH$$
$$^4CHOH \qquad ^4CHOH \qquad ^4CHOH \qquad ^4CHOH$$
$$^5CH_2OH \qquad ^5CH_2OH \qquad ^5CHOH \qquad ^5CHOH$$
$$\qquad\qquad\qquad\qquad\qquad\qquad ^6CH_2OH \qquad ^6CH_2OH$$

$$(2.4) \qquad\qquad (2.5) \qquad\qquad (2.6) \qquad\qquad (2.7)$$

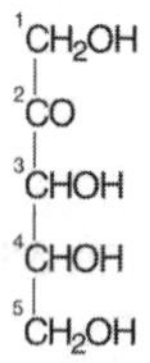

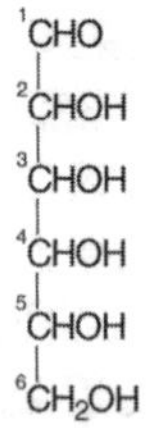

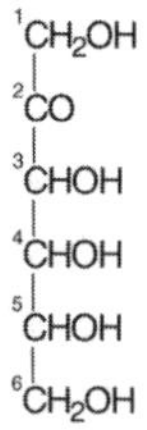

The carbon atom of each CHOH unit carries four different groups and is therefore asymmetrically substituted. This asymmetry is the source of the optical isomerism characteristic of carbohydrates. Almost all naturally occurring monosaccharides belong to the so-called D-series. That is to say their highest numbered asymmetric carbon atom (*e.g.* carbon 4 in a pentose or carbon five in a hexose), the one furthest from the carbonyl group, has the same configuration as D-glyceraldehyde (2.8) rather than its isomer L-glyceraldehyde (2.9).

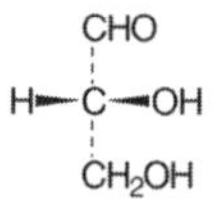

$$CHO \qquad\qquad\qquad CHO$$
$$H-C-OH \qquad\qquad HO-C-H$$
$$CH_2OH \qquad\qquad\qquad CH_2OH$$

$$(2.8) \qquad\qquad\qquad (2.9)$$

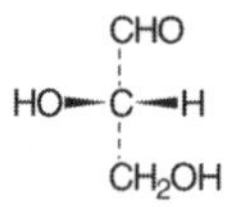

All this isomerism results in there being a very large number of possible monosaccharide structures. Although only a few are relevant to food or even common in nature it is helpful to understand their structural relationships, as shown for the aldose monosaccharides in Figure 2.2.

For simplicity in this diagram and most subsequent structural formulae the carbon atoms within the chain are indicated by the intersections of the vertical and horizontal bonds. The optical configuration corresponds to the conventional representation of the asymmetric carbon atom of glyceraldehyde shown above. A corresponding table of ketose sugars may be drawn up, but with the obvious exception of D-fructose (2.10) none of the ketoses are of much significance in food. Where names of ketose sugars are required they are obtained by changing the '-ose' of the corresponding aldose sugar to '-ulose', as in xylulose (2.11). Dihydroxyacetone and fructose are the exceptions to this convention.

It is important to remember that the monosaccharides of the L-series are

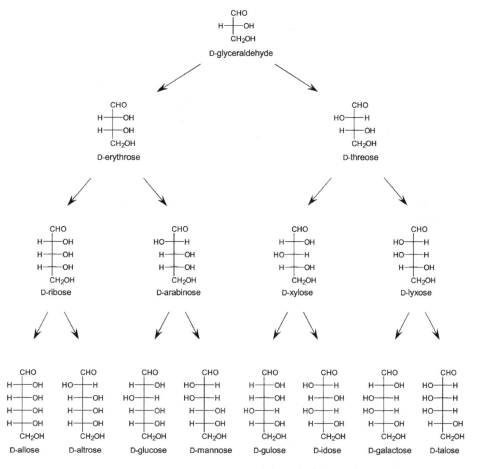

Figure 2.2 *The configurations of the D-aldoses. For anyone who needs to learn them the sequence of names of the aldoses, reading downwards and left to right, will best be remembered by use of the traditionally dreadful mnemonics 'Get Raxl!' and 'All altruists gladly make gum in gallon tanks'. (The author, and innumerable students, would be delighted to hear of any improved versions.)*

related to L-glyceraldehyde (2.9) and have the mirror image configurations to the corresponding D-series sugars. Thus L-glucose is (2.13) rather than (2.12), which is in fact L-idose. It is not surprising, in view of their asymmetry, that the monosaccharides are optically active, *i.e.* their solutions and crystals rotate the plane of polarised light. The symbols (+) and (−) can be used to denote rotation to the right and left, respectively. The old-fashioned names of dextrose and laevulose stem from the dextrorotatory and laevorotatory properties, respectively, of D(+)-glucose and D(−)-fructose.

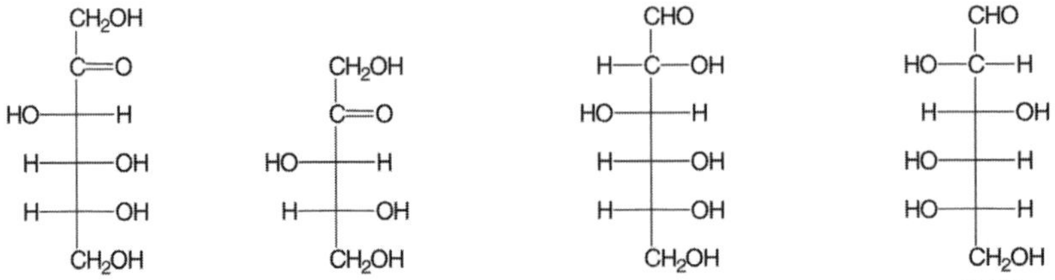

$$(2.10) \qquad (2.11) \qquad (2.12) \qquad (2.13)$$

The straight-chain structural formulae that have been used so far in this chapter do not account satisfactorily for many monosaccharide properties. In particular their reactions, while showing them to possess a carbonyl group, are not entirely typical of carbonyl compounds generally. The differences are explained by the formation of a ring structure when the carbonyl group reacts reversibly with one or other of the hydroxyl groups at the far end of the chain, giving the structure known as a hemiacetal (Figure 2.3).

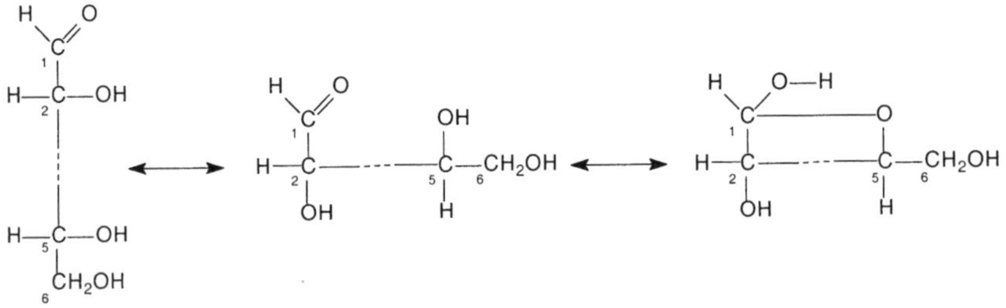

Figure 2.3 *Mutarotation. The natural flexibility of the backbone of the monosaccharide allows the head and tail to come together so that a hydroxyl group at the tail of the molecule can react with the reducing group to form a ring.*

Figure 2.4 shows how a number of different structural isomers can be generated. Rings containing six atoms (five carbon and one oxygen) result when the carbonyl group of an aldose sugar (*i.e.* at carbon 1) reacts with a hydroxyl on carbon 5. The same size ring results when the carbonyl of a ketohexose sugar (*i.e.* at carbon 2) reacts with a hydroxyl on carbon 6 (*i.e.* the $-CH_2OH$, hydroxymethyl, group). Five-membered rings result when the carbonyl of a ketose sugar reacts with a hydroxyl on carbon 5. Six- and five-membered rings are referred to respectively as pyranose and furanose from the structures of pyran (2.14) and furan (2.15).

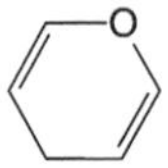

$$(2.14) \qquad (2.15)$$

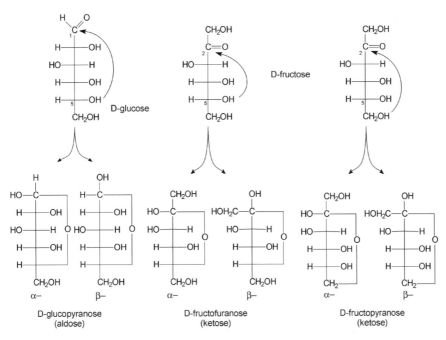

Figure 2.4 *The formation of pyranose and furanose ring forms of* D-glucose. *Only the pyranose forms of* D-glucose *occur in significant amounts. This representation of the ring forms has its shortcomings; C–O bonds are actually about the same length as C–C bonds and certainly do not turn through 90° half way along. Translating the formulae of other monosaccharides from the open chains of Figure 2.2 into the more realistic Haworth rings shown in Figure 2.4 is best achieved by comparison with glucose rather than by resorting to the three-dimensional mental gymnastics recommended by some classical organic chemistry texts.*

Another result of hemiacetal formation, also shown in Figure 2.3, is that what was the carbonyl carbon now has four different substituents, *i.e.* it has become a new asymmetric centre. This means that the ring form of a monosaccharide will occur as a pair of optical isomers, its α- and β-anomers.

The α- and β-anomers of a particular monosaccharide differ in optical rotation. For example, α-D-glucopyranose, which is the form in which D-glucose crystallises out of aqueous solution, has a specific rotation* of

---

* The specific rotation at a temperature of 20°C using light of the D-line of the sodium spectrum is given by the expression

$$[\alpha]_D^{20} = \frac{100\alpha}{l \times c}$$

where α is the rotation observed in the polarimeter tube of length l dm and at a sugar concentration of c g$(100$ ml$)^{-1}$.

+112° whereas that of β-D-glucopyranose, which is obtained by crystallisation from pyridine solutions of D-glucose, is +19°. There are elaborate rules used to decide which of a pair of anomers is to be designated α and which β. However, in practice we observe that α-anomers have their reducing group (*i.e.* the C-1 hydroxyl of aldose sugars and the C-2 hydroxyl of ketose sugars) on the opposite face of the ring to carbon-6; β-anomers have their reducing group on the same face. Careful inspection of the formulae in Figures 2.4 and 2.5 will show how this works out in the potentially tricky L-sugars and in furanose rings.

When crystals of either α- or β-D-glucose are dissolved in water, the specific rotation is observed to change until, regardless of the form one started with, the solution finally gives a value of +52°. This phenomenon is

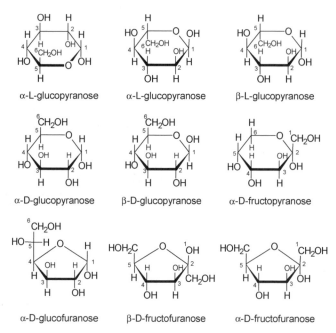

Figure 2.5 *Isomers of glucose and fructose. The obvious shortcomings of the formulae used so far in this chapter as representing molecular structure can be partially resolved, as here, by the adoption of the Haworth convention for ring structures. The ring is treated as planar and drawn to appear perpendicular to the plane of the page. These structures should be compared with those in Figure 2.4. The numbering of carbon atoms may be correlated with that in the open chain structures (2.6) and (2.7) shown on page 10. It should be remembered that only a few of the isomers shown here actually occur naturally in significant amounts and also that not all the possible isomers have been included. Two versions of α-L-glucopyranose are shown to illustrate the result of different approaches to the manipulation of the formula of β-D-glucopyranose.*

Box 2.1   Reducing group reactions

Textbooks of organic chemistry give details of numerous chemical reactions of sugars, particularly involving the carbonyl group. Many of these have little or no relevance to modern food chemistry even though our present understanding of the structure of sugars was initially dependant their study. Until the advent of sophisticated chromatographic methods, notably high-performance liquid chromatography (HPLC) the most important application for these reactions were in food analysis, to identify and perhaps quantify the sugars present in a particular food material.

The carbonyl group of monosaccharides is readily oxidised, hence the reference to *reducing* sugars, *i.e.* sugars that can reduce *oxidising* agents. The reaction of reducing sugars with Cu^{2+} ions complexed in alkali (*e.g.* Fehling's solution or Benedict's reagent) forms the basis of the most popular chemical methods:

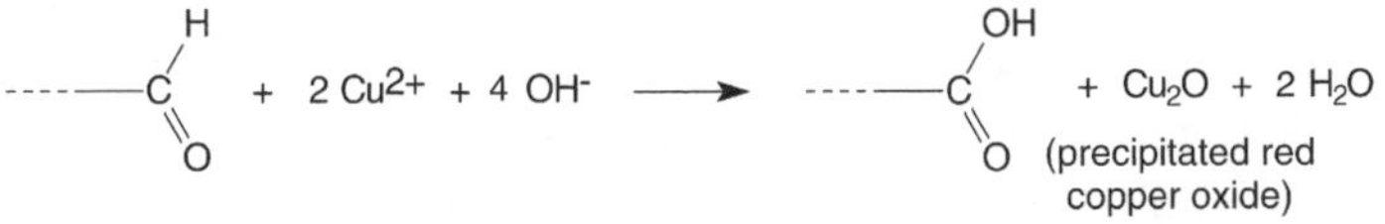

$$----C\genfrac{}{}{0pt}{}{H}{\diagdown O} + 2\ Cu^{2+} + 4\ OH^- \longrightarrow ----C\genfrac{}{}{0pt}{}{OH}{\diagdown O} + Cu_2O + 2\ H_2O \text{ (precipitated red copper oxide)}$$

Although in the reaction above the product is shown as the corresponding '-onic' acid in fact this is not the only product. Depending on the exact conditions a number of other shorter chain carboxylic acids are also produced. Ketose sugars, such as fructose, would not be expected to be oxidised but are, in reactions involving the neighbouring hydroxyl group. The variability in the outcome of these oxidation reactions means that they are non-stoichiometric and are difficult to apply to quantitative sugar analysis. The Lane and Eynon (titrimetric) and Munson and Walker (gravimetric) methods, using Fehling's solution, give data that have to be referred to published tables to obtain results; necessarily expressed as 'reducing sugars as glucose'. The reaction of reducing sugars with 3,5-dinitrosalicylic acid (DNS) to give a red/brown product:

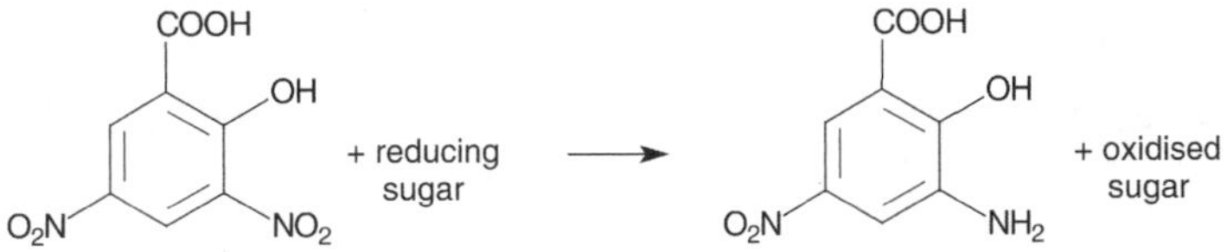

3-amino-5-nitrosalicylic acid

provides the basis of a simple spectrophotometric assay that is sufficiently accurate for use in college laboratory experiments.

known as mutarotation. The transition of one anomer to another proceeds through the open-chain, *aldehydo*, form, and it is clear that it is this isomer which is involved in the sugar reactions that are typical of carbonyl compounds, even though in aqueous solutions only 0.02% of D-glucose molecules are in this form.

The structural relationships between L- and D-isomers, α- and β-anomers, and pyranose and furanose rings in both aldose and ketose sugars are by no means easily mastered. If available, the use of a set of molecular models will help to clarify the issues, but the structural formulae set out in Figure 2.5 illustrate the essential features of terminology in this area.

Over the years chemists have synthesised innumerable derivatives of monosaccharides, but only a few occur naturally or have particular significance to food. Oxidation of the carbonyl group of aldose sugars leads to the formation of the '-onic' series of sugar acids and, depending on the exact details of the conditions, various other products. The enzyme glucose oxidase catalyses the oxidation of the α-anomer of D-glucose to D-gluconolactone (2.16), which hydrolyses spontaneously to D-gluconic acid (2.17) with hydrogen peroxide as a by-product. In the popular analytical technique (*see* Figure 2.6) a second enzyme, peroxidase, is included to provide a measure of the hydrogen peroxide and thus the concentration of the glucose. Besides its obvious applications in food analysis the same reaction system is used in the indicator strips used by diabetics for the measurement of blood glucose concentrations. Glucose oxidase is also used in process for removing traces of glucose from the bulk liquid egg-white used in commercial bakeries and elsewhere to prevent the Maillard reaction – *see* page 33.

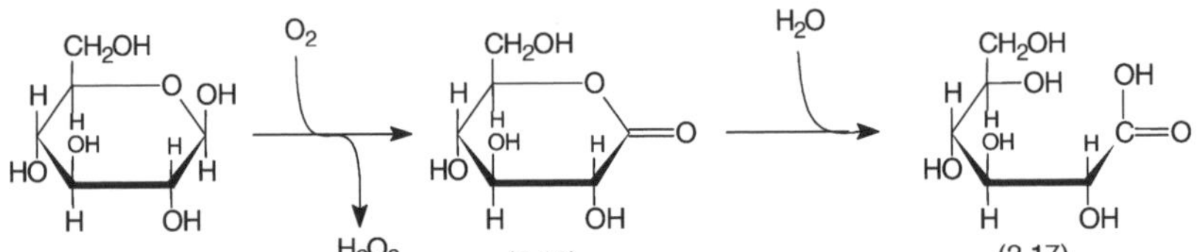

The '-uronic' series of sugar acids are aldohexoses with a carboxyl group at C-6, such as D-galacturonic acid (2.18) and L-guluronic acid (2.19). These are important as constituents of polysaccharides such as pectins and alginates but are of little interest in their own right.

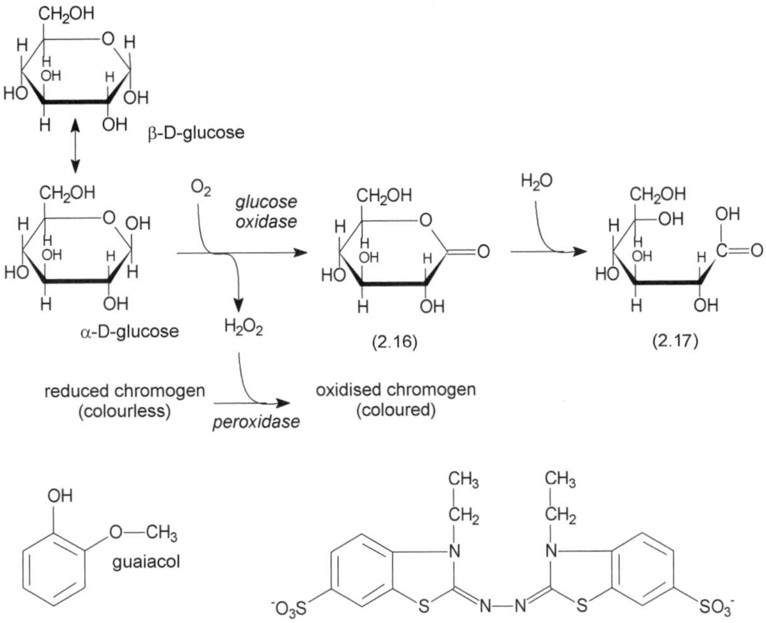

2,2 azino-bis(3-ethylbenzethiazoline-6-sulfonate)

Figure 2.6 *The enzymic determination of glucose. The original colour reagent (chromogen) used with this method was guaiacol but nowadays 2,2 azino-bis(3-ethylbenzthiazoline-6-sulfonate), more commonly known as Perid* ™ , *is the most popular.*

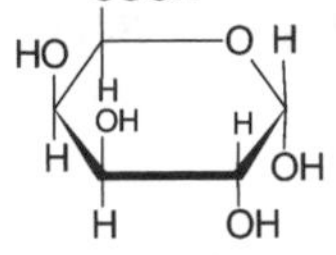

(2.18)

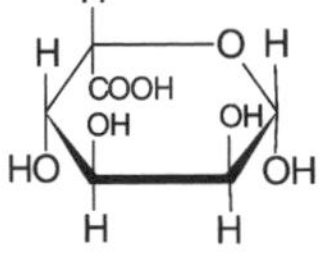

(2.19)

Reduction of the carbonyl group to a hydroxyl gives sugar alcohols such as xylitol (2.20) and sorbitol (2.21). Small amounts of sorbitol occur widely in fruit (although it is notably absent from grapes) but otherwise sugar alcohols are extensively synthesised on an industrial scale from the corresponding aldose sugar by reduction with hydrogen. Sorbitol and other hydrogenated sugars are also used as low-calorie, *i.e.* non-metabolised, sweeteners and are considered in more detail in Chapter 7.

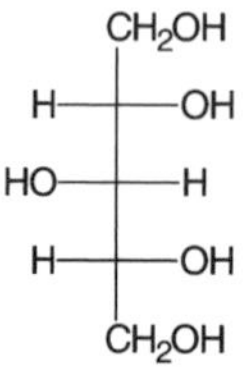
CH2OH
H
OH
HO
H
H
OH
CH2OH

(2.20)

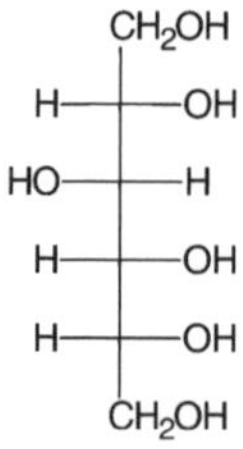
CH2OH
H
OH
HO
H
H
OH
H
OH
CH2OH

(2.21)

Sorbitol is particularly important as the starting point for the industrial synthesis of ascorbic acid (vitamin C, *see* page 284). Sugar alcohols are used to replace sugars in diabetic and other 'calorie-reduced' food products. Though sweet, these are not absorbed from the small intestine and do not reach the bloodstream. Unfortunately their eventual arrival in the large intestine can, if large amounts have been consumed, provoke 'osmotic diarrhoea' when the normal transfer of water from the colon's contents is impaired. Reduction at other positions in the sugar molecule gives deoxy sugars such as L-rhamnose (6-deoxy-L-mannose) (2.22), an important minor constituent of pectins, and 2-deoxy-D-ribose (2.23), the sugar component of DNA.

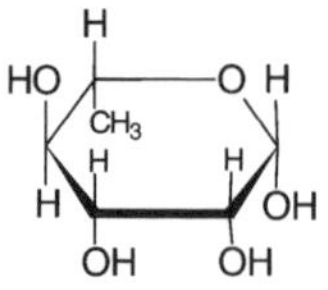
H
HO
O H
CH3
H
H
H
OH
OH
OH

(2.22)

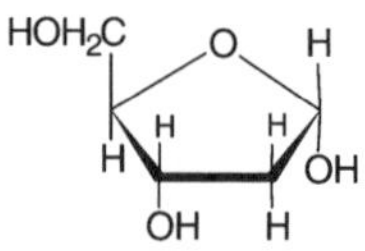
HOH2C
O
H
H
H
H
OH
OH
H

(2.23)

The most important derivatives of monosaccharides are those in which the 'hemiacetal' or 'reducing' group forms an 'acetal' or 'glycosidic' link with a hydroxyl group of another organic compound (2.24).

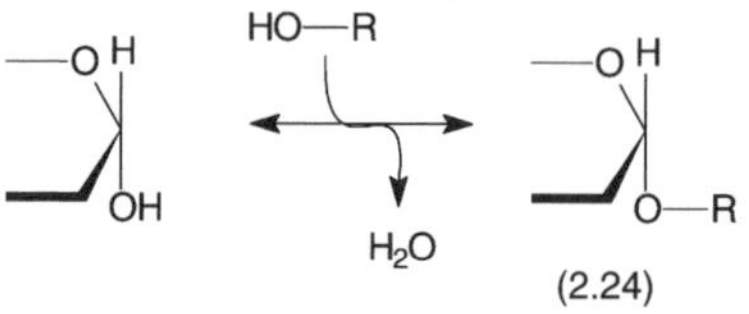
O H
OH
HO—R
H2O
O H
O—R

(2.24)

Glycosidic links are stable under ordinary conditions but are readily hydrolysed in acid conditions or in the presence of appropriate hydrolytic enzymes. The formation of the glycosidic link has the effect of fixing the hemiacetal structure in either the α or β configuration and, of course, abolishing mutarotation. As we will see later in the description of oligosaccharides and other compounds having glycosidic links, it will be necessary when describing a particular structure to specify whether the link in a particular compound is in the α or β configuration.

Although any compound containing a glycosidic link is, strictly speaking, a glycoside, the term is usually reserved for a class of compounds which occur naturally in plants. These particular glycosides have a sugar component linked to a non-sugar component, termed the aglycone. Flavonoids, found widely in plants, normally occur as aglycones with a sugar linked to one of their phenolic hydroxyl groups. The anthocyanin pigments of plants, considered in Chapter 6, are good examples. Many glycosides are best-known as naturally occurring plant toxins (*see* Chapter 10) such as solanine, and the cyanogenic glycosides such as amygdalin (2.25) (in bitter almonds) and the glucoside of α-hydroxyisobutyronitrile (in cassava, *see* Chapter 10). The phenolic hydroxyl groups of flavonoids can react spontaneously with proteins (through the hydroxyl groups of tyrosine residues). Blocking the phenolic hydroxyls with sugar residues maintains their contribution to the biological value of the flavonoid structure while avoiding their potentially destructive effect on the activity of enzymes *etc.* Similarly the presence of the sugar prevents the cyanogenic glycosides from releasing their cyanide until the circumstances are appropriate.

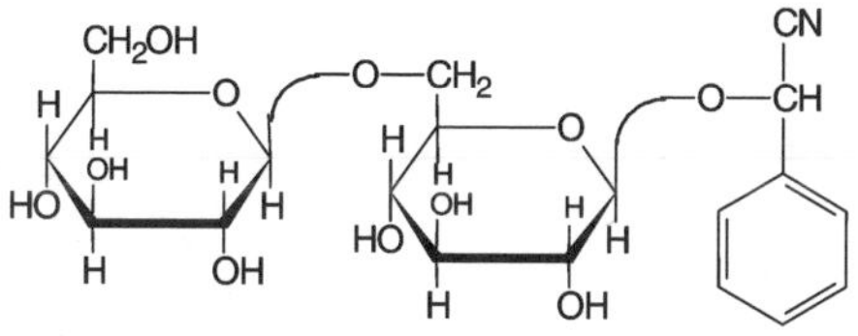

(2.25)

OLIGOSACCHARIDES

When a glycosidic link connects the reducing group of one monosaccharide to one of the hydroxyl groups on another, the result is a disaccharide. Further such linkages will give rise to trisaccharides, tetrasaccharides *etc.*, the oligosaccharides*, and ultimately polysaccharides. The polysaccharides, where hundreds or even thousands of monosaccharide units may be combined in a single molecule, are considered in Chapter 3. In this section we will be considering the di-, tri-, and tetra-saccharides that occur commonly in our food.

---

* Chemists do not recognise a fixed numerical boundary between 'oligo-' and 'poly-'. The Greek origins of these prefixes are usually translated as 'few' and 'many', respectively, and chemists apply these prefixes on this basis.

Even when a disaccharide is composed of two identical monosaccharide units, there are numerous possible structures. This is illustrated in Figure 2.7, which shows five of the more important glucose/glucose disaccharides. Although the configuration of the hemiacetal involved in the link is no longer free to reverse, it should not be overlooked that the uninvolved hemiacetal (*i.e.* that of the right-hand ring of the reducing sugars as they are portrayed in Figure 2.7) is still subject to mutarotation in aqueous

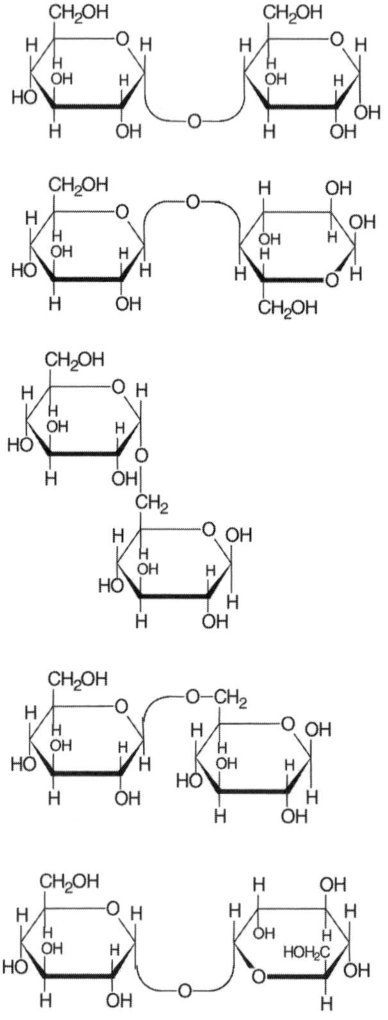

α–D-glucopyranosyl-(1→4)-α–D-glucopyranose

Maltose

β–D-glucopyranosyl-(1→4)-α–D-glucopyranose

Cellobiose

α–D-glucopyranosyl-(1→6)-β–D-glucopyranose

Isomaltose

β–D-glucopyranosyl-(1→6)-β–D-glucopyranose

Gentiobiose

α–D-glucopyranosyl-(1→1)-α–D-glucopyranose

Trehalose

Figure 2.7 *Disaccharides of glucose. The trivial and systematic names are given. In the cases of the reducing oligosaccharides, maltose, cellobiose, isomaltose and gentiobiose, only one of the two possible anomers is shown.*

solution. Thus four of these sugars (trehalose is the exception) occur as pairs of α- and β-anomers. Furthermore, these four sugars continue to show similar reducing properties, associated with the carbonyl group, to those of monosaccharides. The failure of trehalose and sucrose to reduce Cu^{2+} ions in alkaline solutions such as 'Fehling's' provides a useful laboratory test to distinguish these 'non-reducing sugars' from the 'reducing sugars' – the mono- and oligosaccharides that do possess a free reducing group.

Not all of the five disaccharides shown in Figure 2.7 occur in nature. Maltose, together a small proportion of isomaltose, occurs in syrups obtained by the partial hydrolysis of starch, where the same patterns of glycosidic links are found. As was shown in Figure 2.1 maltose also occurs in some fruit, notably grapes. Similarly cellobiose is formed during the acid hydrolysis of cellulose. Gentiobiose commonly occurs as the sugar component of glycosides and trehalose is found in yeast.

The scientific names of oligosaccharides are inevitably rather clumsy. The formally correct procedure is to name a disaccharide as a substituted monosaccharide. Thus lactose (2.26), which consists of a glucose unit carrying a D-galactose unit (in the β configuration) on its C-4 hydroxyl group, is 4-*O*-β-D-galactopyranosyl-D-glucopyranose. This system gets very difficult when applied to higher oligosaccharides (three or more monosaccharide units) and most food scientists prefer to place the details of the link between the monosaccharides where it belongs, between them. Lactose is most frequently described as β-D-galactopyranosyl-(1→4)-D-glucopyranose. In sucrose (2.27), the glycosidic link is between the reducing groups of *both* monosaccharide components, glucose and fructose, so that the usual formal name, α-D-glucopyranosyl-(1→2)-β-D-fructofuranose could be reversed to β-D-fructofuranosyl-(2→1)-α-D-glucopyranose, but no one ever bothers.

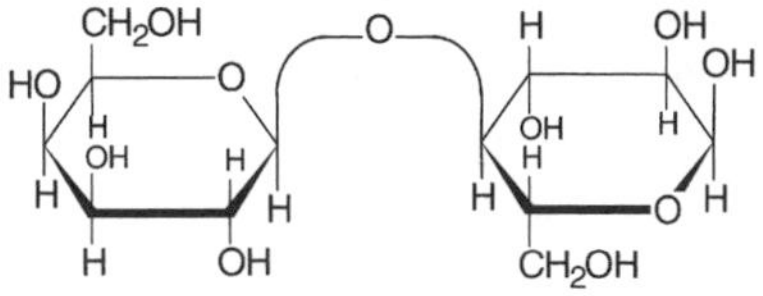

(2.26) 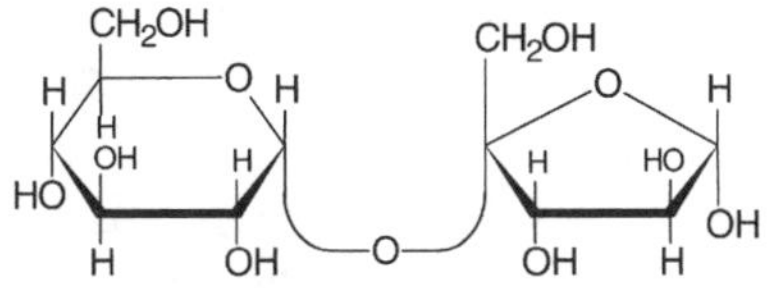   (2.27)

Lactose and sucrose are two of the most important food sugars. Lactose, is the sugar of milk* (approximately 5% w/v in cow's milk) and is of

---

* As secreted by female mammals for feeding their young, neither soya milk nor coconut milk contains lactose.

Box 2.2   Sugar cane, sugar beet and tequila

The fundamental reaction of photosynthesis in all green plants is the reaction, catalysed by the enzyme ribulose-1,5-bisphosphate carboxylase, of carbon dioxide with ribulose-1,5-bisphosphate to form 2 molecules of 3-phosphoglyce-rate. After a complex series of reactions these are converted into fructose-6-phosphate and eventually glucose and sucrose. Although this part of the process, known as the Calvin cycle, is common to all plants two specialist groups of plants have evolved special sequences of reactions that carry out the initial trapping of the carbon dioxide from the atmosphere before it is transferred to the carboxylase. One group is consists of the so-called tropical grasses, including maize, *Zea mays,* ('corn' in North America) and sugar cane, *Saccharum officinarum*, the traditional source of sucrose. These use the Hatch and Slack pathway for initial carbon dioxide fixation. Another pathway, known as Crassulacean Acid Metabolism (CAM) is used by the succulents, of which only two species, the pineapple, *Ananas comosus*, and the blue agave, *Agave tequilana*, are of interest. The leaves of the agave are the source of the sugary syrup that is fermented and then distilled to give tequila, the famous Mexican spirit. Both these variants of normal photosynthesis enable improved efficiency under tropical conditions. In contrast, sugar beet, a variety of the common beet, *Beta vulgaris*, uses the normal Calvin cycle. This grows in a wide range of temperate climates and has been an alternative source of sugar, especially in Europe, since the middle of the eighteenth century.

   In terms of chemistry, biochemistry, human nutrition and health, there is no difference between cane and beet sugars once they have been purified to the colourless crystalline materials we recognise as 'granulated' and 'caster' sugar. Although in isolation the distinction between cane and beet sugar is purely academic it is important from commercial and regulatory points of view to be able to discriminate between them. The method depends upon the 1% of all

course a reducing sugar. Sucrose, is the 'sugar' of the kitchen and commerce and occurs widely in plant materials. The sucrose we buy as 'sugar' has been extracted from sugar cane or sugar beet, but sucrose is also abundant in most plant materials, particularly fruit. Although glucose, as its 6-phosphate derivative, is produced in photosynthesis it is normally converted into sucrose before transport from the leaves to the remainder of the plant. Most plants convert the sucrose into starch in the form of insoluble granules for long term storage as an energy reserve but some, such as sugar beet and sugar cane, retain sucrose in solution. The sucrose, and other sugars (*see* Figure 2.1), in fruit are resynthesised from starch as the fruit ripens.

   As the glucose and fructose units are joined through both of their hemiacetal groups, sucrose is not a reducing sugar. Under mildly acid natural carbon atoms (in organic compounds, as atmospheric carbon dioxide, and minerals such as calcium carbonate) that occur as the non-radioactive isotope ^{13}C. Almost all the rest is the ^{12}C isotope. Most chemical and biological systems, including human metabolism, cannot distinguish between the two isotopes but the enzymes of the Calvin cycle do. As a result plant materials have very slightly different proportions of ^{12}C and ^{13}C depending on the photosynthetic pathway involved in their synthesis. These proportions, shown in Table 2.2, can be measured with sufficient precision by a mass spectrometric technique, stable isotope ratio analysis (SIRA), to discriminate between cane and beet sugar and to detect when these have been used to supplement the natural sugar content of a fruit product dishonestly.

Table 2.2 *The proportion of the stable carbon isotope ^{13}C in the total carbon of plant materials compared with an arbitrary international standard of fossil calcium carbonate (belmenite).*

	International standard	Tropical grasses, incl. sugar cane and maize	Crassulaceae, including agave and pineapple	All other plants, including sugar beet and grapes
% ^{13}C	1.1112	1.0991	1.0975	1.0837

This method has been successful in demonstrating the illegal addition of cane sugar to grape juice prior to the fermentation and in detecting the addition of corn syrups and cane sugar to maple syrup and citrus juices. Unfortunately cane sugar finding its way into pineapple juice or the tequila fermentation is much harder to detect.

conditions or the action of the enzyme invertase sucrose is readily hydrolysed to its component monosaccharides. This phenomenon is termed inversion and the resulting mixture invert sugar, due to the effect of the hydrolysis on the optical rotation properties of the solution. The specific rotation values for sucrose, glucose, and fructose are +66.5°, +52.7°, and −92.4°, respectively, so that we can see that a dextrorotatory solution of sucrose will give a laevorotatory solution of invert sugar, specific rotation −39.7°.

Traditionally invert syrup is manufactured by acid hydrolysis of sucrose, using citric acid, to give a syrup that will not crystallise, with a solids content as high as 80%. Boiled sweets and other confectionery products depend upon the non-crystallising property of invert syrups. Sucrose is dissolved in invert syrup and the mixture is then boiled to reduce the total water content. The result is a glass structure, a supercooled liquid that is malleable when hot but sets to form a glassy solid as it cools. In this state the solute molecules do not form the stable ordered associations character-istic of a crystal but remain totally disordered. However, the water content is too low for the system to remain liquid although it remains, in strict chemical terms, a solution. In the manufacture of boiled sweets flavours and colours are blended in as it cools and the mass is kneaded by machine before being rolled out and moulded into the finished shape.

In modern confectionery all or some of the invert sugar is often replaced by glucose syrups, otherwise known as corn syrups as corn, *i.e.* maize, starch is the usual raw material for their manufacture. In times past glucose syrups were made by the acid hydrolysis of starch. Maize starch or other starchy raw materials such as potatoes were heated with dilute sulfuric acid to hydrolyse the $\alpha1\rightarrow4$ and $\alpha1\rightarrow6$ glycosidic links between the glucose units (*see* Chapter 3). One might expect complete hydrolysis to glucose to be the result but in fact the reaction does not go to completion. As the reaction proceeds the concentration of sugars gets very high and there is a tendency for the reverse reaction to occur. However, the reverse reaction is in no way constrained to give back the $\alpha1\rightarrow4$ and $\alpha1\rightarrow6$ glycosidic links of the original starch. In consequence there is an accumulation of oligosaccharides based on the full range of possible linkages, α or β, $1\rightarrow1$, $1\rightarrow2$, $1\rightarrow3$, *etc.*, especially since many of these are much more resistant to acid hydrolysis. These oligosaccharides are unlikely to be utilised by the human digestive process but their presence does ensure that these syrups, known commercially as liquid glucose*, do not crystallise. Modern commercial cake recipes also take advantage of the refusal of glucose syrups to crystallise. They help the crumb of the cake to retain moisture and thereby remain soft and palatable for longer.

Nowadays enzymic hydrolysis using thermostable α-amylases and other enzymes has superseded acid hydrolysis, making available syrups with a highly controllable compositions, ranging from almost pure glucose to mixtures dominated by particular oligosaccharides. These enzymic meth-ods, which are described in greater detail in Chapter 3, can also be used to produce syrups resembling invert syrup known as 'isosyrups'. These are manufactured by using the enzyme glucose isomerase to convert some of the glucose into fructose.

The inversion of sucrose is also an essential part of the process of making jam and other preserves. In making jam the fruit, together with a

---

* In the food industry the terms 'glucose' or 'liquid glucose' refer to the mixed products of starch hydrolysis. This may still have been produced by acid hydrolysis but with infinitely more care and post-hydrolysis clean up than a hundred years ago. The old fashioned term 'dextrose' is used to indicate D-glucopyranose in the crystalline state.

large amount of sugar, is boiled for a prolonged period. During this period
the fruit is softened and its pectin (*see* Chapter 3), plus any extra pectin
being added, is brought into solution. At the same time a considerable
proportion of the fruit's original water content is removed. The mild
acidity leads to the hydrolysis of up to around half the added sucrose to
glucose and fructose. This has various beneficial effects. Firstly, the total
number of sugar molecules is increased by up to 50%. This gives a modest
increase in the sweetness (*see* Chapter 7), but most importantly it increases
the proportion of the water that is bound to sugar molecules and therefore
unavailable to support the growth of microorganisms (*see* Chapter 12).
The invert sugar is much more soluble in water than the original sucrose
(even syrups containing 80% invert sugar, *i.e.* 20% water, do not crystallise) and the presence of invert sugar actually represses the crystallisation
of sucrose. It is therefore easy to reach the very high solids levels (and
correspondingly low water activity levels) that are needed for long term
stability.

The prolonged high temperatures, reaching 105 °C, of the boiling period
can have a drastic effect on the colour and flavour of the fruit. Commercial
jam makers often reduce this by carrying out the boil at reduced
atmospheric pressure, bringing the water off at lower temperatures. It is
also common to replace some of the sucrose with glucose syrups prepared
from starch (*see* Chapter 3). These contain significant proportions of
glucose oligosaccharides such as maltose, maltotriose (*i.e.* three glucose
units linked $\alpha,1\rightarrow4$), isomaltose and maltotetrose. The presence of these
oligosaccharides represses glucose crystallisation to enable high total sugar
contents to be reached without excessive boiling.

One group of higher oligosaccharides deserves particular attention.
These are the galactose derivatives of sucrose.

raffinose:
  α-D-galactopyranosyl-$(1 \rightarrow 6)$-α-D-glucopyranosyl-$(1 \rightarrow 2)$-β-D-fructofuranose

stachyose:
α-D-galactopyranosyl-$(1 \rightarrow 6)$-α-D-galactopyranosyl-$(1 \rightarrow 6)$-α-D-
                                   glucopyranosyl-$(1 \rightarrow 2)$-β-D-fructofuranose

verbascose:
α-D-galactopyranosyl-$(1 \rightarrow 6)$[-α-D-galactopyranosyl-$(1 \rightarrow 6)$]$_2$-α-D-
                                   glucopyranosyl-$(1 \rightarrow 2)$-β-D-fructofuranose

They are well known for their occurrence in legume seeds such as peas
and beans and some other vegetables (*see* Figure 2.1) and present
particular problems in the utilisation of soya beans. These are neither

hydrolysed nor absorbed by the human digestive system, so that a meal containing large quantities of, for example, beans, becomes a feast for the bacteria that inhabit the large intestine. They produce large quantities of hydrogen and some carbon dioxide as by-products of their metabolism of sugars. Much of the carbon dioxide and hydrogen diffuses through the walls of the colon into the bloodstream and ultimately leaves *via* the lungs but a substantial amount remains to cause flatulence.

A similar fate befalls the lactose of milk consumed by individuals who do not secrete the enzyme lactase (β-galactosidase) in their small intestine. Humans, like other mammals, have this enzyme as infants, but the majority of Orientals and many Negroes no longer secrete the enzyme after weaning and are therefore unable to consume milk without risk of 'stomach upset'. The absence of dairy products from the menu of Chinese restaurants should cause no surprise.

SUGARS AS SOLIDS

Sugars in their crystalline state make important contributions to the appearance and texture of many food products, particularly confectionery, biscuits, and cakes. The supermarket shelf usually offers the customer a choice of two crystal sizes in granulated (400–600 μm) and caster sugar (200–450 μm) and a powdered form for icing (average 10–15 μm). The relative proportions of undissolved sugar crystals and sugar solution (syrup) control the texture of the soft cream centres of chocolates *etc.* but the special problem is that of sealing the soft centre inside the chocolate coating. The answer is the use of the enzyme invertase. This is an enzyme that is obtained commercially from yeast (*Saccharomyces cerevisiae*) that catalyses the hydrolysis of sucrose to glucose and fructose. In the manufacture of soft-centred chocolates the mixture for the centre contains a high proportion of finely milled sucrose suspended in just sufficient glucose syrup (*see* Box 3.2), together with appropriate colourings and flavourings, to produce a very stiff, mouldable paste. Also included is a small amount of invertase. After the paste has been stamped out into suitable shapes it is coated with melted chocolate. During the ensuing weeks, and before the chocolates are sold the public, the invertase hydrolyses a proportion of the sucrose. The glucose and fructose are considerably more soluble in water than their parent sucrose and by dissolving alter the solid/liquid balance of the centre, giving it the familiar soft creamy texture.

The usual crystalline forms of most sugars are anhydrous and consist of only one anomer. However, at temperatures below 50 °C glucose crystallises from aqueous solutions as the monohydrate of the α-pyranose anomer. Likewise, D-lactose crystallises as the monohydrate of the α-anomer. Although quite soluble (about 20 g per 100 cm^3 at room temperature), α-D-lactose monohydrate crystals are slow to dissolve and occasionally form a gritty deposit in evaporated milk. The texture of sweetened condensed milk is also highly dependent on the size of the lactose monohydrate crystals.

These considerations apart, the food chemist is primarily interested in sugars when they are in aqueous solution.

SUGARS IN SOLUTION

The chemical reactions of sugars, and chemists' ability to create a bewildering array of chemical derivatives almost none of which had much relevance outside the chemical laboratory, dominated our knowledge until a few decades ago. There is now much more interest in the behaviour of sugars in the apparently simple environment of an aqueous solution. Studies using nuclear magnetic resonance (NMR) spectroscopy and predictions based on thermodynamic calculations have now revealed a great deal to us. Although the Haworth representation of the configuration of sugar rings always had obvious limitations – after all, van't Hoff proposed the tetrahedral arrangement of carbon's valencies in 1874 – it is only slowly being superseded in textbooks. The 'chair' structure (2.28) is a great improvement.

This is the most thermodynamically favourable arrangement for most six-membered aliphatic ring structures. Using β-D-glucose as an example, it will be seen that there are in fact two possible ring structures, or *conformations* (2.29 and 2.30), for any single anomer:

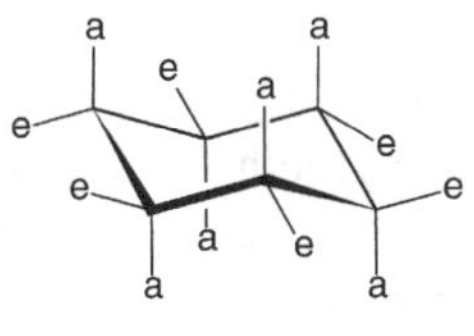

(2.28) 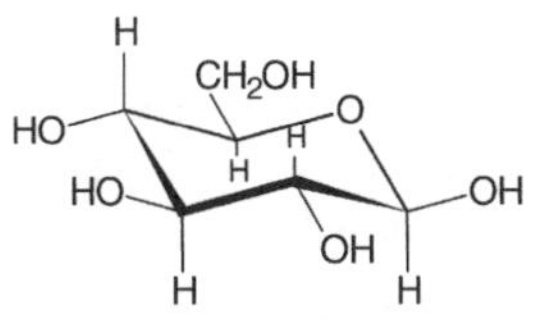   *C*1 (2.29) 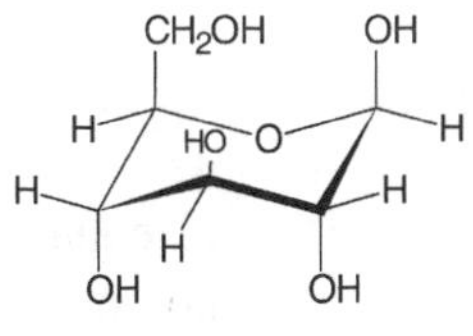   1*C* (2.30)

The designations *C*1 or 1*C* were introduced by Reeves as part of a system for naming these different conformations. For most purposes the designation of a particular structure as *C*1 or 1*C* is most easily achieved by comparison with the two structures shown above. If the sugar ring is portrayed as usual with the ring oxygen towards the right and 'at the back', then the conformation is *C*1 when carbon-1 of the ring is below the plane of the 'seat' formed by carbons 2, 3, and 5 and the ring oxygen. The conformation is obviously $1C$ when carbon 1 is above this plane. Theoretical studies have shown that the favoured conformation will be the one which has the greater number of bulky substituents such as OH or CH$_2$OH in equatorial positions and where neighbouring hydroxyls are as far apart as possible. This has been borne out by experimental work, and it immediately explains why less than 1% of D-glucose molecules in solution are in the $1C$ conformation and why the β-anomer is preferred for glucose. The correct description of a furanose ring is more difficult. There are two most probable shapes, described as the envelope (2.31) and twist (2.32) forms.

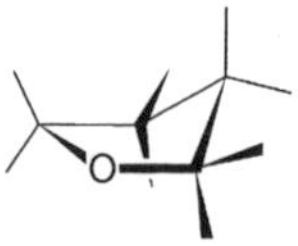

(2.31) 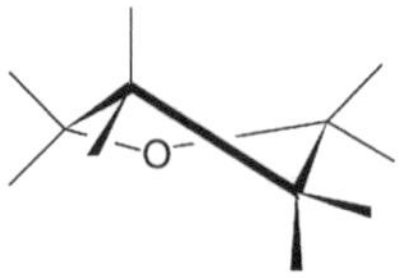   (2.32)

It seems best to assume that furanoses in solution are an equilibrium of many rapidly interconverting forms, such as these, that do not differ greatly from each other and that no one form is dominant. In fact D-fructose occurs predominantly in the $1C$ conformation of the β-pyranose form (2.33), which has the bulky CH$_2$OH group in the equatorial position.

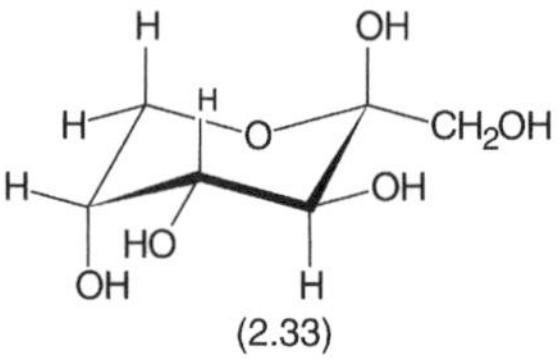

(2.33)

Of the eight D-aldohexoses only D-glucose has a possible conformation where all of its hydroxyl groups, and the hydroxymethyl (*i.e.* the $-$CH$_2$OH), are equatorial. This is almost certainly the explanation for the dominant role of glucose in living systems, both as a metabolic intermediate and as a structural element. The apparent 'untidiness' of its configuration when displayed in straight chain or Haworth formulae is obviously irrelevant to its actual state in living systems.

The conformation and anomer proportions of a particular sugar in solution are not only a function of the intramolecular non-bonding interactions between atoms that have been mentioned so far. They are also dependent on interactions between the sugar molecule and the solvent in foodstuffs, usually water. The fact that there is 37% of the α-anomer of D-glucose at 25 °C in aqueous solution but 45% in pyridine is just one

Box 2.3 Monosaccharide conformations

An analysis of the spatial distribution of the hydroxyl groups of the D-hexoses when in the C1 conformation is given in Table 2.3, along with data showing the proportions of the different anomers and ring types. The C1 (pyranose) ring is the only significant conformation except for D-idose and D-altrose. It is clear that with these two sugars no one conformation offers a predominance of equatorial hydroxyl groups. The same explanation can account for the behaviour of aqueous solutions of D-ribose, which have been shown to contain a broadly similar distribution of anomers and ring sizes to that of fructose except that both the α- and β-pyranose forms occur in roughly equal proportions of the C1 and 1C conformers.

Table 2.3 *The conformation of the aldohexoses. The percentages of each anomer and ring size in aqueous solution at 40 °C are shown, based on the data of S. J. Angyal, Angew. Chem., Int. Edn. Engl., 1969, 8, 157. The positions of the hydroxyls at C−2, C−3, and C−4 are also shown for the C1 conformation. The CH$_2$OH group (C−6) of a D-aldohexose is always equatorial in the C1 conformation, as is the anomeric hydroxyl (on C−1) of the β-configuration.*

	Hydroxyl at			% Furanose		% Pyranose	
	C−2	C−3	C−4	α	β	α	β
D-Allose	eq	ax	eq	5	7	18	70
D-Altrose	ax	ax	eq	20	13	28	39
D-Glucose	eq	eq	eq	<1	<1	36	64
D-Mannose	ax	eq	eq	<1	<1	67	33
D-Gulose	eq	ax	ax	<1	<1	21	79
D-Idose	ax	ax	ax	16	16	31	37
D-Galactose	eq	eq	ax	<1	<1	27	73
D-Talose	ax	eq	ax	<1	<1	58	42
	C−3	C−4	C−5				
1C-D-Fructose	eq	eq	ax	<1	25	8	67

Two factors appear to control the relative proportions of α- and β-anomers for a particular sugar: the usual tendency for a hydroxyl to seek an equatorial position has to be balanced against the unfavourable character of the interaction that can occur between the ring oxygen and the anomeric hydroxyl in the equatorial position.

indication of the existence of strong interactions between water and sugar molecules. As we shall see, the binding of water to sugars and polysaccharides is an important contributor to the properties of many foodstuffs.

The most obvious indication of the binding of water by sugars is that concentrated solutions of sugars do not obey Raoult's Law (which states that 'the relative lowering of the vapour pressure of the solvent is equal to the molecular fraction of the solute in solution'). The observed vapour pressure can be as much as 10% below the expected figure. Comparison of the observed with the expected values permits a fairly straightforward calculation of the number of water molecules that are no longer effectively part of the bulk of the water of the system, *i.e.* those that are apparently bound to the sugar molecules directly by hydrogen bonds. The resulting *hydration numbers* approach 2 and 5 molecules of water per molecule of sugar for glucose and sucrose, respectively.

In recent years NMR and dielectric relaxation techniques have given us alternative probes for examining water binding. They have revealed a primary hydration layer containing an average number of water molecules rather higher than that deduced from departures from Raoult's Law, for example ribose 2.5, glucose 3.7, maltose 5.0, and sucrose 6.6. The water that is involved in hydration is not necessarily part of a permanent structure. NMR techniques will reveal as 'bound' water molecules that remain hydrogen-bonded for as little as one microsecond.

Further aspects of the binding of water by sugars, and other food components, are considered in Chapter 12. The sweetness that sugars impart to our food is considered in Chapter 7.

DECOMPOSITION

It is not only intact sugar molecules that make a positive contribution to the properties of our diet: the products of their thermal decomposition are also important food components in their own right. When sugars are heated to temperatures above 100 °C, a complex series of reactions ensues, known as caramelisation, which give rise to a wide range of flavour compounds as well as the brown pigments that we particularly associate with caramel. A related series of reactions, referred to as the *Maillard reaction* involves amino compounds as well is considered on pages 33–39. Since both of these groups of reactions are particularly associated with the generation of brown pigments in food they are often collectively known as the *non-enzymic browning reactions*. This distinguishes then from the *enzymic browning reactions* encountered in fruit and vegetables and dealt with in Chapter 6.

The first stage of the high-temperature breakdown is a reversible isomerisation (memorably known as the Lobry de Bruyn–Alberda van Eckerstein transformation) of an aldose or ketose sugar, *via* their open chain forms, to a 1,2-*cis*-enediol intermediate (Figure 2.8). If glucose is gently heated in the presence of dilute alkali this reaction leads to the formation of a mixture of glucose, mannose and fructose.

The enediol is readily dehydrated in a neutral or acidic environment *via* the sequence of reactions shown in Figure 2.9. Hydroxymethylfurfural (HMF) is by no means the only product of this pathway. Highly reactive α-dicarbonyl compounds (*i.e.* compounds like 3-deoxyaldoketose in which pairs of carbonyl oxygen atoms are carried by neighbouring carbon atoms) occur as intermediates in these sequences and can give rise to range of cyclic products as well as reacting with amino compounds in the important Strecker degradation reactions discussed below.

HMF can be readily detected in sugar-based food products that have been heated, such as boiled sweets, honey that has been adulterated with invert syrup, and 'golden syrup'.* The brown pigments that characterise caramel and other foods arise from a poorly defined group of polymerisation reactions. These involve both HMF and its precursors. The pigment molecules have very high molecular masses and in consequence are apparently not absorbed from the intestine. Burnt sugar has always been a popular brown colouring because it can be easily made in the home and therefore has a 'natural' image. Commercially manufactured caramels for

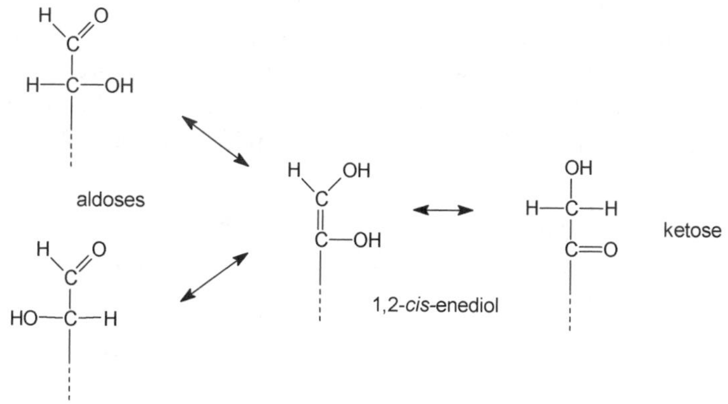

Figure 2.8 *The Lobry de Bruyn-Alberda van Eckerstein transformation.*

---

* Commercial invert syrup is produced by acid hydrolysis of sucrose at high temperatures and is marked as 'golden syrup'.

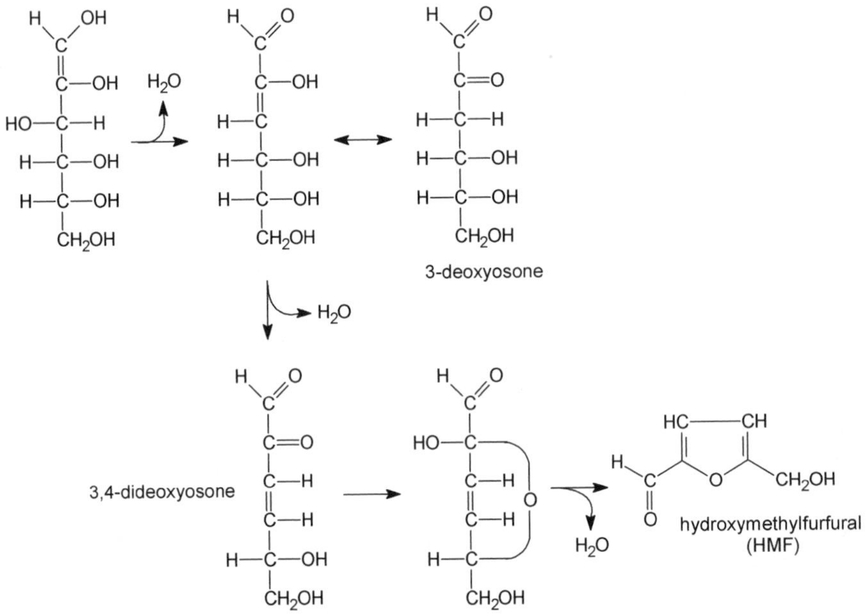

Figure 2.9 *The formation of hydroxymethyl furfural by the sequential dehydration of a hexose enediol. Variations in the sites at which dehydration occurs is the source of related products such as hydroxyacetyl furan (2.34).*

use as food colours are made in systems based on the Maillard reaction, which is to be considered shortly.

Small amounts of a vast range of other breakdown products give burnt sugar its characteristic acrid smell: acrolein (propenal, 2.35), pyruvaldehyde (2-oxopropanal, 2.36), and glyoxal (ethanedial, 2.37) are typical. The caramel flavour itself is reportedly due to two particular cyclic compounds, acetylformoin (2.38) and 4-hydroxy-2,5-dimethylfuran-3-one (2.39).

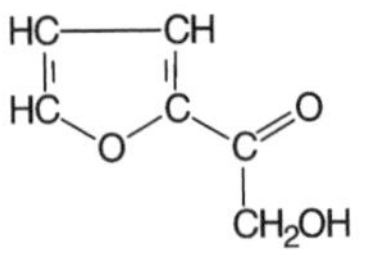

(2.34)

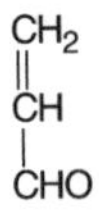

(2.35)

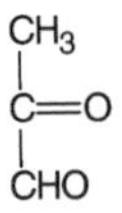

(2.36)

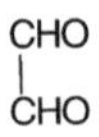

(2.37)

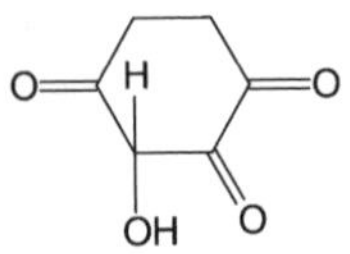

(2.38)

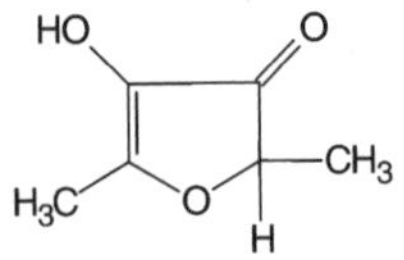

(2.39)

In the presence of amino compounds the browning of sugars occurs much more rapidly, particularly in neutral or alkaline conditions, in a sequence of events known as the Maillard reaction. Low water concentrations, when the reactants are more concentrated, favour the reaction, so that it is implicated in the browning of bread crust and the less welcome discoloration that occurs from time to time in the production of powdered forms of egg and milk. These foodstuffs are ones in which sugars are heated in the presence of protein, the source of the amino groups. The amino groups most often involved are those in the side chains of lysine and histidine, although the α-amino group of any free amino acids will participate. The scheme in Figure 2.10 shows that the initial outcome of the Maillard reaction is essentially the same as for the other caramelisation reactions in that HMF is produced.

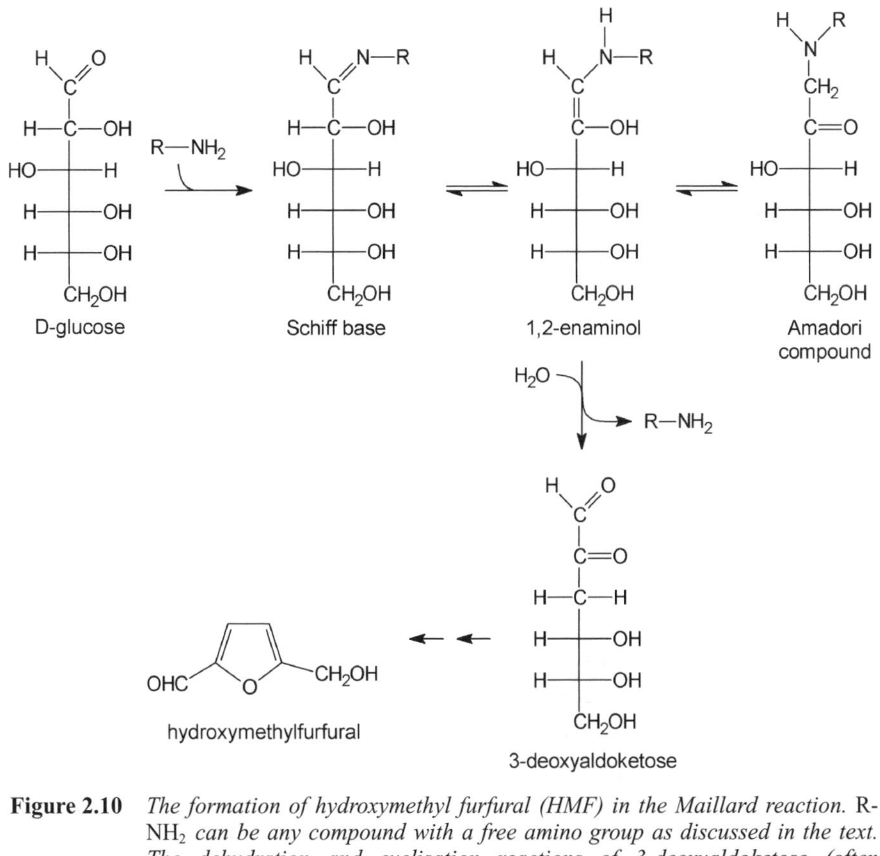

Figure 2.10 *The formation of hydroxymethyl furfural (HMF) in the Maillard reaction. R-NH$_2$ can be any compound with a free amino group as discussed in the text. The dehydration and cyclisation reactions of 3-deoxyaldoketose (often referred to as 3-deoxyhexulose or 3-deoxyosone) to give (HMF) are exactly the same as those in caramelisation shown in Figure 2.8.*

In the more complex situation of complete food materials there are large numbers of alternative pathways besides the simple production of HMF. The *N*-substituted 1-amino-1-deoxy-2-ketoses (the so-called Amadori compounds) also isomerise to the 2,3-enaminol form, and this can give rise, *via* similar sequences of dehydrations, rearrangements and cyclisation reactions to various cyclic compounds, such as maltol (2.40) and isomaltol (2.41) as shown in Figure 2.11, as well as acetylformoin (2.38) and sotolon, 4,5-dimethyl-3-hydroxyfuranone (2.42). Sotolon has been identified as the key compound in the aroma of raw cane sugar.

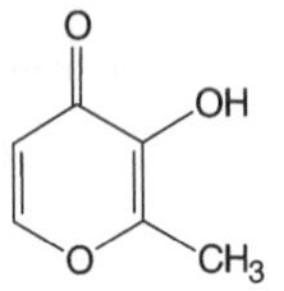

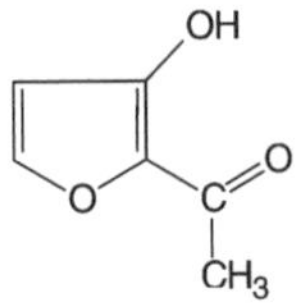

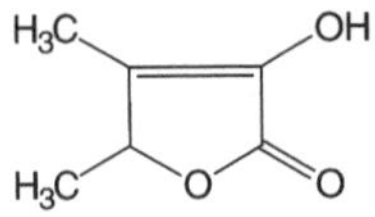

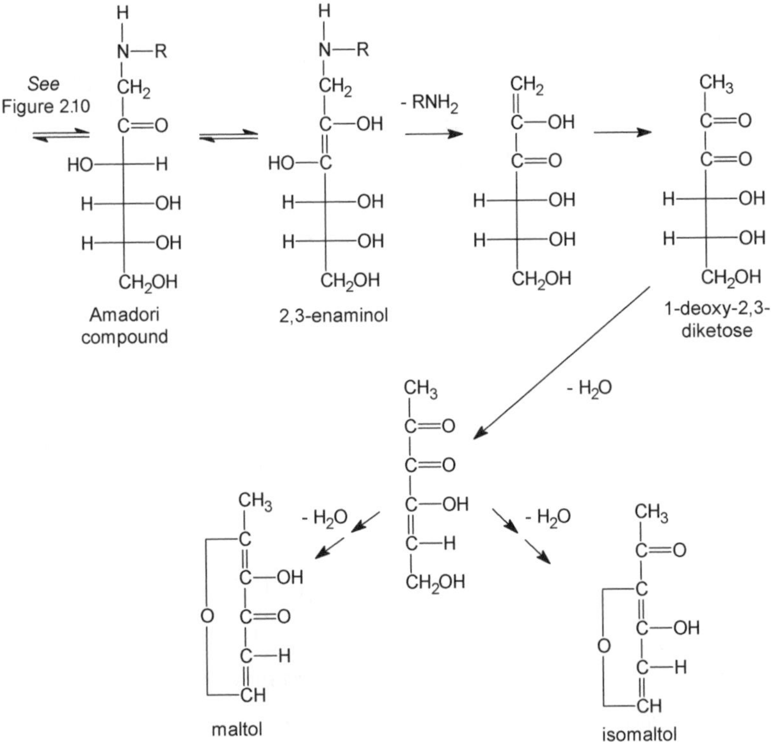

Figure 2.11 *Likely routes for the formation of maltol and isomaltol from the Amadori compound in the Maillard reaction. Compare their structures as drawn here with those in the text, 2.40 and 2.41 respectively.*

Although sotolon's burnt sugar aroma, detectable at parts per billion levels, is regarded as highly desirable in many foods, particularly confectionery, it can be responsible for off-flavours in others, especially at higher levels. Maltol and isomaltol are contributors to the flavour of many cooked foodstuffs.

Disintegration of the α-dicarbonyl molecules also contributes to the flavour of foodstuffs through the formation of substances such as diacetyl (butanedione, 2.43) and acetol (hydroxypropanone, 2.44), as well as pyruvaldehyde and glyoxal, which have already been mentioned.

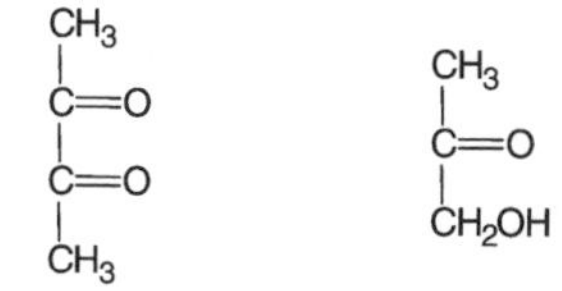

$$
\begin{array}{cc}
\begin{array}{c}
CH_3 \\
| \\
C{=}O \\
| \\
C{=}O \\
| \\
CH_3
\end{array}
&
\begin{array}{c}
CH_3 \\
| \\
C{=}O \\
| \\
CH_2OH
\end{array} \\
(2.43) & (2.44)
\end{array}
$$

A most important class of volatile flavour compounds arises by an interaction at elevated temperatures of α-dicarbonyl compounds with α-amino acids, known as the Strecker degradation, shown in Figure 2.12. The volatile aldehydes derived from amino acids in this reaction make a major contribution to the attractive odour of bread, cake and biscuit making. The pyrazines that are often the other end product of the Strecker degradation also make important contributions to food flavours, *e.g.* those of chocolate and roasted meat. Complex reaction sequences starting with 2,3-enaminols and involving a second amino acid in a Strecker degradation sequence have been suggested as the source of the range of pyrroles and pyridines that can also arise in the Maillard reaction. These are important as they are believed to form the monomers of the polymerised brown pigments (melanoidins) that characterise the Maillard reaction. Actual food situations are far more complex than the model systems of purified starting materials used in current experiments but the structures shown in Figure 2.13 may well turn out to be a close approximation of the brown pigments we find in our food.

While most of the effects of the Maillard reaction are regarded as favourable, there can be adverse nutritional consequences. When food materials containing protein and reducing sugars are heated the free amino groups in the side chains of amino acids can participate in the Maillard reaction. The reaction does not go all the way to coloured end products and may be undetected by the consumer. However the resulting *glycosy-lated* amino acids are unavailable to human metabolism. If a sufficient proportion of the protein's surface is blocked by glycosylation the entire molecule may become unavailable because the proteolytic digestive

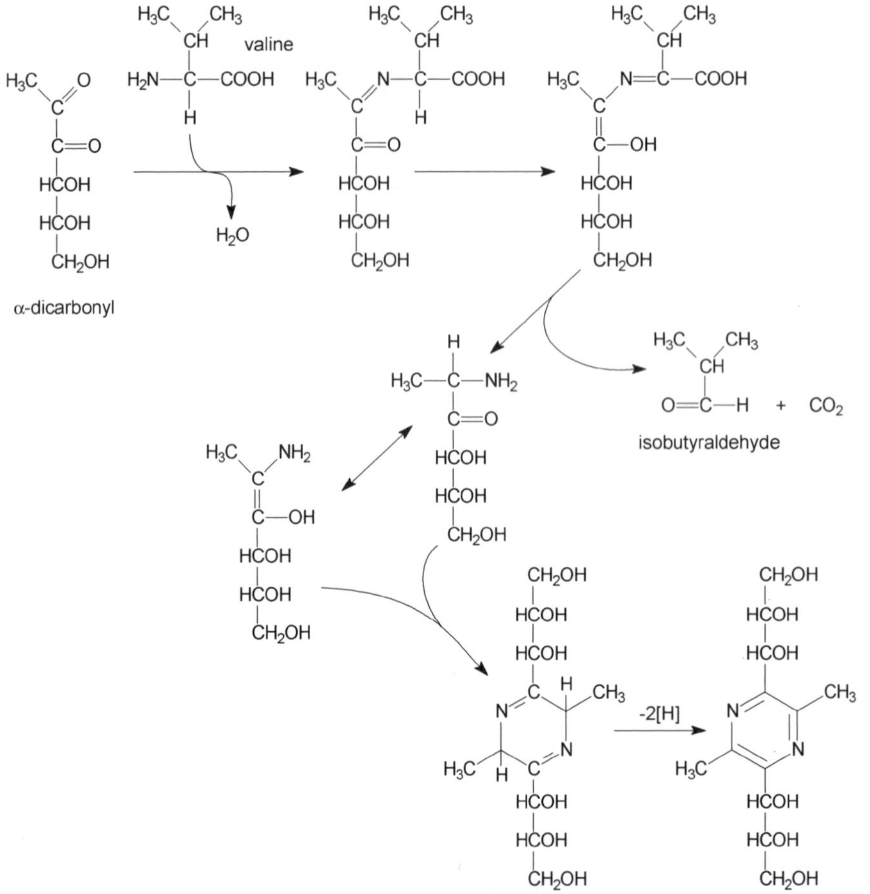

Figure 2.12   *The Strecker degradation. A free amino acid (in this case illustrated by valine) reacts with an α-dicarbonyl compound generated elsewhere in the Maillard reaction. The reactions bears a superficial resemblance to transamination (see Figure 8.2). After rearrangement the carbon skeleton of the amino acid is liberated as a volatile aldehyde. After two ex-sugar fragments (not necessarily, as shown here, identical) condense the product is readily oxidised to give the pyrazine derivative.*

enzymes cannot gain access to the peptide bonds (*see* page 130). Lysine is the most reactive of the vulnerable amino acids, followed by arginine and then tryptophan and histidine. It is the fact that lysine, histidine, and tryptophan must all be supplied in the diet *i.e.* they are all *essential* amino acids, (*see* Chapter 5), and that poor vegetarian diets are often deficient in lysine, that makes this such an important issue.

The reaction does not normally go beyond the formation of a stable

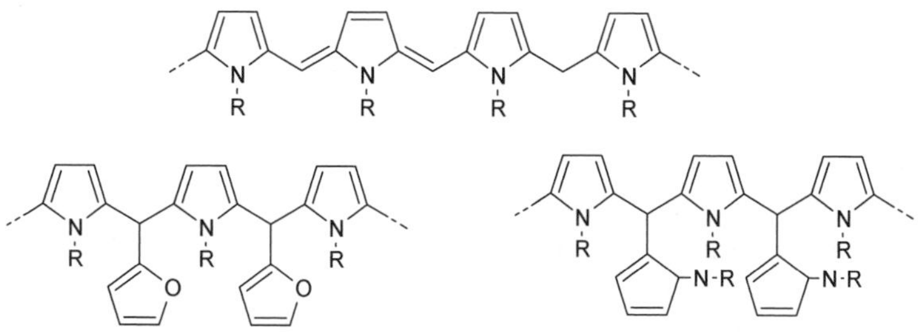

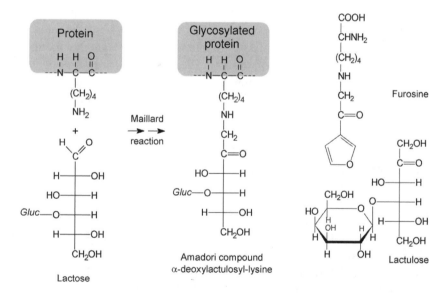

Figure 2.13 *Postulated structures for the brown, melanoidin pigments formed in the Maillard reaction. In these structures R represents any of the possible sugar fragments such as* −CH₃, −(CHOH)₂CH₂OH, *etc. that arise during the course of the reaction. All of the structures shown here are likely to found in a single melanoidin molecule. It has been estimated that around 95% of molecules initially participating in the Maillard reaction eventually end up in melanoidin polymers. (Based on the proposals of R. Tressl et al. pp 69– 75, in 'The Maillard Reaction in Foods and Medicine' – see Further Reading.)*

Figure 2.14 *Lactose reactions during milk processing. The extent of glycosylation in a food protein can be estimated from the furosine content of the hydrolysate after it has been subjected to acid hydrolysis. Also at high temperatures small amounts of lactose undergo the Lobry de Bruyn–Alberda van Eckerstein transformation (see Figure 2.8) to give lactulose, i.e. the glucose component of the disaccharide is converted from glucose to fructose, shown here for clarity in the 'open-chain' structure. Measurements of furosine and lactulose levels are becoming important in establishing the processing history of commercial milk products.*

Amadori compound, such as α-deoxylactulosyl-lysine, as shown in Figure 2.14. Pasteurisation or UHT treatment of milk causes modest losses of lysine (up to around 2%) but the more aggressive heating involved in some milk processes, such as sterilisation, evaporation or roller drying, can result in as much as 25% of the lysine becoming unavailable to human digestion. A mixture of glucose and casein (the principal protein of milk) has been shown to lose as much as 70% of its lysine over a period of five days at temperatures as low as 37 °C. Conditions such as these may well occur during the transport of powdered milk for famine relief in the tropics.

The disease *diabetes mellitus* provides a very important example of protein glycosylation taking place at a 'low' temperature, *i.e.* blood heat, that is sufficiently food related to be included in this book. Semi-automatic devices based on the enzymes glucose oxidase and peroxidase (*see* Figure 2.6) provide an 'instant' measurement of patients' blood glucose concentration at the time of the test but do not provide a long term overall view. This is obtained by measuring the proportion of the patients' haemoglobin that has become glycosylated by the Maillard reaction with glucose in the blood. Normally some 3–7% of haemoglobin is in the glycosylated form (known as HbA1C) but where diabetes has resulted in persistently higher levels of blood glucose higher percentages of HbA1C are found. Many of the complications suffered by long term victims of this disease, including the degeneration of the peripheral vascular system, sight loss (retinopathy) and kidney failure are now known to be the result of crucial proteins in these tissues becoming glycosylated .

The Maillard reaction can pose health problems not strictly related to nutrition. More than 20 complex heterocyclic amines, undoubtedly the products of Maillard type reactions, have been identified in cooked (particularly grilled or fried) meat, fish and poultry as well as in model systems. Many of these have been shown to be mutagenic in tests with bacteria (in the well known Ames test) and must be regarded as *potential*, if not actual, carcinogens. While the concentration of these in grilled food is frequently very low (less than 1 ppb) one, 2-amino-1-methyl-6-phenyli-midazo[4,5-*b*]pyridine (abbreviated to 'PhIP') (2.45), has been detected at 500 ppb in flame cooked chicken. High temperatures (200–300 °C) and fairly long cooking times are required for the formation of heterocyclic amines; ordinary stewing, roasting and baking do not cause their formation. It is very difficult to estimate the risks of human cancer from these compounds in real food, as oppose to model systems, but one current risk estimate is 1 in 10 000 for a normal lifetime and a typical consumption of food cooked at high temperature. Although the humble burger is grilled at high temperatures the short cooking times make it less of a hazard than might be expected.

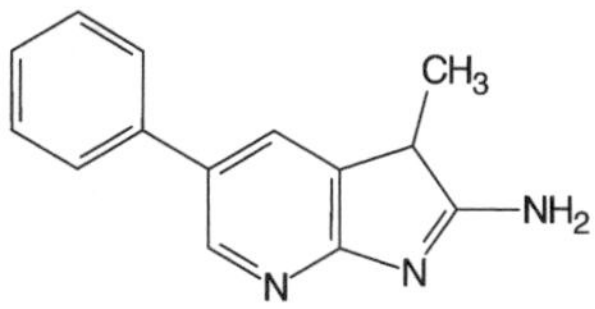

(2.45)

Caramel is produced commercially for use as a food colour, for use in vast range of food products besides the 'gravy browning' found in the domestic kitchen. Around 10 000 tons are used annually in Britain, accounting for all but 5% or so of the artificial colours used. Commercially caramels are manufactured by heating sucrose with catalytic amounts (a few percent) of acid or alkali (Class I caramel), sulfite (Class II), ammonia (Class III) or ammonia plus sulfite (Class IV). The different classes of caramel differ slightly in properties and consequently find different applications especially in drinks. For example Class I and II caramels are used in spirits such as whisky, Class III caramels in beer and Class IV caramels in soft drinks, notably cola.

FURTHER READING

D. A. Southgate, 'Determination of Food Carbohydrates', 2nd Edn., Elsevier, London, 1991.

R. S. Shallenberger, 'Advanced Sugar Chemistry', Ellis Horwood, Chichester, 1982.

W. P. Edwards, 'The Science of Sugar Confectionary', The Royal Society of Chemistry, Cambridge, 2000.

'Food Carbohydrates', ed. D. P. Lineback and G. E. Inglett, AVI, Westport, 1982.

J. F. Kennedy and C. A. White, 'Bioactive Carbohydrates', Ellis Horwood, Chichester, 1983.

'The Maillard Reaction in Foods and Medicine', eds. J. O'Brien, H. E. Nursten, M. J. C. Crabbe and J. M. Ames, The Royal Society of Chemistry, Cambridge, 1998.

R. V. Stick, 'Carbohydrates. The Sweet Molecules of Life', Academic Press, San Diego, CA., 2001.

W. P. Edwards, 'The Science of Sugar Confectionery', The Royal Society of Chemistry, Cambridge, 2000.

Chapter 3

Polysaccharides

In Chapter 2 we considered the properties of the mono- and oligo-saccharides. We can now turn to the high-molecular-weight polymers of the monosaccharides: the polysaccharides. Polysaccharide structures show considerable diversity and are often classified according to features of their chemical structure, for example whether the polymer chains are branched or linear or whether there is more than one type of monosaccharide residue present. Only recently have we begun to discern a general basis for the relationship between the chemical structure of a polysaccharide and its physical properties. In the course of this chapter it is to be hoped that the nature of this relationship will be demonstrated.

The one common feature of the polysaccharides that is important to food scientists is that they occur in plants. Polysaccharides play a relatively minor role in the physiology of animals. Glycogen, which is structurally very similar to amylopectin, acts as an energy/carbohydrate reserve in the liver and muscles*. The other animal polysaccharides are chitin, a polymer of *N*-acetylglucosamine (3.1), the fibrous component of arthropod exoskeletons, and the proteoglycans and mucopolysaccharides, which have specialised roles in connective tissues. The molecules of these animal polysaccharides include considerable proportions of protein.

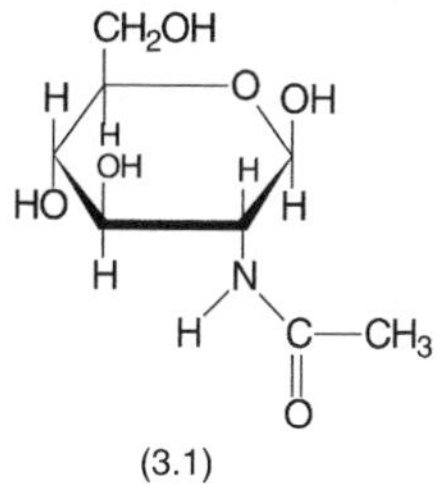

(3.1)

---

* Muscle tissue loses its glycogen in the on-going *post mortem* metabolism of the tissues long before it reaches the plate as meat but fresh liver still contains 1–2% glycogen when it is sold.

41

Polysaccharides have two major roles to play in plants. The first, as an energy/carbohydrate reserve in tissues such as seeds and tubers, is almost always filled by starch. The data in Table 3.1 shows that leafy plant foods, as exemplified by broccoli, have only a trace of starch and that there some important food plants the do not use starch as a major energy store. The second role of polysaccharides is to provide the structural skeleton of both individual plant cells and the plant as a whole, is filled by a wide range of different structural types which offer a correspondingly wide range of physical characteristics. The contribution of polysaccharides to food is no less diverse. From a strictly nutritional point of view starch is the only one of the major plant polysaccharides that is readily digested in the human intestine and thereby utilised as a source of carbohydrate. In fact a high proportion of the human requirement for energy is met by the starch of cereal grains, wheat, rice and maize, and tubers such as potatoes and cassava. It is now widely recognised that polysaccharides make a much greater contribution to human health than the mere provision of energy. Cellulose, pectins, and hemicelluloses, with others, commonly referred to as 'fibre' but more correctly termed 'non-starch polysaccharides' (NSP), have essential roles in the healthy function of the large intestine and in moderating the absorption of nutrients from the small intestine.

In spite of their physiological importance it is through their influence on the texture of food that polysaccharides often make the most immediate impact on the consumer. Furthermore, the relationship between a food's

Table 3.1 *The starch contents of a variety of foods. Most of these figures are taken from 'McCance and Widdowson' (see Appendix II) and in all cases refer to the edible portion. They should be regarded as typical values for the particular food rather than absolute figures which apply to any samples of the same food.*

Food	Starch (%)	Food	Starch (%)
Wheat flour (wholemeal)	61.8	Peas (petit pois)	trace
White bread	46.7	Broad beans	10.0
White rice (uncooked)	73.8	Beetroot	0.6
Spaghetti (cooked)	70.8	Broccoli	0.1
Maize (corn) flour	92.0	Carrots	0.2
Corn flakes	77.7	Chestnuts	29.6
Soya beans (raw)	4.8	Peanuts (raw)	6.3
Cassava (dry)	22.0	Walnuts	0.7
Potatoes (boiled, new)	16.7	Apples	trace
Peas (canned, processed)	14.7	Bananas	2.3

texture and the underlying molecular structure of its components is being revealed most clearly in the case of polysaccharides, and this relationship therefore provides the theme of the accounts of the more important food polysaccharides that follow.

STARCH

Reflecting its role as a carbohydrate reserve, starch is found in greatest abundance in plant tissue such as tubers and the endosperm of seeds. It occurs in the form of granules which are usually an irregular rounded shape, ranging in size from 2 to 100 μm. Both the shapes and sizes of the granules are characteristic of the species of plant and can be used to identify the origin of a starch or flour.

Starch consists of two types of glucose polymer: amylose, which is essentially linear, and amylopectin, which is highly branched. They occur together in the granules, but amylose may be separated from starch solutions since it is much less soluble in organic solvents such as butanol. Most starches are 20–25% amylose, but there are exceptions: pea starch is around 60% amylose and the so-called 'waxy' varieties of maize and other cereals have little or none.

Amylose consists of long chains of α-D-glucopyranosyl residues linked, as in maltose, between their 1- and 4-positions (*see* Figure 3.1). There is

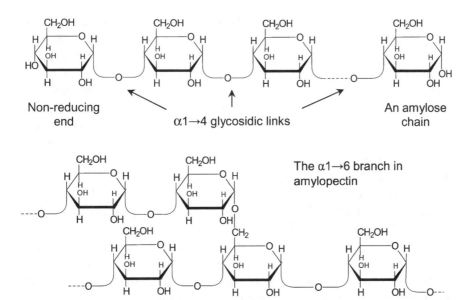

Figure 3.1 *The chemical structures of amylose and amylopectin.*

no certainty about the length of the chains, but they are generally believed to contain many thousands of glucose units so that typical, average molecular weights are between 2×10^5 and 2×10^6. It is becoming clear that amylose chains are not entirely linear, they do contain a very small amount of branching of the type which is characteristic of amylopectin.

Amylopectin is a much larger molecule, having about 10^6 glucose units per molecule. As in amylose the glucose units are joined by α 1→4 glycosidic links, but some 4–5% of the glucose units are also involved in α 1→6 links, creating branch points as shown in Figure 3.1. This proportion of branch points results in an average chain length of 20–25 units. Over the years a great many hypotheses have been put forward to describe the arrangement of the branching of chains of amylopectin, and for many years the tree-like arrangement first proposed by Meyer and Bernfeld in 1940 was accepted. Since then the dissection of the molecule into its individual component chains using the enzyme pullulanase, which is specific for the 1→6 linkages, has provided new insights. Although the chains do indeed have the expected overall average chain length the molecule has been found to consist of two types of chain, the more abundant type has around 15 glucose units, the other around 40, referred to as the A and B chains respectively. The so-called 'cluster model' based on this data, and other supporting evidence, was proposed by Robin in 1974 and is now widely accepted (Figure 3.2A). Its essential feature is a skeleton of mostly singly branched A chains carrying clusters of B chains. As shown in Figures 3.2B and C the amylopectin molecules are believed to be radially orientated in the starch granule, with the terminal reducing group towards the centre, possibly close to the central spot sometimes visible by light microscopy, the *hilum*. The clusters of short chains are believed to be distributed on the B chain skeleton as shown in Figure 3.2B, leading to the formation of concentric domains of regularly orientated (crystalline) polysaccharide chains alternating with amorphous domains, shaded in Figure 3.2B, where most of the α 1→6 branch points are located. In the crystalline regions pairs of neighbouring chains are believed to form short double helices with 6 glucose residues (in each chain) per turn (Figure 3.2D).

When observed in the polarising microscope starch granules show the 'Maltese cross' pattern which is characteristic birefringent materials. This confirms that there is a high degree of molecular orientation. The crystallinity is confirmed by their X-ray diffraction patterns, which, incidentally, reveal that, in general, root starch granules are more crystalline than those of cereals. Another indication of the importance of the amylopectin fraction as the source of starch granules' crystallinity is that waxy starches, which lack amylose, give X-ray diffraction patterns very similar to those

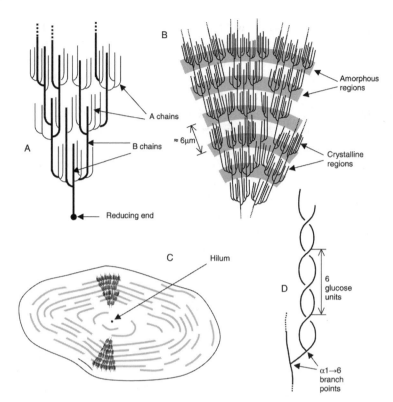

Figure 3.2 *Amylopectin and the starch granule. (A) The essential features of the cluster model first proposed by Robin in 1974. (B) The organisation of the amorphous and crystalline regions (or domains) of the structure generating the concentric layers that contribute to the 'growth rings' that are visible by light microscopy. (C) The orientation of the amylopectin molecules in a cross section of an idealised entire granule. (D) The likely double helix structure taken up by neighbouring chains and giving rise to the extensive degree of crystallinity in a granule.*

of a normal starch. Surprisingly little is known of the arrangement of the amylose molecules within the starch granule. They are assumed to meander amongst the amylopectin molecules, orientated more or less radially. The granules of cereal starches contain approximately 1% of lipid material, mostly in the form of lysophopholipids (*see* Chapter 4). The hydrocarbon chains of the lipid molecules are believed to lie down the axis of helical amylose chains, forming a complex not dissimilar in structure from the iodine/amylose complex (*see below*).

Undamaged starch granules are insoluble in cold water owing to the collective strength of the hydrogen bonds binding the chains together, but, as the temperature is raised to what is known as the initial gelatinisation temperature, water begins to be imbibed. The initial gelatinisation temperatures are characteristic of particular starches but usually lie in the range 55–70 °C. As water is imbibed, the granules swell and there is a steady loss of birefringence. Studies with X-ray diffraction show that complete conversion to the amorphous state does not occur until temperatures around 100 °C are reached. As swelling proceeds and the swollen granules begin to impinge on each other, the viscosity of the suspension/solution rises dramatically. The amylose molecules are leached out of the swollen granules and also contribute to the viscosity of what is best described as a paste. If heating is maintained together with stirring, the viscosity soon begins to fall again as the integrity of the granules is destroyed by the physical effects of the mixing. If the paste is allowed to cool, the viscosity rises again as hydrogen-bonding relationships between both amylopectin and amylose are re-established to give a more gel-like consistency. These changes will be familiar to anyone who has used starch to thicken gravy or to prepare desserts such as blancmange. These changes can be followed in the laboratory with viscometers such as the Brabender Amylograph which apply continuous and controlled stirring to a starch suspension during a reproducible temperature regime. Some typical Brabender Amylograph results for a number of different starches are shown in Figure 3.3. Although our understanding is increasing, we are still some way from being able to explain all the distinctive features of a

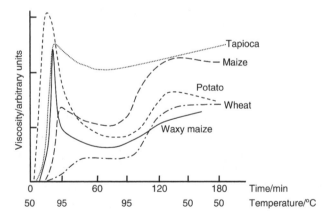

Figure 3.3 *Brabender Amylograph traces of the gelatinisation of a number starches. The temperature is raised steadily from 50 to 95 °C over the first half hour period. It is then maintained at 95 °C for one hour and then brought down back to 50 °C over the final half hour period.*
(Reproduced by permission from Dr D. Howling and Applied Science Publishers.)

particular Brabender trace in terms of the molecular characteristics of the starch.

Solutions of starch have a number of characteristic properties, of which the reaction with iodine is the best known. Solutions of iodine in aqueous potassium iodide stain the surface of starch granules dark blue, but the reaction is most apparent with the amylose component of dissolved starch. In the presence of iodine the amylose chains are stabilised in a helical configuration with six glucose residues to each turn of the helix and the iodine molecules complexed down its axis. The blue colour of the complex is dependent on the length of the chain involved: with 100 residues λ_{max} is at 700 nm, but with shorter chains λ_{max} falls until with only 25 residues it is down to 550 nm. In view of this it is not surprising that amylopectin and glycogen both give a reddish brown colour with iodine.

When starch solutions or pastes are allowed to stand for a few hours, they begin to show changes in their rheological properties. Dilute solutions lose viscosity, but concentrated pastes and gels become rubbery and exude water. Both types of change are due to a phenomenon termed retrogradation, which involves just the amylose molecules. Over a period of time these associate together and effectively 'crystallise out'. If one recalls that the role of amylose in starch solutions is to bind together the expanded granule structure of amylopectin molecules, then these effects of retrogradation on their rheological properties are obviously inevitable.

Special efforts have to be made to avoid the problems that retrogradation could pose for a number of food products. For example, starch is frequently used to thicken the juice or gravy of commercial pies. If after manufacture the pie is frozen, the amylose will undergo retrogradation rapidly. On thawing, the starch paste will be completely liquid, revealing just how much, or how little, of the filling was actually meat or fruit, as well as making the pastry all soggy. One answer to this has been the use of starch from the 'waxy' varieties of maize*, which consist only of amylopectin and therefore give pastes that will survive freezing and thawing.

The manufacturers of 'instant' desserts must also be wary of retrogradation. These starch-based dry mixes are designed to be blended with milk to give a blancmange-like dish without the need for cooking. A slurry of raw starch in water is passed over steam-heated rollers which results in rapid gelatinisation immediately followed by drying to a film which is scraped off the roller and subsequently milled to a fine powder. The water is thus removed before the amylose molecules have had the chance to get

---

* The term 'waxy' is derived from the appearance of the starch endosperm of the maize kernel when it is broken in two.

Box 3.1   Chemically Modified Starches

The use of waxy starches and pregelatinisation widen the range of behavioural characteristics that are available to food processors but the most significant changes can obtained by chemical modification of the starch molecules. Three types of modification are used.

(i)   Depolymerisation. Starch granules are treated with hydrochloric acid followed by neutralisation. Only a very small proportion of the glycosidic links are broken but the fragmentation of the amylopectin molecules is still sufficient for the granule structure to be lost much more easily on heating and gelatinisation to be much faster. Known as thin-boiling starches they give stronger clearer gels and are used in many sorts of jelly confectionery. Maltodextrins are produced by much more extensive starch hydrolysis, with DP values (Degree of Polymerisation, the average the number of glucose units per molecule) in the range 5–100.

(ii) Derivatisation. Esterification of a small proportion (normally less than 1%) glucose units with organic acids or phosphates gives 'stabilised' starches. Depending on the particular groups attached one can achieve gels with excellent freeze-stability and resistance to retrogradation (acetyl or hydroxypropyl groups), high paste viscosity (phosphates) or emulsifying capabilities (1-octenyl-succinates).

(iii) Crosslinking. Less than one crosslink per 1000 glucose units is sufficient to produce significant changes in starch properties. Crosslinks between chains tend to strengthen the granule structure. This makes for pastes that are much more resilient in the face of low pH, extended cooking and physical shear forces than those made from unmodified starch. Not surprisingly crosslinked starches require much more cooking to achieve gelatinisation. They are particularly valuable in canned products. Their slow gelatinisation allows the can's contents to remain liquid during heating, so that heat transfer is not impaired by solidification but the final product has a very firm texture.

Commercial modified starch products often have both stabilisation and crosslinking used in the same product, quite often applied to a waxy starch

organised. The resulting pregelatinised starch will disperse in water fairly easily, but no one would claim that this gives as good a texture as that of a proper blancmange.

In the manufacture of bread and other bakery products the behaviour of starch is obviously very important. When flour is milled it is inevitable that a significant proportion of the starch granules will be physically damaged, scratched, cracked or broken. When a dough is first mixed, these damaged starch granules absorb some water. During bread-making the rather than the normal type. Some of the chemical reactions involved in derivatisation are shown in Figure 3.4. The extent of these reactions is restricted by food legislation but is always far too small to have any significant effect on the digestibility and consequent nutritive value.

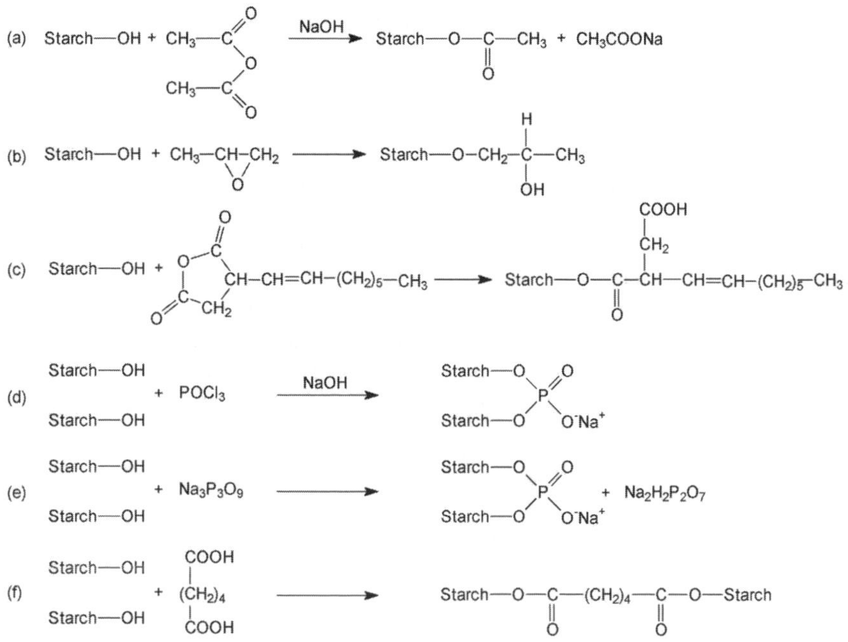

Figure 3.4 *Chemical modification of starch. Reactions (a)–(c) are stabilising deriviti-sations, (d)–(f) are crosslinking reactions. (a) Acetylation with acetic anhydride. (b) Formation of a hydroxypropyl ether by reaction with propylene oxide. (c) Formation of the 2-(octenyl-1)succinyl ester. (d) Phosphate diester formation with phosphorus oxychloride. (e) Phosphate diester formation with sodium trimetaphosphate. (f) Adipate diester forma-tion with adipic acid.*

kneading and proving stages prior to the actual baking allow time for the α- and β-amylases that are naturally present in flour to break down a small proportion of the starch to maltose and other sugars. These are fermented by the yeast to give the CO_2 which leavens the dough. Other baked goods rely on the CO_2 from baking powder together with air, whipped into the mixture, for the same purpose.

Once in the oven the starch granules gelatinise and undergo varying degrees of disruption and dispersion. The α-amylase activity of flour

---

Box 3.2   Syrups from starch

The conversion of starch into glucose, and other sugars is a major industry. The old fashioned process based on acid hydrolysis has been almost totally superseded by processes based on enzymic catalysed hydrolysis. These processes take advantage of the great specificity of enzymes for particular types bonds in particular substrates. Although enzymes, being proteins, are usually thought of as being inactivated at high temperatures nowadays many enzymes for industrial use are obtained from thermophilic bacteria that have thermo-stable enzymes.

In the usual two-phase process starch, mostly from maize, and cold water are mixed to form a slurry with up to 40% solids content. A thermostable α-amylase from a thermophilic strain of a bacillus (*Bacillus licheniformis, B. stearothermophilus* or *B. amyloliquefaciens*) is then incorporated into the slurry which is then brought rapidly up to a temperature of 105–110 °C and held there for a few minutes. The starch granules gelatinise and as they do so the enzyme is able to gain access to the $\alpha1 \rightarrow 4$ glycosidic links of the amylose and amylopectin. As a result both polymers are fragmented to oligosaccharides (sometimes referred to as dextrins) averaging around 10 glucose units in size. Many of the fragments will include $\alpha1 \rightarrow 6$ branches derived from the branched structure of amylopectin but these bonds are not attacked by the α-amylase. Unlike the impossibly stiff starch paste that would have resulted from simply gelatinising a 40% starch slurry the solution of oligosaccharides is now manageably fluid.

---

persists until temperatures around 75 °C are reached, so that some enzymic fragmentation of starch granules also occurs, particularly in bread. Amongst the factors that affect the degree of starch breakdown are the availability of water and the presence of fat. Obviously, in a shortbread mixture containing only flour, butter, and sugar the coating of fat over the starch granules will limit the access of the small amount of water present, and we find that, after baking, the granule structure is still visible. A *roux* is another example of fat being used to control the gelatinisation behaviour of starch.

After one or two days cakes with a low fat content and bread start to go stale. Staling is often thought of as a simple drying-out process, but of course it occurs just as readily in a closed container. What we are actually seeing is the retrogradation of the starch. The current view of bread staling is that it is a two-stage process. The first stage, marked by the transition from the soft 'new-bread' state takes place only a few hours after baking and is believed to involve the retrogradation of the unbranched amylose fragments. The second stage, staling proper, involves retrogradation of a proportion of the amylopectin fragments. The crystallinity of the retro-

Having gone through this initial 'liquefaction' phase the oligosaccharides are subjected to further enzymic hydrolysis in the 'saccharification' phase of the process. The temperature is brought down to around 55–60 °C and preparations of the amyloglucosidase (otherwise known as glucoamylase) from the fungus *Aspergillus niger* are are added. Amyloglucosidase is remarkable for its ability to catalyse the hydrolysis of both the $\alpha1\rightarrow4$ and $\alpha1\rightarrow6$ glycosidic links and given time would break down the oligosaccharides almost entirely to glucose. However the amyloglucosidase is often supplemented at this stage with the enzyme pullulanase, from *Bacillus acidopullulyticus*, which is specific for the $\alpha1\rightarrow6$ links. Although the saccharification stage may take as long as 90 hours the results are spectacular. After suitable clean-up procedures yields of pure glucose of 95% of the theoretical result are obtained.

This need not be the end of the story. The high purity glucose syrup may be passed through a reactor in which the enzyme glucose isomerase has been immobilised on solid carrier particles. This enzyme is obtained from a number of different fungi where its primary role is the interconversion of xylose and xylulose, *i.e.* interconversion of an aldose sugar and its ketose equivalent. This reaction is completely analogous to the interconversion of glucose and fructose and enables syrups with fructose:glucose ratios of 55:45 to be produced. Known as 'high fructose corn syrups' these have a sweetness comparable with invert sugar and are becoming increasingly popular, especially with soft drink manufacturers.

graded starch gives the crumb of stale bread its extra 'whiteness', and the increased rigidity of the gelatinised starch causes the lack of 'spring' in the crumb texture. Retrograded starch can be returned to solution only by heating, *i.e.* regelatinisation; stale bread can be 'revived' by moistening and then returning it to a hot oven for a few minutes.

The phenomenon of retrogradation is a reminder that not all the starch in a foodstuff is necessarily available to us, *i.e.* readily digested and absorbed from the small intestine. In partially milled cereal products such as muesli that are eaten uncooked some of the starch remains undigested simply because the amylases cannot gain access to it. The starch in raw potatoes and unripe bananas is in a crystalline form that is highly resistant to intestinal hydrolysis. Another type of what is known as 'resistant starch' is formed when starchy materials are heated in the absence of abundant water, for example during the manufacture of breakfast cereals. It also appears during the canning of beans or other legumes. There may be plenty of water in the can but little penetrates into the beans. These prolonged high-temperature treatments damage the granule structure so that it becomes immune to the effects of digestive enzymes and very reluctant to gelatinise. It must therefore be classed as an insoluble fibre component even though it is not easy for analytical procedures to distinguish it from available starch.

Over 10% of the total starch in canned beans, and 2% of that in bread, comes into the resistant category, a factor which may need to be borne in mind in calculations of their nutritional value.

PECTINS

Pectins are the major constituents of the middle lamella of plant tissues and also occur in the primary cell wall. They make up a substantial proportion of the structural material of soft tissues such as the parenchyma of soft fruit and fleshy roots. As the structure of pectins can only explored after they have been separated from the rest of the plant tissue we cannot be certain as to the exact nature of pectins as they occur in plant tissues and the term protopectin is applied to the water-insoluble material in the plant that yields water-soluble pectin on treatment with weak acid. Most elementary accounts of pectins describe them as linear polymers of α-D-galacturonic acid linked through the 1- and 4-positions, with a proportion of the carboxyl groups esterified with methanol as shown in Figure 3.5. Their molecular weights can best be described as uncertain but high, probably around 100 000. The degree of esterification in unmodified pectins, *i.e.* those that have not been deliberately de-esterified during extraction or processing, can vary from around 60%, *e.g.* those from apple pulp and citrus peel, to around 10%, *e.g.* that from strawberries.

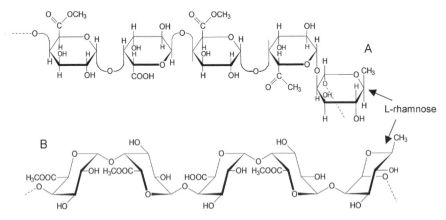

Figure 3.5 *The polygalacturonan backbone of pectin. (A) shows four galacturonic acid residues, three of which are methylated, and a rhamnose residue. (B) shows the same structure in their chair configuration to illustrate the zig-zag shape of the chain. Ring hydrogen atoms have been omitted for clarity. The effect of the rhamnose residue on the configuration of the chain is clear.*

Although this simple structure was always recognised by carbohydrate chemists to be an over-simplification, it is only in recent years that food scientists have begun to appreciate that the behaviour of pectins in food systems such as jam and confectionery could be properly understood only if minor components and structural features were also taken into account. As many as 20% (but usually rather less) of the sugars in a pectin may be found to be neutral sugars such as L-rhamnose, D-glucose, D-galactose, L-arabinose, and D-xylose. L-Rhamnose residues are found in all pectins, bound by a 1,2-link into the main chains. The occurrence of the other neutral sugars varies from one plant species to another, but studies of the pectin of apples (an important source of commercial pectin) are likely to provide a useful model for other pectins. It appears that the apple pectin molecule is an extended rhamnogalacturonan* chain that contains regions that are essentially free of L-rhamnose residues. For reasons that will become clear these regions are referred to as 'smooth'. The smooth regions are interspersed with 'hairy' ones characterised by the kinks introduced into the otherwise neat zigzag by the insertion of the L-rhamnose residues (Figures 3.5 and 3.6). The hairy regions get their name from the presence of numerous short branches carried on the galacturonate residues. One type of branch, thought also to occur in citrus peel pectins, consists of a short, possibly branched, chain of D-galactose residues carrying a highly branched cluster of some twenty or so L-arabinose

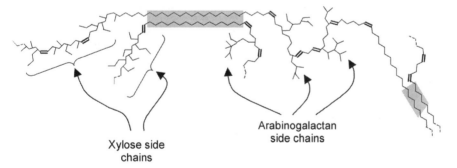

Xylose side chains

Arabinogalactan side chains

Figure 3.6 *A suggested arrangement for the rhamnogalacturonan backbone and neutral sugar side chains based on Pilnik's studies of apple pectin. Each line in the zig-zag of the backbone represents a galacturonate unit, with or without an esterified methyl group, except where a double line indicates the presence of a rhamnose residue. Junction zones formed from paired 'smooth' regions of the backbone are shaded.*

* Terms such as galacturonan, glucan and arabinogalactan describe polysaccharides consisting of, respectively, galacturonic acid, glucose and both arabinose and galactose. The 'an' suffix is derived from 'anhydro', indicating that the links between the monosaccharide units are formed by the departure of the elements of water.

residues. The other type of branch found in apple pectin consists of very short chains (one to three units only) of D-xylose residues. Figure 3.6 is an illustration of this still tentative view of apple pectin structure.

The importance of pectin in food is its ability to form the gels that are the basis of jam and other fruit preserves. Gels consist largely of water and yet are stable and retain the shape given to them by a mould. We have already seen that sugar molecules, and by inference polysaccharides also, are effective at binding water. However, a stable gel also requires a three-dimensional network of polymer chains with water, together with solutes and suspended solid particles, held in its interstices. It has often been assumed that simple interactions, such as single hydrogen bonds, between neighbouring chains were all that was required to maintain the stability of the network. The fact that the lifetime of an isolated hydrogen bond in an aqueous environment is only a tiny fraction of a second was quite overlooked. It is obvious that attractive forces of this character provide an adequate explanation for the viscosity of the solutions of many polymers, such as the polysaccharide gums, but do not explain gel structure.

The concept of the junction zone was first introduced by Rees to explain the structure of the gels formed by carageenans, a group of polysaccharides that will be considered later in this chapter, but the concept can be applied to most gel-forming polymers, including the pectins. Although single hydrogen bonds are not sufficiently durable to provide a stable network, a multiplicity of hydrogen bonds or other weak forces, acting co-operatively, are the basis of the stability of crystal structures. Rees has shown that polymer chains in a gel interact in regions of ordered structure, involving considerable lengths of the chains. These regions, the junction zones, are essentially crystalline in nature, and details of their structure have been revealed by X-ray crystallography.

The detailed structure of the junction zone is dependent on the nature of the polymer involved. In agar and carageenan gels the interacting chains form double helices which further interact to form 'superjunctions'. The pectins have been less studied, but it is clear that the two interacting chains pack closely together as suggested in Figure 3.6. It is now possible to understand the importance to gel structure of the so-called *anomalous* residues, such as L-rhamnose, and the 'hairy' region of the chain. Without them there would be nothing to prevent interaction taking place along the entire length of the chains, resulting in precipitation. This is precisely the phenomenon that is involved when the unbranched, uninterrupted chains of amylose undergo retrogradation. The generalised two-dimensional picture of a polysaccharide gel that emerges from these observations is presented in Figure 3.7.

It may have been inferred from the discussion so far that pectins form

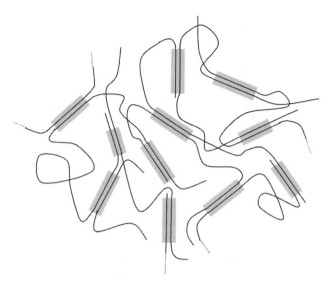

Figure 3.7 *A generalised, two-dimensional view of a polysaccharide gel. Regions of the polymer chains involved in junction zones are shaded. Between them the polymer chains form regions of irregular orientation, so-called random coils. Water, dissolved substances and suspended particles occupy the spaces between the chains. Each end of the junction is terminated by an anomalous residue or the presence of a bulky side chain on at least one of the chains involved. The junction zones of some polysaccharides may involve many chains stacked together, rather than the pairs shown here.*

gels quite readily, but anyone who has been involved in jam-making at home will know how untrue this is. The presence of pectin, solubilised by boiling the fruit, is itself not enough. For successful jam-making we must meet other conditions as well. The use of a highly esterified pectin (at least 70% of the carboxyl groups methylated) is advisable, with a concentration of at least 1% w/v. The acid present, from the fruit or added, must maintain the pH below 3.5. The total sugar content must be at least 50% w/v. While the relationships between these requirements and the procedures of jam-making can be readily discerned, the reasons for them are less obvious. In fact pectin is a reluctant participant in gel formation. The low pH and high degree of esterification ensure that very few galacturonate residues have carboxyl groups in the ionised (*i.e.* $-COO^-$) form:

$$-COO^- + H^+ \leftrightarrow -COOH$$

so that the affinity of the polymer chains for water is minimised. The high sugar content effectively competes for water with the polymer chains, with the result that the amount of bound water which would otherwise disrupt the fragile affinities of the junction zone is reduced and the gel can form.

During the boiling phase of jam making a considerable proportion of the added sucrose is hydrolysed to its parent monosaccharides, glucose and fructose (*see* Chapter 2). This increases the total number of bound water molecules and thereby enhances the gel-forming process.

The breakdown of pectins is an essential feature of the fruit ripening and the enzymes involved occur widely in plant tissues, as well as being secreted by the fungi and bacteria the specialise in attacking and breaking down plant tissues. There are three important classes of enzymes (pectinases) that catalyse pectin breakdown:

(a) Pectin methyl esterases that remove the methanol residues from methylated galacturonate residues.

(b) Polygalacturonases that hydrolyse the glycosidic links between unmethylated galacturonate residues. There are two types of these enzymes. *exo*-Polygalacturonases release small fragments from the non-reducing end of the chain whereas *endo*-polygalacturonases hydrolyse linkages within the chain and generate short fragments. Polygalacturonases only become active after the action of pectin methyl esterases since they cannot work when the degree of methylation of the pectin molecule as a whole is higher than around 60–70%.

(c) Pectin lyases break the glycosidic links between methylated galacturonate residues by an unusual β-elimination reaction where a *trans* double bond is inserted between C-4 and C-5 rather than the usual hydrolysis:

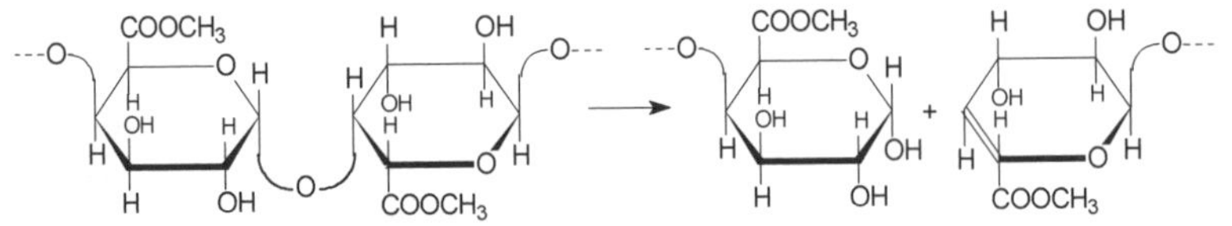

Figure 3.8 *The β-elimination reaction catalysed by pectin lyase.*

As ripening takes place these enzymes will break down the pectin surrounding the cells of the fruit tissue causing it to soften and become more palatable. Of course there are many other changes taking place at the same time including pigment formation and the formation of sugars, pigments and flavour compounds.

Pectinase preparations containing these same enzymes, and others that attack the pectin side chains, are produced by the bacteria and fungi we associate with rotting fruit and vegetables. They are extensively used in wine and cider making and the fruit and vegetable juice industry to increase the degree of tissue damage, and therefore juice yield, when the juice is extracted by pressing. They are also widely used to clarify fruit juices, cider and wines since the large polygalacturonate molecules are very effective at binding protein particles and other cell debris, thereby producing an unwanted haze. Of course consumers do not expect all fruit juices to be clear, particularly tomato. Before tomatoes are fully ripe most of their pectinases are within the cell membrane, kept away from the pectins of the middle lamella that holds the cells together. When the fruit is crushed the cell membranes are disrupted and the pectinases are able to reach their substrate. In the case of the tomato the ensuing breakdown is extremely rapid. Tomato juice is therefore made in a 'hot break' process where the tomatoes are rapidly heated either before or even during the crushing in order to inactivate the enzymes.

Pectinases can be used for stripping off the albedo, or pith, of citrus segments before canning but a curious variation of this process is to introduce the enzyme under the peel. After an hour or so of enzymic breakdown the residues of the peel can be 'washed' off. Further enzyme treatment then separates the fruit into segments which, after another good wash, are ready for canning. With other fruit such as peaches or apricots pectinase treatment is beginning to replace the traditional 'lye peeling' in which the peel is dissolved away by immersion in very strong sodium hydroxide solution.

SEAWEED POLYSACCHARIDES

As we have just seen, the pectins have a particular place in the structure of higher-plant tissues. In the seaweeds this place is taken by either the alginates (in the brown algae, the Pheophyceae) or the agars and carageenans (in the red algae, the Rhodophyceae).

The alginates are linear polymers of two different monosaccharide units: β-D-mannuronic acid (M) (3.2) and α-L-guluronic acid (G) (3.3). These two will be seen to be C-5 epimers of each other. Both occur together in the same chains but linked in three different types of sequence:

-M-M-M-M-M-   -G-G-G-G-G-   -M-G-M-G-M-G-

in each case by 1,4-links. It will be noticed that the pyranose ring conformations that these two sugar acids take up places the bulky carboxyl group in the equatorial position.

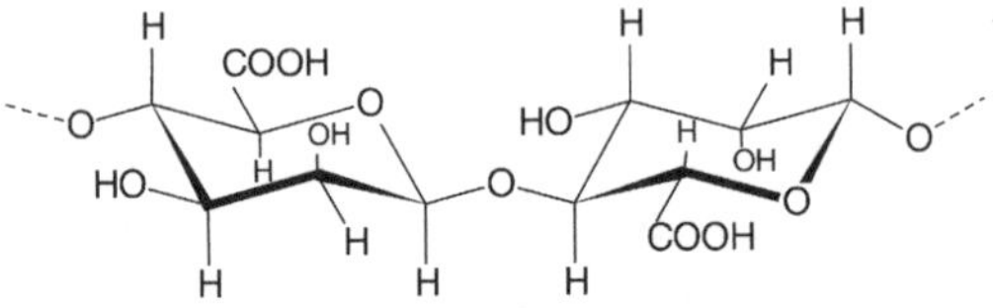

(3.2)

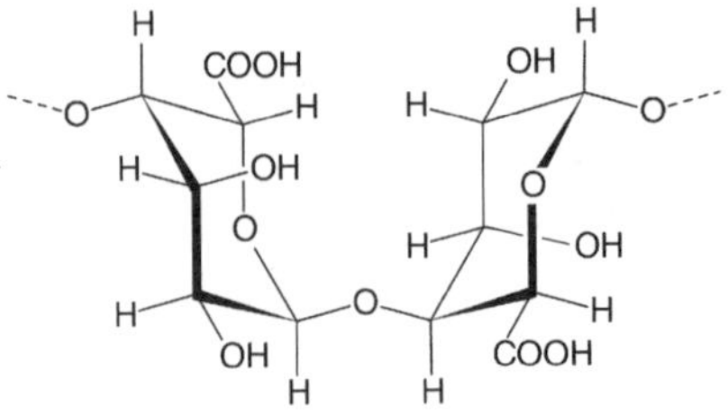

(3.3)

Alginic acid itself is insoluble, but the alkali-metal salts are freely soluble in water. Gels readily form in the presence of Ca^{2+} ions. Rees' group has shown that the junction zones only involve -G-G-G-G- sequences in what is described as an 'egg box' structure with a Ca^{2+} ion complexed by four L-guluronate residues. As might be expected, alginate preparations from different seaweed species vary in their M/G ratios and have corresponding variations in the characteristics of their gels.

Calcium alginate gels do not melt below the boiling point of water, and they have therefore found a number of applications in food products. If a fruit purée is mixed with sodium alginate and is then treated with a calcium-containing solution, various forms of reconstituted fruit can be obtained. For example, if a cherry/alginate purée is added as large droplets to calcium solution, the convincing synthetic cherries popular in the bakery trade can be made. Rapid mixing of a fruit/alginate purée with a calcium solution followed by appropriate moulding will give the apple or apricot 'pieces' that are now sold as pie fillings.

The agars and carageenans are a more complex group than the alginates. Most are linear chains of galactose derivatives with the following generalised structure:

$$\to 3)A\beta(1 \to 4)B\alpha(1 \to 3)A\beta(1 \to 4)B\alpha(1 \to 3)\beta(1 \to 4)B\alpha(1 \to 3)\beta(1 \to$$

where A and B are residues from two different groups as indicated in Table 3.2. Different monosaccharides predominate in particular carageenans but the distinction between then different carageenans in particular seaweeds is not always clear cut. Commercial carageenan preparations sold

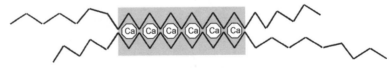

Figure 3.9 *The junction zone of an alginate gel. The junction zone (shaded) consists of two strands of contiguous guluronate residues, terminated by sequences containing mannuronate residues.*

Table 3.2 *The monosaccharides of agar and carageenans. The structures of some of the monosaccharide units listed here are shown in Figure 3.10. There is a steady increase in the sulfate content as one goes down the list from agarose to μ-carageenan, which is matched by a steady decrease in gelling power; λ- and μ-carageenans will not gel at all.*

		Position		
		A	B	
			Major	Minor
Agarose		I	IV	
κ-Carageenan		III	V	VI–VIII
ι-Carageenan		III	VI	VIII
λ-Carageenan		I, II	VIII	
μ-Carageenan		III	VII	V, VIII
I	β-D-Galactose			(3.4)
II	β-D-Galactose 2-sulfate			(3.5)
III	β-D-Galactose 4-sulfate			(3.6)
IV	3,6-Anhydro-α-L-galactose			(3.7)
V	3,6-Anhydro-α-D-galactose			
VI	3,6-Anhydro-α-D-galactose 2-sulfate			(3.8)
VII	α-D-Galactose 6-sulfate			
VIII	α-D-Galactose 2,6-disulfate			(3.9)

for food use as gelling or thickening agents will be blends designed to have properties appropriate for a particular application.

The junction zones that these polysaccharides form are quite unlike the simple 'egg boxes' that we have met so far. Rees' group showed that gelation first involves the formation of double helices of galactan chains wound around each other. The formation of these helices is not sufficient in itself for the gel to form. This occurs only when these rod-like structures associate together to form 'superjunctions', as shown in Figure 3.11. Examination of carageenans by X-ray crystallography has shown that it is the intervention of residues in the B-position that lack the 3,6-anhydro bridge (VII and VIII) that disrupts the initial helix formation. As the sulfate content rises, the resulting negative charge on the helix requires the presence of neutralising cations such as K^+ (for κ-carageenan) or Ca^{2+} (for ι-carageenan) if the helices are to associate into superjunctions. Agar, with its very low sulfate content, gels well regardless of the cations present. The agar superjunctions are sufficiently large to refract light and are the source of slight opalescence shown by even the purest agar gels.

Although agar has been used extensively in microbiology for a very

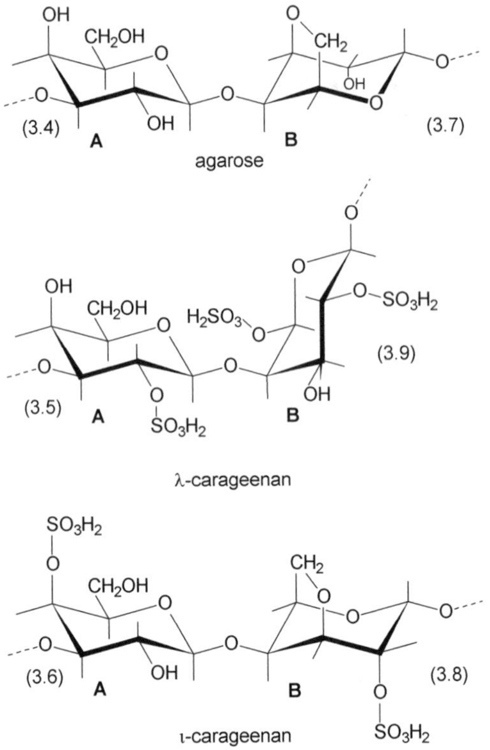

OH
CH₂OH
O
O
CH₂
OH
----O
O
O----
(3.4) A OH B (3.7)

agarose

OH
CH₂OH H₂SO₃
O
O
O-SO₃H₂
----O
O
O
(3.5) A SO₃H₂ B OH (3.9)

λ-carageenan

SO₃H₂
O
CH₂OH
O
CH₂
O
----O
O
O----
(3.6) A OH B O (3.8)

ι-carageenan SO₃H₂

Figure 3.10   *The repeating units of agar and carageenan. The role of the anhydro bridge in maintaining the 'B' residue in the unfavourable (in terms of the axial position of sulfate groups) conformation is clear in these structural formulae.*

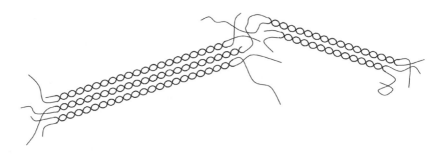

Figure 3.11   *Carageenan or agar 'superjunctions'.*

long time, the development of food applications for this group of polysaccharides has been slow. They are beginning to be used both as gelling agents in milk-containing desserts and as stabilisers in a wide range of products. Concerns over their safety as food additives based on suggestions that carageenans could cause intestinal irritation have not been sustained, owing largely to the apparent freedom from this problem in Japan, where seaweeds are an important part of the diet.

CELLULOSE, HEMICELLULOSES AND FIBRE

Cellulose and hemicelluloses are major components of the fraction of plant foods, particularly cereals, that are commonly known as 'dietary fibre' or 'roughage'. The definition of these terms is rather more problematical than for most food substances as we are dealing with a group that does not have a clear boundary, in terms of either their chemical nature or their contribution to human nutrition. It would be as well to attempt to reach a proper understanding of these terms, and some of the others used in this area, before going on to potential describe members of the group in any detail.

Besides cellulose and the hemicelluloses all of the following substances could be described as fibre: lignin; gums; seaweed polysaccharides; pectins; resistant starch; inulin. What all these have in common is that they are not broken down in the small intestines of mammals by digestive enzymes. However, most are to some degree at least broken down by the bacteria resident in the large intestine of omnivores such as humans. The colon bacteria go on to ferment the resultant sugars to provide the energy for their growth. The various fermentation processes give rise to a number of end products. Besides the hydrogen, carbon dioxide and methane already mentioned on page 26 considerable quantities of short chain fatty acids, (notably ethanoic, propionic and butyric) lactic acid and even ethanol are produced. These by-products are re-absorbed by the lining of the colon and ultimately contribute to the human energy economy – approximately $1.8 \, \text{kcal mol}^{-1}$ of original monosaccharide, about half that of a monosaccharide absorbed directly from the small intestine. The term 'unavailable carbohydrate' that is often applied to these polysaccharides is therefore only true as far as absorption of their component monosaccharides from the small intestine is concerned. Thus stachyose (*see* page 25) could be unavailable carbohydrate but as a sugar it hardly qualifies. Conversely lignin (*see* page 65) should really be included although it is certainly not a carbohydrate.

The term 'fibre', like the old term 'roughage', implies insolubility, but the substances listed above range from the totally insoluble cellulose to the totally soluble gums. Traditionally 'fibre' to the analyst was the insoluble residue that remained after a defatted plant foodstuff had been extracted first with boiling dilute acid and then with boiling dilute alkali, a treatment that solubilises almost all plant substances except the cellulose and lignin.

Nowadays the extractions are carried out using enzymes. Amylases and proteases are used to break down starch and proteins followed by extraction with 76–80% ethanol. This dissolves the amino acids and glucose, derived from the proteins and starch respectively, as well as removing other low molecular weight substances. The resulting insoluble residue is considered to be close in composition to the unabsorbed residue from human digestion in the small intestine.

Food composition tables such as 'McCance and Widdowson' usually quote data obtained by two of the methods. In the slightly earlier one, the Southgate method, the undissolved residue from the enzymic digest is simply dried and weighed. In the more recent method, developed by Englyst and Cummings (usually known as the Englyst method), the residue is subjected to an acid digestion and the resulting monosaccharides identified and quantified (by GLC or HPLC), and related back to their parent polysaccharides. The Englyst method gives a result for what is referred to as the Non-Starch Polysaccharide (NSP) and this method and terminology was officially adopted in Britain for nutritional data. Unfortunately, and with hindsight unsurprisingly, this terminology was never recognised by consumers as indicating what had been understood as dietary fibre. Since 1999 Britain has changed its position and joined the rest of Europe in adopting the Association of Official Analytical Chemists International (AOAC) methodology as the preferred method. This is an updated and refined version of the Southgate method that corresponds to the now internationally accepted definition of dietary fibre developed by the American Association of Cereal Chemists (AACC) in 2000:

> *Dietary fiber is the edible parts of plants or analogous carbohydrates that are resistant to digestion and absorption in the human small intestine with complete or partial fermentation in the large intestine. Dietary fiber includes polysaccharides, oligosaccharides, lignin, and associated plant substances. Dietary fibers promote beneficial physiological effects including laxation, and/or blood cholesterol attenuation, and/or blood glucose attenuation.*

The notable features of this definition are that ignores the issue of laboratory analysis but includes a statement on the physiological activity, in humans, of dietary fibre. Table 3.3 summarizes the relationships between these different approaches. Unless specified otherwise data on food labels currently refer to NSP. Some data on the fibre contents of a variety of foodstuffs is presented in Table 3.4.

The broad distinction between so-called 'soluble fibre' and 'insoluble fibre' goes beyond its considerable nutritional importance. Insoluble fibre is the subject of this section; one form of soluble fibre, the pectins, has

Table 3.3 *The components of the different fibre classes. The descriptions of 'Soluble fibre' and 'Insoluble fibre' refer to earlier usage of these terms rather than suggest what or is not soluble. The soluble non-cellulosic polysaccharides include the pectins, gums and seaweed polysaccharides.*

	Lignin	Cellulose (insoluble)	Hemi-celluloses	Soluble non-cellulosic polysaccharides	Resistant starch
Southgate	+	+	+	+	+
Englyst (NSP)		+	+	+	
Insoluble fibre		+	+		
Soluble fibre				+	
AACC	+	+	+	+	+

Table 3.4 *The fibre components of different foods.*[1]

	Grams per 100 grams				
	Cellulose	Other insoluble fibre[2]	Soluble fibre[3]	Lignin	Total
Wheat flour					
wholemeal	1.6	6.5	2.8	2.0	12.9
white	0.1	1.7	1.8	trace	3.6
Oatmeal	0.7	2.3	4.8	~2.0	9.8
Broccoli	1.0	0.5	1.5	trace	3.0
Cabbage	1.1	0.5	1.5	0.3	3.4
Peas	1.7	0.4	0.8	0.2	3.1
Potato	0.4	0.1	0.7	trace	1.2
Apple	0.7	0.4	0.5	–	1.6
Gooseberry	0.7	0.8	1.4	–	2.9

[1] Data quoted by D. A. T. Southgate in 'Determination of Food Carbohydrates', 2nd Edn., Elsevier Applied Science, London, 1991; *see* Further Reading. [2] Hemicellulose. [3] Including pectin.

already been considered but another, the gums, and the dietary role of soluble fibre generally, will be given further attention in the next section.

The insoluble-fibre polysaccharides are divided into totally insoluble cellulose and the sparingly soluble hemicelluloses. Cellulose, said to be the most abundant organic chemical on earth, is an essential component of all plant cell walls. It consists of linear molecules of at least 3000 $\beta1{\rightarrow}4$ linked glucopyranose units. This linkage leads to a flat-ribbon arrangement maintained by intramolecular hydrogen bonding, as shown in Figure 3.12.

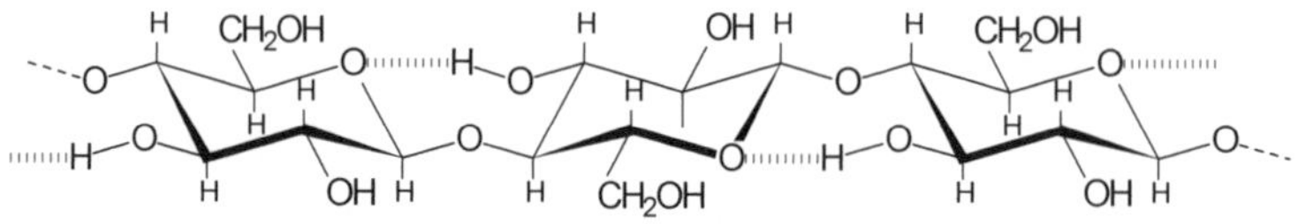

Figure 3.12   *The flat-ribbon arrangement of the cellulose chain. Hydrogen bonds are shown between the ring oxygen and the C4 hydroxyl group of the neighbouring glucose residue.*

Within plant tissues cellulose molecules are aligned together to form microfibrils several micrometres in length and about 10–20 nm in diameter. Although doubt still exists as to the exact arrangement of the molecules within the microfibrils, it is clear that the chains pack together in a highly ordered manner maintained by intermolecular hydrogen bonding. It is the stability of this ordered structure that gives cellulose both its insolubility in virtually all reagents and also the great strength of the microfibrils. Herbivorous animals, particularly the ruminants, are able to utilise cellulose by means of the specialised microorganisms that occupy sections of their alimentary tracts. These secrete cellulolytic enzymes that release glucose; they then obtain energy by fermenting the glucose to short-chain fatty acids, such as butyric acid, which can be absorbed and utilised by the animal. Even in these specialised animals cellulose digestion is a slow process, as evidenced by the disproportionately large abdomens of herbivores compared with carnivores.

The hemicelluloses were originally assumed to be low molecular-weight (and therefore more soluble) precursors of cellulose. Although this idea has long been discarded, we are left with a most unfortunate name to describe this group of structural polysaccharides. They are closely associated with the cellulose of plant cell walls, from which they can be extracted with alkaline solutions (*e.g.* 15% KOH). Three types of hemicellulose are recognised: the xylans, the mannans and glucomannans, and the galactans and arabinogalactans. Although the leaves and stems of vegetables provide a high proportion of the indigestible polysaccharide of our diet, very little is known about the hemicelluloses of such tissues. We know much more about the hemicelluloses of cereal grains, where the xylans are the dominant group. The essential feature of these xylans is a linear or occasionally branched backbone of $\beta1\rightarrow4$ linked xylopyranose residues with single L-arabinofuranose and D-glucopyranosyluronic acid residues attached to the chain, as indicated in Figure 3.13. There is every reason to suppose that the conformation of the xylan backbone resembles the flat ribbon of cellulose but of course the side chains will prevent the tight packing and thereby allow some solubility.

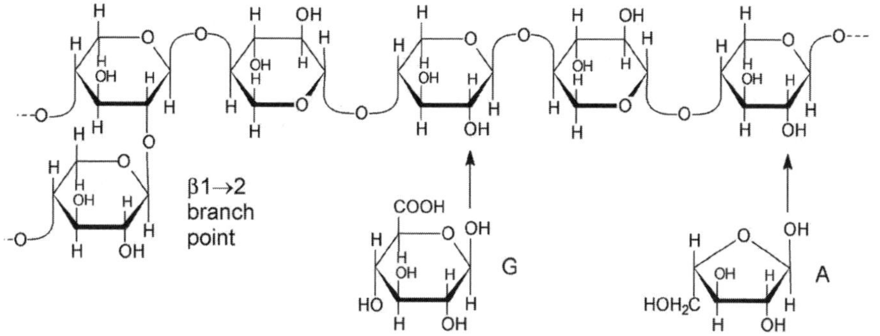

Figure 3.13 *Typical structural features of a wheat xylan. The more soluble endosperm xylans are richer in L-arabinose whereas bran xylans less soluble, more highly branched, and richer in glucuronic acid. The arrows show the points of attachment of any L-arabinose (A) or glucuronic acid (G) residues.*

Xylans are major constituents of the seed coats of cereal grains, which on milling constitute the bran. A typical wheat bran composition (on a dry basis) is: lignin* 8%, cellulose 30%, hemicellulose 25%, starch 10%, sugars 5%, protein 15%, lipid 5%, and inorganic and other substances making up the remainder. Much less hemicellulose (about 5%) is found in the endosperm, which is the basis of white flour. Nevertheless various effects of the hemicelluloses have been noted on the milling and baking characteristics of particular flours.

The nutritional benefits of high-fibre diets are highly publicised in relation to many of the diseases that characterise modern urban man and his highly refined food. There are two quite distinct areas of human health where fibre is believed to exert a beneficial influence. The benefits of a high-fibre diet to the functions of the large intestine have been recognised for a long time. Significantly lower incidences of a number of bowel disorders from constipation and diverticulosis to cancer are associated with high-fibre diets. These benefits appear to be derived from the greater bulk that fibre brings to the bowel contents. Besides improving the musculature of the bowel wall, this also reduces the time that potential carcinogens derived from our food spend in the bowel. Recent research suggests that a quite different mechanism may also be important. Some fibre components are broken down and fermented by the resident bacteria of the bowel to give rise to high concentrations of the short-chain fatty acids acetic,

---

* Lignin is the characteristic polymer of woody tissues. The detailed structure is not well understood but it is based on the condensation of aromatic aldehydes such as vanillin, syringaldehyde and *p*-hydroxybenzaldehyde.

propionic, and butyric. These have been shown to induce the death of bowel tumour cells in a process known as apoptosis.

It is becoming clear that the relationship between dietary fibre intake and the incidence of arterial disease is a complex one and that not all forms of fibre are equally effective in lowering the levels of the different types of blood lipids. However, what does seem to be agreed is that it is soluble rather than insoluble fibre that is more effective, and this aspect will therefore be considered in the next section rather than here.

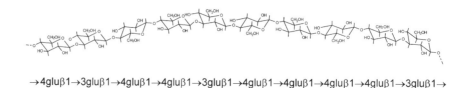

→4gluβ1→3gluβ1→4gluβ1→4gluβ1→3gluβ1→4gluβ1→4gluβ1→4gluβ1→4gluβ1→3gluβ1→

Figure 3.14   *Oat β-glucan.*

On the boundary between hemicelluloses and gums are the β-glucans, that in oats being the best known. Oat β-glucan consists of long chains of glucose units (at least 5000) joined by both β1→3 and β1→4 glycosidic links. As Figure 3.14 shows the chain is dominated by cellulose-like β1→4 links but interrupted every third or fourth glucose unit by a β1→3 linkage. This has the effect of disrupting the ordered structure associated with cellulose and imparting solubility in water. Solutions of oat β-glucan extremely viscous, as are rolled oats cooked to give porridge. There are strong indications that oat β-glucans are beneficial in relation to heart disease by lowering the levels of cholesterol in the blood. The most likely mechanism is that its viscous character impedes the reabsorption of cholesterol and bile acids from the small intestine thereby placing a drain on the body's cholesterol economy. Sadly, as in so many cases, the amount of porridge oats one would need to consume on a regular basis to have a useful effect (several bowls-full daily) is somewhat prohibitive.

One should be cautious of automatically regarding a high dietary fibre level as beneficial. Phytic acid, inositol hexaphosphate (3.10) is an important component of bran, a valued source of dietary fibre in many diets. Phytic acid is the phosphate reserve for germinating seeds and its phosphate groups make it a potent complexer of divalent cations, implicating it in calcium and zinc deficiency diseases amongst children who consume very little milk and large quantities of baked goods made from wholemeal flour.

$$\text{(3.10)}$$

Inulin is another polysaccharide that could be described as a gum. It is the most important member of a group of plant polysaccharides known as fructans, *i.e.* polymers of fructose. Although fructans occur widely in small amounts in leaves, inulin, whose structure is shown in Figure 3.15, is notable for its starch like role, in terms of plant physiology, in the tuberous roots of a number of food plants, especially chicory (*Cichorium intybus*) and Jerusalem artichokes (*Helianthus tuberosus*). These both contain around 18 g per 100 g fresh weight. Many other food plants, including asparagus, onions, garlic, bananas and cereal flours such as wheat contain just a few percent. The rather different levels of wheat consumption, compared with chicory and Jerusalem artichokes, means that wheat is actually the dominant source of dietary inulin in most diets.

Inulin as not a large polymer like starch or cellulose, only a small proportion of the polymer chains have more than 50 fructose units. A significant proportion are sufficiently short to qualify as oligosaccharides rather than polysaccharides.

Until recently inulin was little more than a phytochemical curiosity but its consumption has now been shown to have a range of effects similar to other soluble fibre components. It resists digestion in the human stomach and small intestine but is readily hydrolysed and fermented by the bacteria

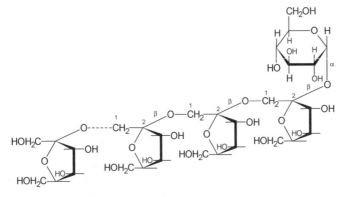

Figure 3.15 *The β-(2→1) linked fructofuranose chain of inulin. A sucrose residue is usually found providing a non-reducing terminus for what would otherwise be the reducing end of the chain.*

of the colon. The fermentation generates considerable amounts of short chain fatty acids such as acetic, propionic and butyric acids and promotes the growth in the colon the bifidobacteria, important members of the colon flora. These bacteria are believed to impart considerable health benefits to their hosts but whether these are sufficient to justify the manipulation of one's diet that consuming as much as 40 g of inulin (extracted from sources such as chicory) daily would require remains to be seen.

GUMS

There are many substances in cereal products and vegetables that could properly be described as gums. However, the term is usually reserved for a group of fairly well defined substances that are obtained as plant exudates or as 'flours' made from certain non-cereals seeds. Two other gums that are secreted by bacteria, xanthan gum and gellan, have also become important food ingredients in recent years.

The distinctive feature of the gums is their great affinity for water and the high viscosity of their aqueous solutions. However, they will not form gels and their solutions always retain their plasticity even at quite high concentrations. The reason for this becomes clear when the molecular structure of some representative gums is examined. The key structural features of gum tragacanth (an exudate from the tree *Astralagus gummifer*) and guar gum (the storage polysaccharide in the endosperm of the seeds of a leguminous shrub, *Cyamopsis tetragonoloba*) are illustrated in Figure 3.16. The essential structural feature of all gums, shown clearly by both these examples, is extensive branching. This leaves no lengths of ordered backbone available to form the junction zones characteristic of polysaccharide gels. The branches are nevertheless capable of trapping large amounts of water, and interactions between them ensure that even quite dilute solutions will be viscous. If one examines the solution properties of polysaccharides in the sequence cellulose, amylose, agar, pectin, amylopectin, and the gums, there is a clear parallel with the extent of branching and/or the diversity of monosaccharide components.

In the past the usual application for gums in food products was as thickening agents as a substitute for starch. Nowadays they are often used at quite low concentrations to assist in the stabilisation of emulsions (*see* Chapter 4) and to 'smoothen' the textures of products such as ice-cream. However, in recent years the realisation that it is the soluble components that are responsible for the beneficial effects of a high-fibre diet on the incidence of arterial disease has focused attention on a nutritional role for gums, as was discussed on page 66 in the context of oat glucan.

Another valuable effect, particularly associated with guar gum but

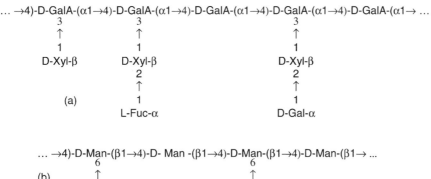

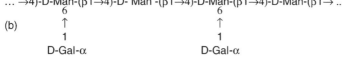

Figure 3.16 (a) *Gum tragacanth. Shown here is the structure of its principal component, tragacanthic acid. The backbone of* D-*galacturonic acid residues has occasional insertions* L-*rhamnose. Almost all the backbone residues carry one or other of the* D-*xylose based side chains shown. (*L-*Fuc, i.e.* L-*fucose, is 6-deoxy-*L-*galactose.) (b) Guar gum. Alternate backbone residues carry single* D-*galactose residues. Another galactomannan, locust bean gum (from Caratonia siligua), differs from guar gum only having a lower frequency of* D-*galactose side chains.*

clearly involving other soluble fibre substances, concerns the behaviour of starch and sugars in the intestine. The presence of guar gum in the intestine appears to retard the digestion and absorption of carbohydrates and thereby flatten the peak of glucose flow into the bloodstream that normally follows a carbohydrate-rich meal. The effect appears to be purely physical in that sugars and starch fragments get caught up in the three dimensional network of fibre molecules, making them less accessible to digestive enzymes and slowing down their diffusion towards the absorptive surface of the small intestine. The application of this effect is believed to have considerable potential in the management of diabetes, particularly the non-insulin dependent type.

Xanthan gum is produced by the bacterium *Xanthomonas campestris*. The gum encapsulates the cell and helps it to stick to the leaves of its favoured hosts, cabbages, as well as helping to protecting it from it from dehydration and other environmental stresses. Commercially the gum is prepared by growing the microorganism in large-scale fermentation processes. Steps are taken to ensure that final product contains no viable cells of the bacterium. The enormous molecules of the polymer contain between 10000 and 250000 sugar units. As shown in Figure 3.17 the molecule has a backbone of glucose units linked by β-1,4 linkages, as in cellulose. However, every fifth glucose unit carries a trisaccharide side

(a)

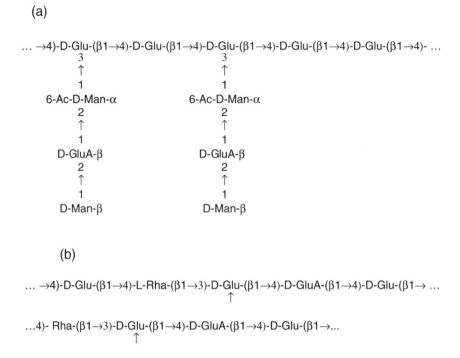

... →4)-D-Glu-(β1→4)-D-Glu-(β1→4)-D-Glu-(β1→4)-D-Glu-(β1→4)-D-Glu-(β1→4)- ...

(b)

... →4)-D-Glu-(β1→4)-L-Rha-(β1→3)-D-Glu-(β1→4)-D-GluA-(β1→4)-D-Glu-(β1→ ...

...4)- Rha-(β1→3)-D-Glu-(β1→4)-D-GluA-(β1→4)-D-Glu-(β1→...

Figure 3.17   (a) *Xanthan gum. The trisaccharide side chains are carried on approximately every other glucose unit of the backbone. The mannose residues next to the backbone have acetyl groups (CH$_3$CO-) attached to the C-6 hydroxyl. Approximately half the terminal mannose residues have a pyruvate residue bridged across their C-4 and C-6 hydroxyl groups. (b) Gellan gum. In its native form the glucose residues arrowed here carry an acetyl group on the C-6 hydroxyl and a glyceryl group (CH$_2$OH.CHOH.CO-) on the C-2 hydroxyl. If they are left in place gellan forms very elastic gels. These substituents are normally removed during the manufacture of gellan.*

chain which includes one, and sometimes two, carboxyl groups. These charged groups help to ensure xanthan gum's high affinity for water.

The special properties of xanthan gum solutions result from the ease with which the polymer chains associate together, and the ease with which these associations are reversibly broken. When a xanthan gum solution (typically in food products at around 0.1–0.3%) is both cool and stationary, *i.e.* not flowing or being stirred or shaken, the polymer chains take on a rigid zig-zag shape (see Figure 3.12) of sufficient length for them to come together in junction zones (*see* page 54) and form a gel-like network. What makes these junction zones special is that the slightest hint of physical stress is sufficient to disrupt the junctions. This stress, or shear, can result from any flow imposed on the gel. This means that xanthan gum solutions are thixotropic, a property they share with non-drip paint. Another term often applied to these solutions is 'shear thinning'.

The viscosity of a xanthan gum solution in a bottle at rest can be sufficiently high to ensure that even quite large suspended particles do not settle out. However, as soon the bottle is inverted the solution instantly liquefies so that it can flow readily out of the bottle. The most obvious applications of xanthan gum to food are salad dressings with suspended particles of herbs *etc.*, and ketchups. The concept of a tomato ketchup that flows easily from the bottle but does not dribble off one's chips has a certain appeal. Xanthan gum is also very effective at stabilising emulsions.

Gellan gum comes from the bacterium *Pseudomonas elodea*, manufactured by similar processes to those for xanthan gum. Although it has now been in use around the world for some years it was only cleared for use in Europe in 1994. Gellan gum has an unbranched polymer chain consisting of a repeating tetrasaccharide unit (Glucose-Rhamnose-Glucose-Glucuronic acid), with a total length of around half a million monosaccharide units (Figure 3.17). Individual polymer chains form double helices around each other. Gels are formed when hot gellan solutions cool in the presence of cations. Divalent cations, such as Ca^{2+}, associate with the carboxyl groups of the glucuronic acid residues cross-linking double helical sections together to form the junctions zones of the gel. Monovalent cations such as Na^+ give the same result by simply neutralising acid carboxyl groups and allowing the chains to come together. The levels of sodium and calcium normally found in found systems are usually sufficient to promote gelation.

The striking feature of gellan gum is that it forms gels at 0.1%, a much lower concentration than other gelling agents. The gels it forms are rather brittle so that they crumble in the mouth, giving the impression of melting.

FURTHER READING

'Plant Polymeric Carbohydrates', eds. F. Meuser, D. J. Manners and W. Seibel, The Royal Society of Chemistry, Cambridge, 1997.

'Modified Starches: Properties and Uses', ed. O. B. Wurzburg, CRC Press, Boca Raton, FL, 1986.

'Handbook of Hydrocolloids', eds. G. O. Phillips and P. A. Williams, CRC Press, Boca Raton, FL, 2000.

'Complex Carbohydrates in Foods', eds. S. S. Cho, L. Prosky and M. Dreher, Dekker, New York, 1999, 1986.

'Starch: Properties and Potential', ed. T. Galliard, Wiley, London, 1987.

'Starch Structure and Functionality', eds. P. J. Frazier, P. Richmond and A. M. Donald, The Royal Society of Chemistry, Cambridge, 1997.

D. A. Rees, 'Polysaccharide Shapes', Chapman and Hall, London, 1977.

'Food Gels', ed. P. Harris, Elsevier Applied Science, London, 1990.

British Nutrition Foundation Task Force, 'Complex Carbohydrates in Food', Chapman and Hall, London, 1990.

D. A. T. Southgate, 'Dietary Fibre Analysis', The Royal Society of Chemistry, Cambridge, 1995.

S. Cho, J. W. Devries and L. Prosky, 'Dietary Fibre Analysis and Applications', Aspen, Gaithersburg, Maryland, 1998.

Chapter 4

Lipids

The term 'lipid' refers to a group of substances that is even less clearly defined than the carbohydrates. It generally denotes a heterogeneous group of substances, associated with living systems, which have the common property of insolubility in water but solubility in non-polar solvents such

Table 4.1 *The total fat contents of a variety of foods. These figures are taken from 'McCance and Widdowson'. (see Appendix II) and in all cases refer to the edible portion. They should be regarded as typical values for the particular food rather than absolute values to which all samples of the food comply. The definition of 'fat' that applies here covers all lipids but in fact only in the case of egg yolk, and products containing egg, are polar lipids quantitatively significant.*

Food	Total fat (%)	Food	Total fat (%)
Wholemeal flour	2.2	Low fat spread	40.5
White bread	1.9	Lard	99.0
Rich tea biscuits	16.6	Vegetable oils	99.9
Shortbread	26.1	Bacon, streaky, fried	42.2
Madeira cake	16.9	Pork sausage, grilled	24.6
Flaky pastry	40.6	Beef, sirloin, roast raw	21.1
Skimmed milk (cow's)	0.1	Beefburger,	20.5
Whole milk (cow's)	3.9	Lamb chop, grilled	29.0
Human milk	4.1	Chicken, roast	5.4
Soya milk	1.9	Turkey breast, roast	1.4
Clotted cream	63.5	Cod, raw	0.7
Cheese (Brie)	26.9	Cod, battered & fried	10.3
Cheese (Cheddar)	34.9	Mackerel, smoked	30.9
Dairy ice cream	9.8	Salmon, smoked	4.5
Egg yolk	30.5	Taramasalata	46.4
Egg white	trace	Brazil nuts	68.2
Scotch egg	17.1	Peanuts, dry roasted	49.8
Butter	81.7	Plain chocolate	29.2
Margarine	81.6	Milk chocolate	30.3

as the hydrocarbon paraffins. Included in the group are the oils and fats[*] of our diet together with the so-called phospholipids associated with cell membranes. While food materials like the oils and fats clearly consist of very little other than lipids their presence in many other foods is less obvious, as indicated by the data in Table 4.1. What the lipids in all these foods have in common is that they are esters of long-chain fatty acids, but there are many other lipids that lack this structural feature. They include the steroids and terpenes, but, with the exception of cholesterol, its long chain fatty acid esters, and the phytosteroids (all covered later in this chapter) where these substances do have significance to food they will be found under the more relevant headings of vitamins, pigments, or flavour compounds.

FATTY ACIDS – STRUCTURE AND DISTRIBUTION

Monocarboxylic fatty acids (*i.e.* fatty acids that have only one carboxyl group) are the structural components common to most of the lipids that interest food chemists, and, since many of the properties of food lipids can be accounted for directly in terms of their component fatty acids, they will be considered in some detail. Almost without exception the fatty acids that occur in foodstuffs contain an even number of carbon atoms in an unbranched chain, *e.g.* lauric, or dodecanoic, acid (4.1). (It is worthwhile to master both the trivial names, usually derived from an important source, and the systematic names of the principal fatty acids.)

$$CH_3-CH_2-CH_2-CH_2-CH_2-CH_2-CH_2-CH_2-CH_2-CH_2-CH_2-COOH \quad (4.1)$$

Besides the saturated fatty acids, of which lauric acid is an example, unsaturated fatty acids having one, two, or sometimes up to six double bonds are common. The double bonds are almost invariably *cis* (*see* Figure 4.1), and when the fatty acid has two or more double bonds they are 'methylene interrupted':

$$-CH_2-CH=CH-CH_2-CH=CH-CH_2-$$

rather than conjugated:

$$-CH_2-CH=CH-CH=CH=CH-CH_2-$$

---

[*] There is no formal distinction between oils and fats. The former are liquid and the latter solid at room temperatures.

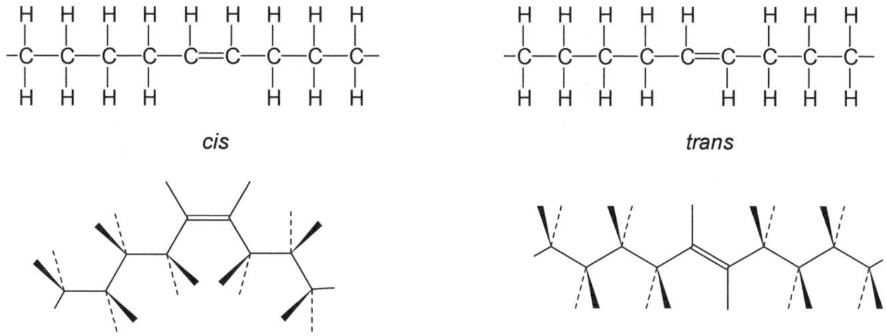

Figure 4.1 Cis *and* trans *double bonds and their effects on the conformation of the fatty acid chain. 'Cis' and 'trans' are Latin for 'on this side of' and 'across' respectively. Unlike single bonds, double bonds cannot rotate, so that the four atoms involved in the double bond, two carbon atoms and two hydrogen atoms, remain fixed in a flat plane.*

Thus α-linolenic acid* (systematically all-*cis*-9,12,15-octadecatrienoic acid) has the structure:

$$CH_3-CH_2-CH=CH-CH_2-CH=CH-CH_2-CH=CH-(CH_2)_7-COOH$$

The system used for the identification of double-bond positions will be apparent by comparison of this structure with its systematic name. The structure of a fatty acid can be indicated by a convenient shorthand form giving the total number of carbon atoms followed by a colon and then the number of double bonds; the positions of the double bonds can be shown after the symbol Δ. Thus for example, linoleic acid would be written simply as '18:3Δ9,12'.

Table 4.2 gives the names and structures of many of the fatty acids commonly encountered in food lipids.

The oils and fats are obviously the lipids that most interest the food chemist. These consist largely of mixtures of triglycerides, *i.e.* esters of the trihydric alcohol glycerol (propane-1,2,3-triol) (4.2), and three fatty-acid residues which may or may not be identical. Triglycerides are frequently referred to as triacylglcerols. 'Simple' triglyceride molecules have three identical fatty-acid residues; 'mixed' triglycerides have more than one species of fatty acid. Thus a naturally occurring fat will be a mixture of quite a large number of mixed and simple triglycerides. It is important to remember that organisms achieve a desirable pattern of physical properties for the lipids of, say, their cell membranes or adipose

---

* A rare isomer with double bonds at the 6-, 9- and 12- positions is known as γ-linolenic acid.

Table 4.2 *The fatty acids commonly found in foodstuffs. The meanings of the term 'cis' and its opposite 'trans' that are used in the names of the unsaturated fatty acids is explained in Figure 4.1.*

Systematic name	Common name	'Structure'
Saturated acids		
n-Butanoic	Butyric	4:0
n-Hexanoic	Caproic	6:0
n-Octanoic	Caprylic	8:0
n-Decanoic	Capric	10:0
n-Dodecanoic	Lauric	12:0
n-Tetradecanoic	Myristic	14:0
n-Hexadecanoic	Palmitic	16:0
n-Octadecanoic	Stearic	18:0
n-Eicosanoic	Arachidic	20:0
n-Docosanoic	Behenic	22:0
Unsaturated acids		
cis-9-Hexadecenoic	Palmitoleic	16:1Δ9
cis-9-Octadecenoic	Oleic	18:1Δ9
cis, *cis*-9,12-Octadecadienoic	Linoleic	18:2Δ9,12
all-*cis*-9,12,15-Octadecatrienoic	α-Linolenic	18:3Δ9,12,15
all-*cis*-6,9,12-Octadecatrienoic	γ-Linolenic	18:3Δ6,9,12
all-*cis*-5,8,11,14-Eicosatetraenoic	Arachidonic	20:4Δ15,8,11,14
all-*cis*-7,10,13,16,19-Docosapentaenoic	Clupanodonic	22:5Δ7,l0,13,16,19

tissue by utilising an appropriate, and possibly unique, mixture of a number of different molecular species, rather than by utilising a single molecular species which alone has the desired properties, as is the usual tactic with proteins or carbohydrates.

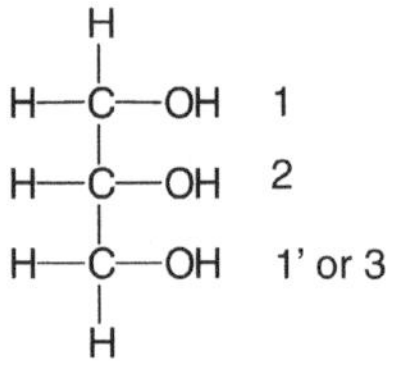

$$
\begin{array}{l}
\text{H} \\
\text{H---C---OH}\quad 1 \\
\text{H---C---OH}\quad 2 \\
\text{H---C---OH}\quad 1' \text{ or } 3 \\
\text{H}
\end{array}
$$

(4.2)

The relative abundance and importance of the various fatty acids will be appreciated from an inspection of Table 4.3, where the molar proportions of the fatty acids in a number of fats and oils are presented. The origins of the names of some fatty acids will also be apparent.

The relationship between the fatty-acid composition of a fat and its

Table 4.3 *The fatty-acid composition of fats and oils. The figures here are typical values expressed in mol%. There is often considerable variation in composition between different samples of the same fat due to, for example, variations in the animal's diet or the conditions for plant growth. Trace components, i.e. those representing less than 0.5%, have been ignored.*

Fatty acid	Beef fat	Lard	Cow's milk fat	Human milk fat	Herring oil	Maize (corn) oil	Cocoa butter[a]	Coconut oil	Palm oil	Olive oil
4:0			9							
6:0			5							
8:0			2					12		
10:0			4	1				8		
12:0			3	3				49		
14:0	4	2	10	5	7			16	1	
14:1	1		2	1						
15:-[b]	2		1	1						
16:0	28	26	23	26	13	14	29	7	48	11
16:1	5	4	2	5	9					1
17:-[b]	1		1							
18:0	20	15	12	7	1	2	35	2	6	3
18:1	34	44	23	37	12	34	32	5	34	79
18:2	3	9	2	11	2	48	3	1	11	5
18:3	2		1	1		1				1
20:0						1	1			
20:1					19					
20:5					8					
22:1					25					
22:5					2					
22:6					2					

[a] The principal component of chocolate. [b] The sums of straight and branched chain, saturated and unsaturated fatty acids.

melting properties will be apparent in Table 4.3. The low melting temperature that characterises the oils is associated with either a high proportion of unsaturated fatty acids, *e.g.* corn oil and olive oil, or a high proportion of short-chain fatty acids, *e.g.* milk fat and coconut oil. The melting points of the fatty acids are shown in Table 4.4.

The insertion of a *cis* double bond has a dramatic effect on the shape of the molecule, introducing a kink of about 42° into the otherwise straight (overall) hydrocarbon chain, as shown in Figure 4.1. The insertion of a *trans* double bond has very little effect on the conformation of the chain and therefore very little effect on the melting temperature. For example, elaidic acid, the *trans* isomer of oleic acid, has similar physical properties to stearic acid. Although some *trans* fatty acids do occur in nature their primary importance is as by-products of the hydrogenation process in margarine manufacture and they are considered in detail on page 86.

The milk fats of ruminants are characterised by their high proportion of short-chain fatty acids. These are derived from the anaerobic fermentation of carbohydrates such as cellulose by the micro-organisms of the rumen. These micro-organisms are also the source of the very small proportions of branched-chain fatty acids that occur in cow's milk fat. Branched-chain fatty acids usually belong to the *iso* series, which have their hydrocarbon chains terminated:

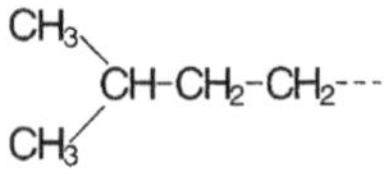

$$CH_3 \diagdown \atop CH_3 \diagup CH-CH_2-CH_2---$$

(4.3)

or to the *anteiso* series:

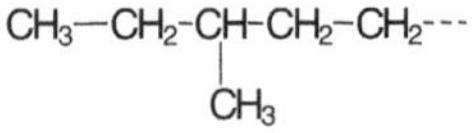

$$CH_3-CH_2-\underset{\underset{CH_3}{|}}{CH}-CH_2-CH_2---$$

(4.4)

Table 4.4 *Fatty acid melting points (M.P.).*

Fatty acid	M.P./°C	Fatty acid	M.P./°C
Butyric	−7.9	Elaidic	43.7
Caproic	−3.4		
Caprylic	16.7	Oleic	10.5
Capric	31.6	Linoleic	−5.0
Lauric	44.2	α-Linolenic	−11.0
Myristic	54.1		
Palmitic	62.7	Arachidonic	−49.5
Stearic	69.6		
Arachidic	75.4		

Human milk fat, like that of other non-ruminant species such as the pig, is rich in linoleic acid. Although no direct benefits of this to human babies have yet been identified, it will be surprising if there are none in view of the importance of linoleic acid in human nutrition (see later in this chapter).

Cod liver oil is most familiar to us as a source of vitamin D, but fish oils are particularly interesting for the diversity of long-chain, highly unsaturated fatty acids they contain. The extremely cold environments of fish such as cod and herring are perhaps one reason for the nature of fish lipids.

There are a number of fatty acids with unusual structures which are characteristic of particular groups of species or plants. For example, petroselenic acid (18:1Δ11*cis*) is found in the seed oils of celery, parsley, and carrots. Ricinoleic acid (4.5) makes up about 90% of the fatty acids of castor oil.

$$CH_3(CH_2)_5CHCH_2CH\!=\!CH(CH_2)_7COOH$$
$$|$$
$$OH$$

(4.5)

Fatty acids containing cyclopropene rings are mostly associated with bacteria, but traces of sterculic acid (4.6) and malvalic acid (4.7) are found in cottonseed oil. They are toxic to non-ruminants, and the residual oil in cottonseed meal is sufficient to have an adverse effect on poultry fed exclusively on the meal. It must be assumed that the low levels man has consumed for many years in salad dressing and margarine manufactured from cottonseed oil have had no deleterious effects.

$$\overset{CH_2}{\overset{/\quad\backslash}{CH_3(CH_2)_7C\!=\!C(CH_2)_7COOH}}\qquad\overset{CH_2}{\overset{/\quad\backslash}{CH_3(CH_2)_7C\!=\!C(CH_2)_6COOH}}$$

(4.6)   (4.7)

Erucic acid, *cis*-13-docosenoic acid (22:1Δ13), is a characteristic component of the oils extracted from the seed of members of the genus *Brassica*, notably varieties of *B. napus* (in Europe) and *B. campestris* (in Canada and Pakistan) which are known as rape. Rapeseed meal is an important animal feeding stuff, and rapeseed oil is important for both human consumption (as a raw material in margarine manufacture) and industrial purposes such as lubrication. Most rapeseed oil used to contain some 20–25% erucic acid, to which was attributed a number of disorders concerning the lipid metabolism of rats fed with high levels of the oil. The problem of possible toxicity to humans has been largely eliminated by plant breeders who have developed strains of rape that give oil with acceptably low erucic acid levels (less than 5%), which is utilised in margarine manufacture. Oils rich in erucic acid remain important for many industrial applications.

Bacteria, especially those that occupy the rumen of cattle *etc.* synthesise a considerable variety of fatty acids that do not fit the patterns associated with most higher animals and plants. Small amounts of some of these fatty acids eventually reach the depot and milk fats of these animals and therefore, ultimately, the human diet. Elaidic acid is one of these that has already been mentioned, vaccenic acid, *cis*-11-octadecenoic acid, is another.

Conjugated Linoleic Acids

The most interesting of all the naturally occurring fatty acids containing *trans* double bonds, largely due to their apparent anti-cancer functionality, are the conjugated linoleic acids (CLAs). In these the two double bonds are not in their usual positions but conjugated (*see* page 74), starting at one of carbons 9, 10 or 11, and most commonly with at least one of the two bonds in the *trans* configuration. The most abundant CLAs are 18:2Δ*cis*9, *trans*11 (4.8) and 18:2Δ*trans*10, *cis*12 (4.9). The CLA content of cow's milk based dairy products is around 5 to 10 mg g^{-1} of total fat. Similar levels are found in the depot fat and milk of other ruminants but much lower levels are found in non-ruminants (monogastrics), including horses, humans and pigs. CLAs arise as by-products of the hydrogenation of polyunsaturated fatty acids by specialist rumen bacteria such as *Butyrovibrio fibrisolvens*. Vegetable oils and seafoods have much lower levels. CLAs, particularly the most abundant *cis*9, *trans*11 type, have been shown to be effective anti-mutagens* and antioxidants (*see* page 98). In experiments with laboratory rodents dietary levels of around 1% (5 to 10 times the level that would be expected in diets containing normal levels of CLA-rich fats) CLAs have proved effective anti-tumour agents. Details of the proposed mechanisms of these effects are outside the scope of this book. Research in this area is some way from clinical studies with humans but this has not prevented CLAs (commercially synthesised by a chemical isomerisation) being taken up enthusiastically as dietary supplements.

---

* In the well-known Ames test for mutagens and potential carcinogens (*see* Chapter 10).

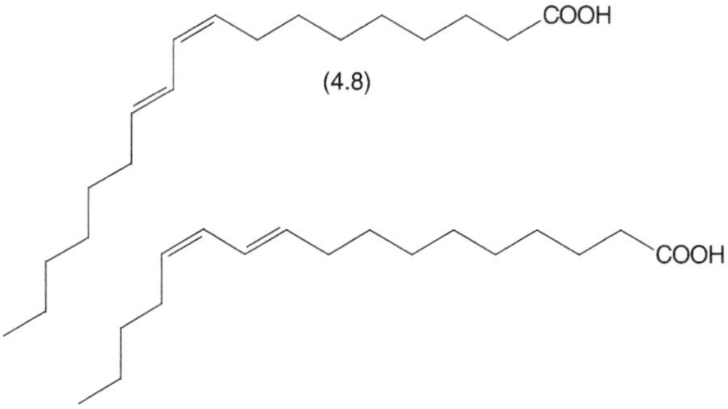

(4.8)

(4.9)

ESSENTIAL FATTY ACIDS

The fatty acids, in the form of the triglycerides of the dietary fats and oils, provide a major proportion of our energy requirements as well as, when in excess, contributing to the unwelcome burden of superfluous adipose tissue that so many of us carry. In recent years we have begun to appreciate that certain dietary fatty acids have a more particular function in human nutrition. Rats fed on a totally fat-free diet show a wide range of acute symptoms affecting the skin, vascular system, reproductive organs, and lipid metabolism. Although no corresponding disease state has ever been recorded in a human patient, similar skin disorders have occurred in children subjected to a fat-free diet. The symptoms in rats could be eliminated by feeding linoleic or arachidonic acids (which in consequence became known for a time as vitamin F), and it is generally accepted that 2–10 g of linoleic acid per day will meet an adult human's requirements.

The identification of these two 'essential fatty acids' in the 1930s preceded by some 25 years their identification as precursors of a group of animal hormones, the prostaglandins*. Although normally regarded as the province of biochemists interested in metabolic pathways, an insight into the various interconversions of the unsaturated fatty acids is useful for food chemists seeking to understand their nutritional status, especially the grouping of unsaturated fatty acids into three series, known as 'ω-3', 'ω-6', and 'ω-9'.

Animals, including humans, are equipped to synthesise saturated fatty acids up to 18 carbon atoms in length. However, desaturation, *i.e.* the insertion of double bonds, can only take place between carbons 9 and 10,

---

* Hormones are defined as substances that affect cellular processes at sites in the body distant from the tissue that produces them. Prostaglandins are not therefore true hormones in that they exert their effects locally and their effects are specific to that tissue, often the mediation or facilitation of the action of other hormones.

as in Figure 4.2, or much more rarely at other positions nearer to the carboxyl group. This means that animals are unable to synthesise linoleic acid but humans are able to convert linoleic acid into arachidonic acid, Figure 4.3. Arachidonic acid is not found in plants. This is of little consequence to humans but of great significance to cats. In the course of evolution they have lost, but not missed, the ability to convert linoleic acid into arachidonic acid. Being carnivores cats have always been able to obtain sufficient arachidonic acid from their diet. This only becomes a serious problem for cats whose owners attempt to make them into vegetarians*. Fortunately most cats should be in a position to supplement such a misguided and unnatural diet with arachidonic acid supplied by small mammals, birds *etc.*

The usual isomer of linolenic acid, sometimes referred to as α-linolenic ($18:3\Delta^{9,12,15}$), is formed in plants by insertion of a third double bond into linoleic acid.

The fatty acids shown in Figure 4.3 are all classified as members of the ω-6 series since their first double bond lies between the sixth and seventh carbon atoms, counting from the terminal methyl group.

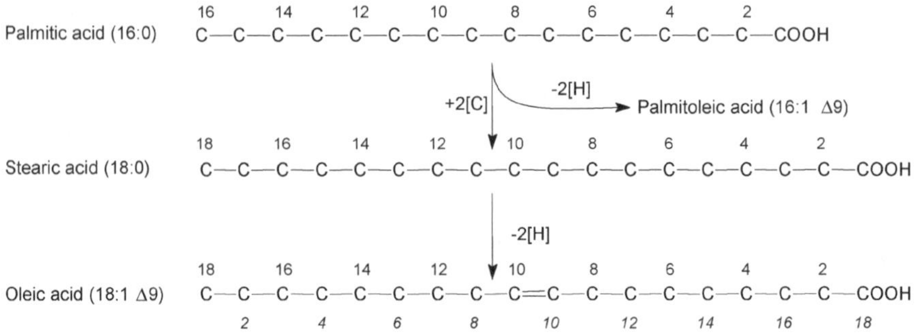

Figure 4.2 *The biosynthesis of oleic acid from palmitic acid. For clarity the hydrogen atoms of the hydrocarbon chain have been omitted. +[2C] Indicates the addition of (CH_2CH_2) at the carboxyl end of the chain, −[2H] an oxidation to form a double bond. A biochemistry textbook should be consulted for details of these reactions, which also feature in Figures 4.3 and 4.4. The numbering of carbon atoms is the usual one, starting from the carboxyl end. However, when the complex structural relationships of unsaturated fatty acids are being considered it is usual to classify them by the distance of the last double bond from the terminal methyl group of the chain, the so-called ω-carbon. Oleic acid, as shown here, has this bond at the ω-9 position making it a member of the 'ω-9 series'. (Some authorities use the letter 'n' instead of 'ω'.)*

* ... and by inference the domestic cat's larger cousins. The author is unaware of any attempts to discover whether lions or tigers could be persuaded to become vegetarians.

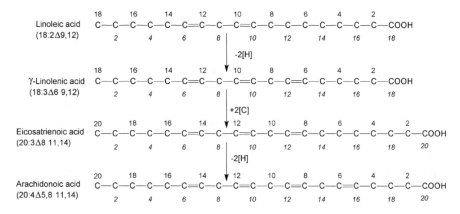

Figure 4.3 *The biosynthesis of arachidonic acid from linoleic acid – the ω-6 series. The two intermediate fatty acids are found in only trace amounts in most species, except that γ-linolenic acid replaces α- in a handful of higher plants, most notably the evening primrose and borage, and the Phycomycete fungi, the class that includes the familiar moulds Mucor and Rhizopus. The theoretical justification for the quasi-pharmaceutical use of evening primrose oil is that our natural synthesis of arachidonic acid may need some encouragement and could benefit from an enhanced supply of its precursor, γ-linolenic acid.*

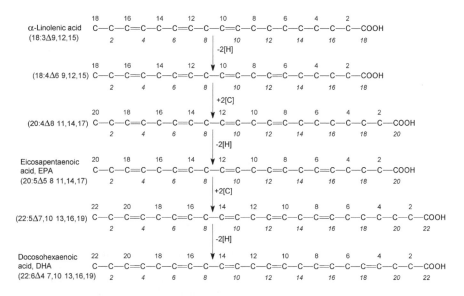

Figure 4.4 *The formation of ω-3 series fatty acids in algae, the origin of those found in fatty fish.*

α-Linolenic acid is the starting point for the formation of the ω-3 series of fatty acids, Figure 4.4. This sequence occurs in unicellular algae and one would be excused for questioning their relevance to food chemistry. However, fish consume algae and the oils of fatty fish such as herring, mackerel, salmon, and trout provide ω-3 fatty acids in human diets.

The prostaglandins are now recognised as being just one class of a group of hormones known as the eicosanoids. Other eicosanoids include the thromboxanes and leukotrienes. There are a great many different prostaglandins but their formation always follows the same pattern, typified by that for prostaglandin E_2, shown in Figure 4.5. Although the list of physiological activities in which prostaglandins are involved continues to grow they are best known for their role in inflammation and the contraction of smooth muscle. Thromboxanes are involved in the aggregation of platelets, part of the process of blood clot formation.

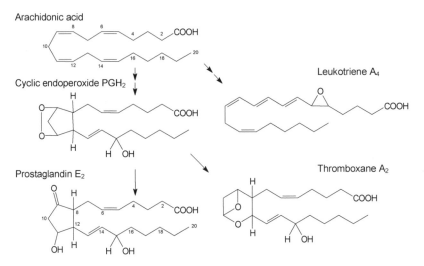

Figure 4.5 *The biosynthesis of eicosanoids from arachidonic acid. The numbering of arachidonic acid carbon atoms will facilitate comparison with the unfolded structure in Figure 4.3. The F series prostaglandins differ in that both the oxygen atoms attached to the ring are in the form of hydroxyl groups. This pathway also shows why the positions of the double bonds in essential fatty acids are critical. Prostaglandin E_1, is derived from eicosatrienoic acid (C20:3Δ8,11,14) (Figure 4.3) and therefore lacks the C5-6 double bond. Similarly prostaglandin E_3, derived from eicosapentaenoic acid (Figure 4.4) has an extra double bond at C17-18.*

Fatty Acids and Coronary Heart Disease

The involvement of dietary fatty acids in the occurrence of atherosclerosis and particularly heart disease is a complex issue. Nutritional guidelines focus on reducing the contribution that fats make to the total energy content of the diet and on lowering the proportion of saturated fatty acids in dietary fat. However, it is becoming increasingly clear that the apparently simple relationships (which were the scientific basis of these guidelines) between dietary lipids, blood cholesterol levels, and the risk of coronary heart disease are not simple at all. A diverse range of other 'risk factors' and refinements are being identified which help us to explain the apparent inconsistencies of the so-called 'Lipid Hypothesis' of coronary heart disease (CHD). While some of these risk factors still have an obvious nutritional element others, such as ethnicity, employment, smoking, and family history, do not.

Thus when the fatty acids in the diet are being considered it is now thought important to discriminate between saturated fatty acids of different chain length as well as between the mono- and poly-unsaturated. Currently the most popular candidates for the true villains of the piece appear to be myristic, lauric and palmitic acids. Stearic acid is considered to be much less effective at raising plasma cholesterol levels. The difficulty is that that fats relatively rich in stearic acid are usually well endowed with palmitic acid as well. The pattern of unsaturation is also important, as indicated by the remarkably low incidence of arterial disease amongst Eskimos, in spite of a diet that appears to break all the usual nutritional rules. Their traditional diet, though extremely fatty, is very rich in polyunsaturated fatty acids of the ω-3 series (*see* page 84). Eicosanoids from ω-3 fatty acids are generally less potent than those from ω-6 fatty acids in promoting the formation of the blood clots that are involved in CHD. This observation, and subsequent studies of the relationship between fatty fish consumption and the incidence of heart disease is the basis of recommendations that we should eat more fatty fish, such as herring and mackerel.

The observation that the occurrence of CHD is linked to the peroxidation (*i.e.* oxidation involving the formation of peroxides) of blood lipids has drawn attention to the possible role of the 'antioxidant vitamins', A, E, and C, as well as other naturally occurring antioxidants and hence fruit and green vegetable consumption. The function of antioxidants, substances that interfere in oxidation of unsaturated fatty acids is discussed in detail later in this chapter. There are now indications that many of the differences in the incidence of CHD between population groups (*e.g.* between the north and south of the British Isles) that defy explanation in terms of dietary fat consumption can be correlated with the intake of fruit and green vegetables.

The evidence that changing one's diet, *e.g.* by replacing butter with margarine, actually lowers the risk of CHD by a significant amount is also less clear cut than the advertisements for margarine 'rich in polyunsaturates' imply. The account of *trans* acids that follows shows that margarine may not be all that it appears to be. Similarly the fatty acid compositions presented in Table 4.3 show that animal fats can be highly unsaturated (*e.g.* fish oils) and plant fats can be the reverse (*e.g.* coconut oil). The observation that the level of cholesterol in the diet has little or no influence on the levels of cholesterol in the blood is also widely ignored by the advertisers of products derived from plant oils.

The superficiality of this brief account of the complexity of the CHD issue is evidence that its proper consideration has now moved well beyond the scope of food chemistry. Readers with a particular interest in this field are strongly recommended to consult the excellent review published by the British Nutrition Foundation (*see* Further Reading).

REACTIONS OF UNSATURATED FATTY ACIDS

Unsaturated fatty acids take part in a number of chemical reactions that are important to the food scientist, the hydrocarbon chain of the saturated fatty acids being essentially chemically inert under the conditions encountered in food. These reactions do not involve the carboxyl group, and since most occur whether or not the fatty acid is esterified with glycerol it is convenient to deal with them at this point rather than when the particular properties of the fats and oils are considered.

Halogens react readily with the double bonds of fatty acids, and the stoichiometry of the reaction has been utilised for many years as a guide to the proportion of unsaturated fatty acids in a fat. The iodine number of a fat, *i.e.* the number of grams of iodine that react with 100 g of fat, is quoted in older literature as a measure of the overall degree of unsaturation. However, for most purposes it has been replaced by gas chromatography of the volatile methyl esters of the fatty acids obtained by saponification and methylation of the fat. This can provide complete data on the fatty-acid composition of an unknown fat.

Hydrogenation, Margarine and *trans* Fatty Acids

In the presence of hydrogen and a suitable catalyst the double bonds of unsaturated fatty acids are readily hydrogenated, a reaction at the heart of modern* margarine manufacture. The aim of hydrogenation is to convert a liquid vegetable or fish oil to a fat with a butter-like consistency by reducing the degree of unsaturation of its component fatty acids.

After preliminary purification to remove polar lipids and other substances which tend to 'poison' the catalyst, the oils are exposed to hydrogen gas at high pressures and temperatures (2–10 atm, 160–220 °C) in the presence of 0.01–0.2% finely divided nickel. Under these conditions the reduction of the double bonds is fairly selective. Those furthest from the ester link of the triglycerides and those belonging to the most highly unsaturated fatty-acid residues are most reactive. This fortunate selectivity thus results in trienoic acids being converted into dienoic acids and dienoic to enoic acids rather than an accumulation of fully saturated acids. Of course the margarine manufacturer regulates the degree of hydrogenation to give the particularly desired characteristics to the finished product.

The hydrogenation reactions are by no means as straightforward as one might imagine. Isomerisation of *cis* to *trans* occurs under the extreme conditions of the hydrogenation process; *trans* is the thermodynamically preferred configuration of the double bond. Double bonds also migrate during the process so that a fatty acid chain starting with a *cis* double bond at carbon 9 may end up with a *cis* or *trans* double bond anywhere between carbons 4 and 16.

The proportions of these isomeric fatty acids vary widely between different samples of margarines and other hydrogenated fats but some typical data are shown in Table 4.5. The total amount of *trans* fatty acids

Table 4.5 *Isomers of C18 monoene and diene fatty acids in margarine. (Derived from data quoted by Craig-Schmidt in 'Fatty Acids in Foods and Their Health Implications', ed. C. K. Chow; see Further Reading.)*

Isomer	% Total fatty acids	Isomer of C18:1 trans	% Total fatty acids
C18:1 *trans*	14.8	$\Delta 7$	0.5
		$\Delta 8$	1.3
		$\Delta 9$	2.5
C18:2 *trans, cis*	1.3	$\Delta 10$	3.1
C18:2 *cis, trans*	1.3	$\Delta 11$	2.8
C18:2 *trans, trans*	0.3	$\Delta 12$	2.1
		$\Delta 13$	1.4
		$\Delta 14$	0.7
		$\Delta 15$	0.4

* The first margarine was invented by a French chemist, Miége Mouriés, in 1869 as a butter substitute. It was a mixture of milk, chopped cow's udder tissue, and a low melting fraction of beef fat. The name margarine was derived from the Greek, *margaritas*, meaning 'pearl-like'!

Box 4.1   Hydrogenation in detail

The hydrogenation process involves a number of reactions (illustrated in Figure 4.6):

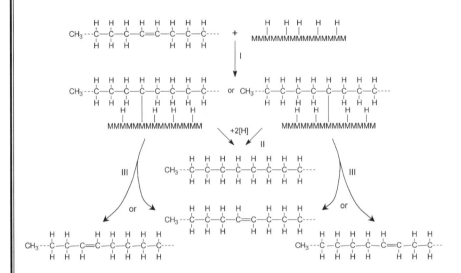

Figure 4.6   *The hydrogenation reactions. Reaction IV, the re-entry of isomerised by-products into the sequence, has been omitted.*

I   Hydrogen is absorbed onto the metal surface of the catalyst and in doing so is effectively converted from H_2 into atomic hydrogen as individual hydrogen atoms are bound to metal atoms. At the site of a double bond the fatty acid chain is bound to the metal surface and one hydrogen atom from the metal surface. This link may be formed at either end of the double bond. From this point there are two possible outcomes:

can range from 5%, in lightly hydrogenated vegetable oils, to more than 40% in commercial frying fats. (Light hydrogenation of cheaper vegetable oils is often done to reduce their polyunsaturated fatty acid content and thereby their tendency to rancidity – see below)

The total consumption of *trans* fatty acids has been quite high. A fairly recent average value (for UK adults) is around 5 g per day but consumption of over 12 g per day has not been uncommon. The contribution of different foods to one's *trans* acid intake differs widely from one individual to another. Spreading fats, used on bread, contribute about one third; shortening fats, used in pastry and biscuits, contribute a similar amount.

II A second hydrogen reacts to fully saturate the double bond and the hydrogenated fatty acid chain dissociates from the catalyst, *or*

III The fatty acid chain dissociates from the catalyst without further reaction so that a double bond is restored. However, there is no special reason why the hydrogen returned to the catalyst should be the original one donated by the catalyst so the resulting double bond can be in the original position or between a neighbouring pair of CH_2 groups. The result is the apparent migration of the double bond, *e.g.* from 9,10 to 10,11 or 8,9. The newly formed bonds are most likely to be have the *trans* configuration, rather than the original *cis* as this form is greatly preferred on thermodynamic grounds.

IV A proportion of these migration products will re-enter the sequence, some ending up with *trans* double bonds back where they started from but others ending up even further away down the chain.

The tendency of linoleic acid residues to be converted to oleic (desirable) compared with the conversion of oleic to stearic (undesirable) is expressed as the Selectivity Ratio (SR), the ratio of the rate constants for the two processes. The SR is affected by a number of operating parameters including the temperature, pressure (*i.e.* H_2 concentration), catalyst level and degree of agitation (stirring rate) in the reactor. Elevation of any of these parameters raises the rate of hydrogenation but high temperature and catalyst concentration both lead to high SR values and high proportions of *trans* acids whereas high pressure and high intensity stirring have the reverse effect on SR and *trans* acid production. If the catalyst surface is densely packed with hydrogen atoms hydrogenation is more likely and selectivity is also low. However, if hydrogen atoms are sparse selectivity will be high but there will be a much greater likelihood of isomerisation as there will not be a conveniently placed second hydrogen atom to complete the hydrogenation reaction.

Although the *trans* acid content of animal fats is low[*] dairy products and meat can contribute around one tenth of total consumption. A typical 'fast food' meal of a cheeseburger, French fries, and a milk shake with a pastry for dessert can supply nearly 4 g of *trans* acids.

In spite of their presence in our diet since the early years of this century the metabolism and possible effects on health of *trans* acids have not received much attention until recently. It is reasonable to conclude that any

---

[*] The fats of ruminants, including the fat in meat and milk, contain traces of these fatty acids, as a by-product of hydrogenation reactions carried out by rumen bacteria.

well defined toxic effects would have shown themselves by now. Humans are equipped with appropriate enzymes to metabolise at least some *trans* acids but it remains to be shown how far we cope with some of the more exotic positional isomers. There are hints from epidemiological studies of a positive correlation between the mortality from arterial disease and the consumption of hydrogenated fats. The most likely explanation appears to be that the body reacts towards *trans* unsaturated fatty acids as if they were saturated. This would certainly apply to their contribution to the physical properties of membrane lipids. If this proves to be the case then it will be necessary to exclude *trans* fatty acids from the totals of mono- and poly-unsaturated ones when the analyses of margarines *etc.* are quoted on labels. At the present time there appear to be no plans for including *trans* acid contents of food products on their labels although their measurement is not too problematical. Besides gas–liquid chromatography they may be detected by their characteristic infrared absorption bands (965–990 cm^{-1}).

The *trans* acids problem is likely to prove short-lived. More recent data (1999) from the oils and fats industry shows that while just a few hydrogenated fat products, all shortenings (in food industry parlance 'white fats') have *trans* acid contents around 10% most, including a wide range of margarines and low fat spreads (yellow fats!) now have 1% or less. This has been achieved by means of the an adaptation of the interesterification process, to be discussed later in this chapter (page 110) after the properties of triglycerides have been considered in more detail.

RANCIDITY

Rancidity is a familiar indication of the deterioration of fats and oils leading to unpleasant smells. There are two completely distinct mechanisms by which this can arise, lipolytic rancidity and oxidative rancidity.

Lipolytic rancidity can be a problem in dairy fats, particularly butter. It results when lipases secreted by the microbial flora catalyse the hydrolysis of the fat's triglycerides, liberating short chain fatty acids (*see* Table 4.3). Although low concentrations of these acids make an important contribution to the desirable flavour of butter too much can lead to flavours resembling over-ripened cheese. However, these enzymic reactions have an essential role in cheesemaking. Lipases from the moulds that develop as cheese matures liberate butyric acid, caproic acid, *etc.* which constitute a major element in the flavour of cheese. Proteolytic enzymes (proteases) from the same moulds liberate peptides and free amino acids (*see* Chapter 5) from the milk proteins which make up the other major component of cheese flavour. For many of us Stilton, Camembert and Gorgonzola and their thousands of country cousins are what we mean by cheese, the yellow plastic slice in the cheeseburger is not. Nevertheless we have to accept that vast numbers of cheeseburgers do get sold every day and for such a large market any reduction in the maturation period of cheese makes financial sense. So-called 'accelerated cheese ripening' preparations containing both lipases and proteases are now used extensively in the manufacture of cheese for applications such as cheeseburgers.

Oxidative rancidity, which has no redeeming features, affects fats and oils and the fatty parts of meat and fish, and is the result of autoxidation of the unsaturated fatty acids. The course of the reaction has been studied in free fatty acids (as shown here) but there is no reason to suppose that the behaviour of fatty acids esterified in triglycerides is significantly different.

The sequence of reactions is traditionally presented in three stages, initiation, propagation, and termination, shown in outline in Figure 4.7. The initiation reactions give rise to small numbers of highly reactive fatty acid molecules that have unpaired electrons, *free radicals*. These are shown as R· in Figure 4.7a. The dot (·) shows the presence of an unpaired electron and the R denotes the remainder of the molecule. Free radicals are very short-lived, and highly reactive, as they seek a partner for their unpaired electron.

In the propagation reactions atmospheric oxygen reacts with these to generate *peroxy* radicals, ROO· (Figure 4.7b). These are also highly reactive and go on to react with other unsaturated fatty acids generating *hydroperoxides*, ROOH and another free radical, R· (Figure 4.7c). [Do not confuse the hydroperoxide group with a carboxyl group; in a hydroperoxide (*see* Figure 4.10c) the hydroxyl group, −OH, is attached directly to the other oxygen atom, not the carbonyl carbon as in a carboxyl group.]

This free radical can then go round and repeat the process, forming a chain reaction, (Figure 4.7b) but the hydroperoxide can also break down to give other free radicals (Figure 4.7d). These too can behave much as

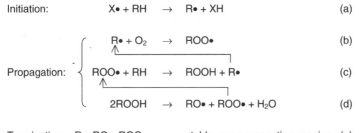

Figure 4.7 *An outline of the reactions involved in the autoxidation of unsaturated fatty acids. The various stages, a–e, are discussed in the text.*

Box 4.2   Singlet and triplet oxygen

These terms relate to the bonding between the two atoms in the oxygen molecule. The oxygen molecule, O_2, has a total of twelve valency electrons originating from the outer shells of the two oxygen atoms. In its normal low energy 'ground' state eight of these are located in seven different molecular orbitals. In five of these the paired electrons spin in opposite directions as shown by the arrows in the first five of these boxes.

↑ ↓	↑ ↓	↑ ↓	↑ ↓	↑ ↓	↑	↑
$(2s\sigma)^2$	$(2s\sigma^*)^2$	$(2p\sigma)^2$	$(2p\pi_x)^2$	$(2p\pi_y)^2$	$(2p\pi_x^*)^2$	$(2p\pi_y^*)^2$

The asterisks indicate antibonding orbitals. The number of bonds between the atoms is obtained by subtracting the number of electrons in antibonding orbitals from those in bonding orbitals and dividing the result by 2.

The two highest molecular orbitals contain single electrons with parallel spin. This gives the molecule a *multiplicity* (S) of 3 from the equation: $S = 2s + 1$ where $s = $ the sum of the spins of all the electrons when $\boxed{\uparrow}$ counts as $+\frac{1}{2}$ and $\boxed{\downarrow}$ counts as $-\frac{1}{2}$. Having a multiplicity of three is more commonly known as the *triplet* state. The triplet state of oxygen molecules in their lowest energy, *ground*, state is unusual. Most molecules have *singlet* ground states in which all electrons' spins are paired. Triplet oxygen is shown symbolically by $^3O_2$. Besides its triplet ground state oxygen molecules can occur in two *excited*, high energy singlet states with their two highest orbitals filled as shown here: $\boxed{\uparrow \downarrow}\boxed{\uparrow \downarrow}$ or $\boxed{\uparrow}\ \boxed{\downarrow}$, written as: $^1O_2$ and $^1O_2{}^*$ respectively. The others are as shown above. These excited states are extremely short-lived.

Only the first of these excited states, $^1O_2$, is relevant to us. This has the two electrons that were unpaired in singlet oxygen in a $2p\pi$ orbital leaving the corresponding antibonding $2p\pi^*$ orbital empty. The resulting molecule is an extremely reactive electrophile as it seeks to fill the empty orbital, for example by reaction with unsaturated fatty acids. The energy difference between this and triplet oxygen is $+ 24$ kcal mol^{-1}.

ROO· did. The result is that ever increasing numbers of free radicals accumulate in the fat, which absorbs considerable quantities of oxygen from the air. Eventually the concentration of free radicals reaches a point when they start to react with each other to produce stable end-products (Figure 4.7e). These are the termination reactions.

In the past there was great uncertainty about the nature of the reactions that give rise to the first free-radical species (X·) that initiates the autoxidation sequence. However, it is now widely accepted that the initiation reaction involves a short-lived but highly reactive, high-energy form of oxygen known as singlet oxygen, $^1O_2$. Singlet-oxygen molecules arise from the normal low-energy, ground-state oxygen (triplet oxygen, $^3O_2$) in a number of reactions, but most relevant to food systems are those in which oxygen reacts with pigments such as chlorophyll, riboflavin (very important in fresh-milk deterioration), or haem, in the presence of light. As shown in Figure 4.8 the singlet oxygen reacts with the double bond of an unsaturated fatty acid. In the reaction the double bond changes its configuration from *cis* to *trans* at the same time as moving along the chain.

Autoxidation proper begins as these first hydroperoxides decompose to hydroperoxy and alkoxy radicals, Figure 4.9. These are sufficiently reactive to abstract H·, *i.e.* a hydrogen atom or more correctly a hydrogen radical, from particularly vulnerable sites in monoenoic (*e.g.* oleic) and polyenoic (*e.g.* linoleic) fatty acid residues in the fat. The most vulnerable sites are the CH_2 groups next to double bonds, known as α-methylene groups. A hydrogen atom of an α-methylene group is particularly labile, *i.e.* easily abstracted. Not surprisingly the most vulnerable α-methylenes are those lying between neighbouring double bonds in a polyunsaturated fatty acid, *e.g.* number 11 in linoleic acid, as shown in Figure 4.10a and 10b.

Once the hydrogen radical has been abstracted the unpaired electron can migrate, leading to the formation of *trans* double bonds. Reaction with oxygen then leads (Figure 4.10c) to the formation of hydroperoxy radicals. These abstract hydrogen from other fatty acids forming hydroperoxides and continuing the chain reaction. Oleic acid gives rise to 8-, 9-, 10-, and 11-hydroperoxides in roughly equal amounts. Linoleic acid, which reacts with oxygen 10 times faster than oleic acid, gives only 9- and 13-

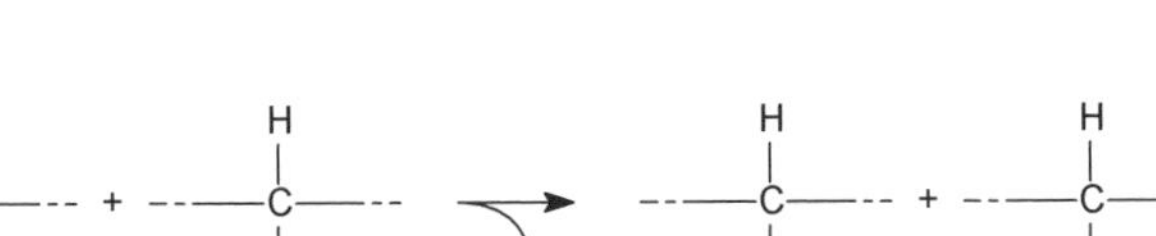

Figure 4.8   *The reaction of singlet oxygen with a double bond to form a hydroperoxide. The curly arrows show serious organic chemists how the electrons migrate during the course of the reaction but they can be ignored by the rest of us.*

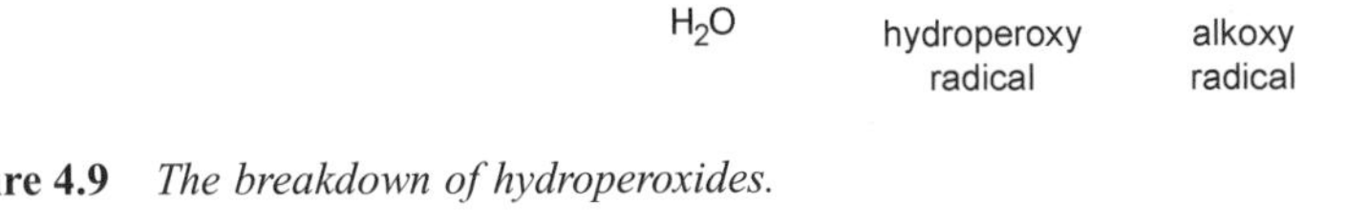

Figure 4.9   *The breakdown of hydroperoxides.*

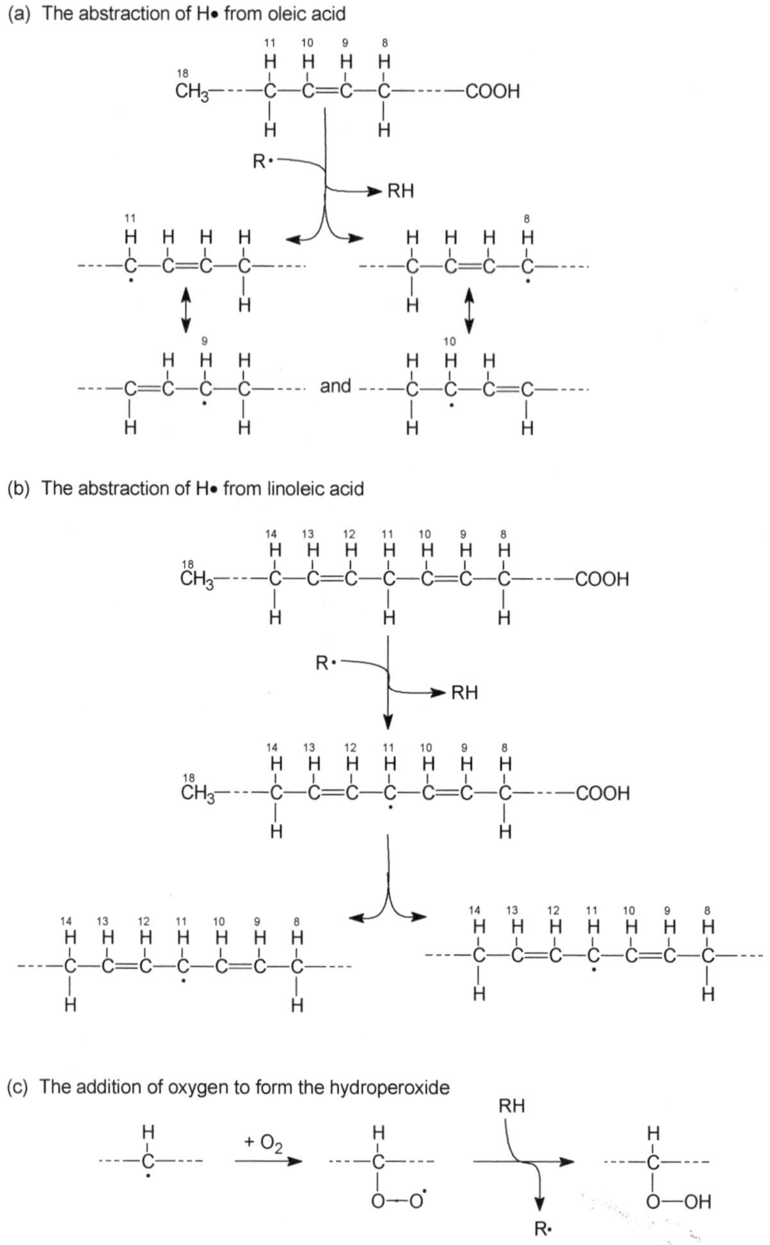

(a) The abstraction of H• from oleic acid

(b) The abstraction of H• from linoleic acid

(c) The addition of oxygen to form the hydroperoxide

Figure 4.10   *The formation of hydroperoxides in the propagation phase of autoxidation.*

hydroperoxides as the stability of the conjugated diene system favours attack by oxygen at the ends of the system rather than in the middle.

As hydroperoxides accumulate in the autoxidising fat, considerable quantities of oxygen are absorbed from the atmosphere, but, as the chain reaction proceeds, the build up of their breakdown products becomes increasingly important. As shown in Figure 4.11, the alkoxy radicals give rise to aldehydes, ketones, and alcohols as well as a continuing supply of free radicals to maintain the chain reaction.

Transition-metal cations (inevitable trace contaminants of oils processed, stored, or utilised in metal vessels) are important catalysts of hydroperoxide breakdown:

$$R\text{-}OOH + M^+ \rightarrow R\text{-}O\cdot + OH^- + M^{2+}$$

$$R\text{-}OOH + M^{2+} \rightarrow R\text{-}OO\cdot + OH^- + M^+$$

$$2R\text{-}OOH \rightarrow R\text{-}O\cdot + R\text{-}OO\cdot + H_2O$$

Aldehydes arising from cleavage of the carbon chain on either side of the alkoxy radical are the source of rancid fat's characteristic odour. For example, the 9-hydroperoxide from linoleate will give either 2-nonenal or 2,4-decadienal, as shown in Figure 4.12.

One particular compound, malonaldehyde (4.10), results from cleavage at both ends of a diene system. It forms a pink colour with thiobarbituric acid which forms the basis of a useful method for assessing the deterioration of fats. Other tests include measurements of the carbonyl content using 2,4-dinitrophenylhydrazine, the hydroperoxide content by

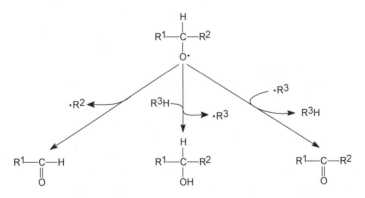

Figure 4.11 *The formation of stable end-products from alkoxy radicals.*

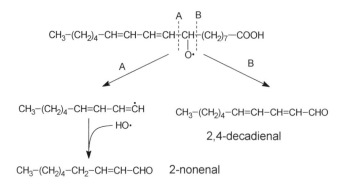

Figure 4.12 *The formation of unsaturated aldehydes during autoxidation. Cleavage can occur on either of the alkoxy group, at A or B.*

reaction with iodine, and characteristic absorbances of conjugated dienes and trienes at 230 and 270 nm, respectively.

$$HC-CH_2-CH$$
$$\parallel \qquad \parallel$$
$$O \qquad O$$
(4.10)

The use of tests such as these, as well as monitoring the oxygen uptake, has shown that the course of oxidation of a fat is marked by an induction period of slow oxygen uptake (while the level of initiating free radicals builds up) followed by a period of rapid oxidation. Usually a pronounced rancid odour is not detectable until the rapid oxidation phase is well established, although some oils do develop off-flavours and -odours even when very little oxidation has occurred. Soya bean oil is particularly susceptible to this phenomenon, termed reversion, apparently because of its content of highly labile linoleic acid. Amongst the compounds that have been implicated in the 'flavour' of reverted soya bean oil are *cis*-3-hexenal, diacetyl (butanedione), 2,3-pentanedione, and 2,4-pentadienal. Reduction in the linolenic acid content of seed oils is an important objective for plant breeders.

Fats and oils exposed to the atmosphere and heating over a long period show the final stage of the oxidation sequence, polymerisation. The highly unsaturated oils used in paint show the phenomenon even more readily! The crosslinking reactions can be of various types as shown in Figure 4.13

Polymerisation reactions are particularly associated with frying oils, where prolonged use leads to high-molecular-weight compounds which cause foaming and increased viscosity. Discarded oils are sometimes found to contain as much as 25% of polymerised material. The chain reaction

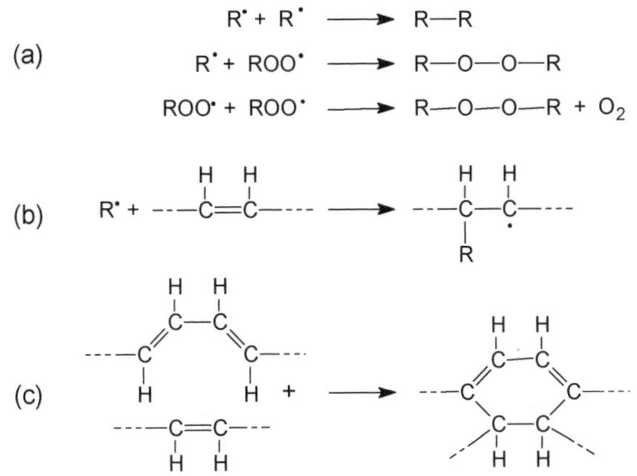

$$R^{\cdot} + R^{\cdot} \longrightarrow R{-}R$$

(a)

$$R^{\cdot} + ROO^{\cdot} \longrightarrow R{-}O{-}O{-}R$$

$$ROO^{\cdot} + ROO^{\cdot} \longrightarrow R{-}O{-}O{-}R + O_2$$

Figure 4.13 *Crosslinking reactions in the formation of end-products. Free radical may react directly together* (a) *or with other alkenic systems* (b). *Diels–Alder type reactions also occur to give cyclic structures* (c).

character of the autoxidation reactions provides the explanation for why a half empty bottle of oil approaching the end of its useful life should be discarded in its entirety rather than topped up with fresh oil.

The possibility that lipid oxidation products are toxic to humans remains unresolved. The most important source in the human diet is deep-fried foods although fatty fish and their oils may be significant in some diets. In order to produce measurable levels of adverse effects experiments with animals use much larger amounts (in relation to body weight and life span) than would be encountered amongst humans. Feeding these large amounts of highly oxidised fats induces a wide range of symptoms, most corresponding to vitamin E deficiency. It is assumed that the high intake of oxidised fats overwhelms the body's natural antioxidant systems (*see* page 85).

In recent years attention has been focused on the potentially harmful involvement of lipid oxidation products (particularly those derived from cholesterol) in arterial disease. The mechanisms of these effects are complex, and beyond the scope of this book (see 'Free Radicals and Food Additives' in Further Reading). As is so often the case the link between experimental results using laboratory animals and the incidence of human disease is a tenuous one and vulnerable to criticism from scientists, and others, with different priorities. Suggestions, not wholly endorsed by other workers, have been made that the elevated levels of cholesterol oxides in clarified butter (ghee) used for cooking may account for the high level of atherosclerosis in Britain's Asian community. Cholesterol oxides have been recorded at 50 p.p.m. in French fries from fast-food outlets.

It is commonly assumed that at least some fat oxidation products will be carcinogenic. Some have definitely been shown to be mutagenic. The observation that many antioxidants have some anti-carcinogenic activity supports this view. However, there is little evidence at the present time to suggest that oxidised fats do cause cancer in humans. Perhaps the adverse effects of a diet over-rich in fats generally swamp these more subtle influences.

Antioxidants

Although the development of rancidity in bulk fats and oils can be retarded by careful processing procedures avoiding high temperatures, metal contamination, *etc.*, such measures are never wholly adequate. Fatty foods such as biscuits and pastry are particularly susceptible to rancidity as their structure necessarily exposes the maximum surface of the fat to the atmosphere. The shelf-life of these types of foods (on the customer's shelf as well as the shopkeeper's) can be massively extended by the inclusion of antioxidants in many of the fats we buy, particularly lard. Antioxidants are substances used as food additives to retard the autoxidation reactions. Some are wholly synthetic products of the chemical industry but antioxidants of natural origins are becoming more significant as consumers and the food industry seek out what are perceived to be healthier alternatives. As was mentioned earlier in this chapter (page 85) the involvement of lipid peroxidation reactions in arterial disease has focussed attention of the role of antioxidants as dietary components in their own right, not merely as protectors of fats prior to consumption.

There are two well-recognised mechanisms of antioxidant action. In the best known, particularly associated with the synthetic antioxidants such as BHA and BHT (*see* Figure 4.14), the antioxidant blocks the propagation phase of the autoxidation process (Figure 4.15).

Many other antioxidants, but most famously β-carotene, lycopene and other carotenoids (*see* Chapter 6), operate by a singlet quenching mechanism. These molecules react with singlet oxygen, $^1O_2$ (*see* page 92), returning it to the relatively unreactive triplet state:

$$^1O_2 + \text{carotenoid} \rightarrow {}^3O_2 + \text{carotenoid}^*(\text{excited})$$

$$\text{carotenoid}^* \rightarrow \text{carotenoid} + \text{heat}$$

In biological systems a number of highly reactive free radical forms of oxygen are important, such as the superoxide anion, $\cdot O_2^-$ (an oxygen molecule carrying an extra unpaired electron giving it a negative charge),

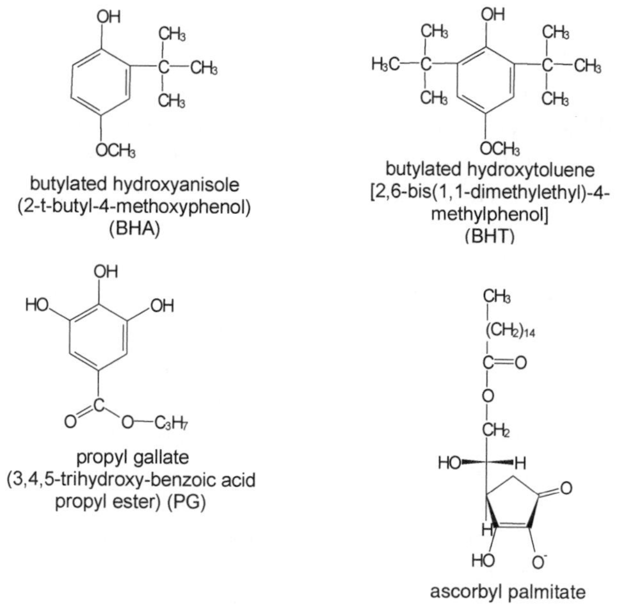

butylated hydroxyanisole
(2-t-butyl-4-methoxyphenol)
(BHA)

butylated hydroxytoluene
[2,6-bis(1,1-dimethylethyl)-4-
methylphenol]
(BHT)

propyl gallate
(3,4,5-trihydroxy-benzoic acid
propyl ester) (PG)

ascorbyl palmitate

Figure 4.14   *Synthetic antioxidants presently permitted for food use.*

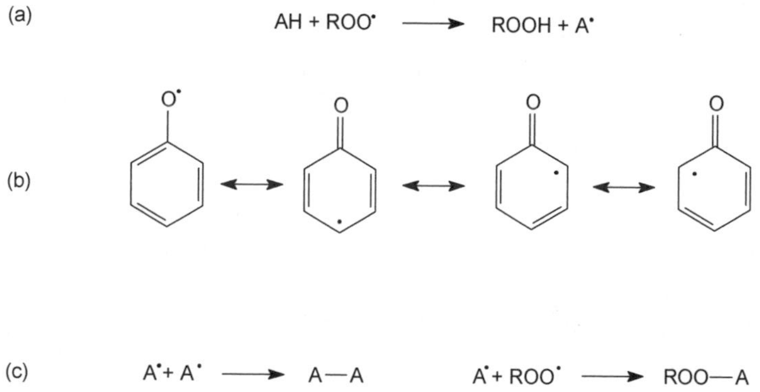

(a) $AH + ROO^\bullet \longrightarrow ROOH + A^\bullet$

(b)

(c) $A^\bullet + A^\bullet \longrightarrow A{-}A$ $A^\bullet + ROO^\bullet \longrightarrow ROO{-}A$

Figure 4.15   *Free radical stabilisation by resonance. (a) A free radical intermediate in the lipid autoxidation is stabilised when an antioxidant molecule, AH, donates a hydrogen atom. (b) The antioxidant free radical is stabilised by resonance and is consequently insufficiently reactive to propagate the autoxidation sequence. (c) Antioxidant radicals react together to terminate the process.*

and the hydroxy radical, ·OH (a hydroxyl ion lacking an electron). These arise as by-products of enzyme and other reactions involving oxygen (*see* Figure 5.13) and it appears likely that singlet quenching antioxidants also react with them.

Antioxidants do not reduce the ultimate degree of rancidity; rather they lengthen the induction period in rough proportion to their concentration. Thus the inclusion of 0.1% BHA in the lard used to make pastry could be expected to add at least a month to the period, otherwise only a few days, before obvious rancidity develops. The three entirely synthetic antioxidants shown in Figure 4.14 (BHA, BHT, and PG) are normally added to fats at rather lower levels than this, up to about 200 p.p.m. There is growing concern over the safety of these synthetic antioxidants: some laboratory studies with animals have hinted at carcinogenicity. Although this issue is far from resolved, there is growing pressure for their withdrawal from use. However, it seems likely that they will not be banned completely until (a) the carcinogenicity question is clarified by more research and (b) more progress is made towards their replacement with more natural antioxidants such as the tocopherols and fatty-acid derivatives of ascorbic acid.

Tocopherols occur in most plant tissues, as much as 0.1% in vegetable oils. Animals require tocopherols in their diet, as vitamin E, and they are therefore discussed more fully in Chapter 8. Two other vitamins, ascorbic acid and retinol, also have antioxidant activity as discussed in Chapter 8. There are indications from both epidemiological studies and the results of experimental dietary supplementation that the 'antioxidant vitamins' can have a collective beneficial effect on arterial disease. The explanation of this effect is that the progress of atherosclerosis involves the peroxidation of the lipid components of lipoproteins. It remains to be seen whether the dramatic differences in the incidence of heart disease between different populations will prove to be as much (or more) a function of vitamin intake as of fat intake.

There are two other sources of antioxidants that are beginning to attract attention, tea and spices. In the Far East consumption of green (*i.e.* unfermented, *see* 196) tea has long been associated with good health. Catechin (*see* Figure 6.11), one of the principal polyphenolic compounds in green tea, has been shown to be a most effective free radical scavenger and inhibitor of lipid peroxidation in laboratory experiments. Theaflavins (6.24), also polyphenolics and the major red pigments in black (*i.e.* fermented) tea are also very effective free radical scavengers. It remains to be seen whether these activities translate from the laboratory to normal human nutrition. Spices have traditionally been used to add to the flavour of foods, or mask off-flavours, but modern research has shown that many spices are rich in antioxidants that have the potential for use in curbing rancidity. Some naturally occurring antioxidants are illustrated in Figure 4.16. They pose a number of practical problems including the high costs of production in useful amounts and in many cases flavours that while desirable in some foods will be unwelcome in others. Rosemary extracts

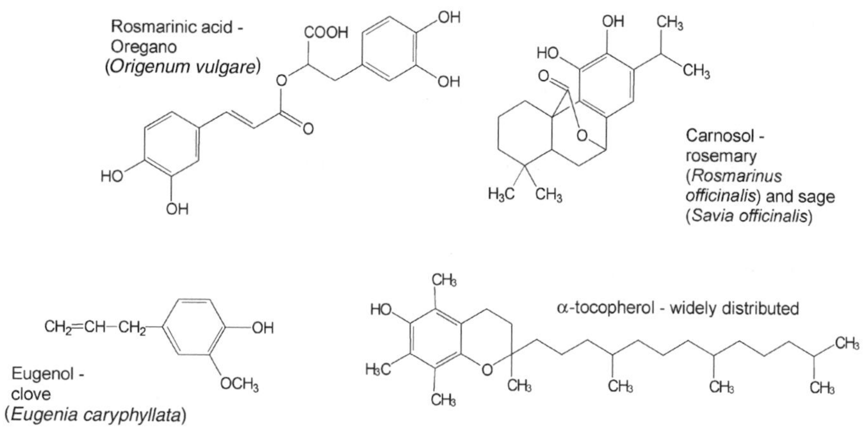

Figure 4.16 *Some naturally occurring antioxidants.*

are already finding applications in meat products where both the flavour and antioxidant activity are of benefit.

TRIGLYCERIDES

Having considered the properties of the fats and oils from the point of view of the chemistry of their component fatty acids, we can now examine them in terms of their component triglycerides. The first descriptions of the glyceride structure of fats assumed that all their component triglycerides were simple. Thus a fat containing palmitic, stearic, and oleic acids would be a mixture of the three triglyceride species tripalmitin, tristearin, and triolein. The first attempts to separate the component glycerides of fats, by the laborious process of fractional crystallisation from acetone (propanone) solutions at low temperatures, made it clear that much greater numbers of species of triglycerides occurred than would be expected from this simple concept. Fats and oils became recognised as clearly defined mixtures of mixed, *e.g.* 1-palmityl, 2-linoleyl, 3-oleyl glycerol (4.11) and simple triglycerides.

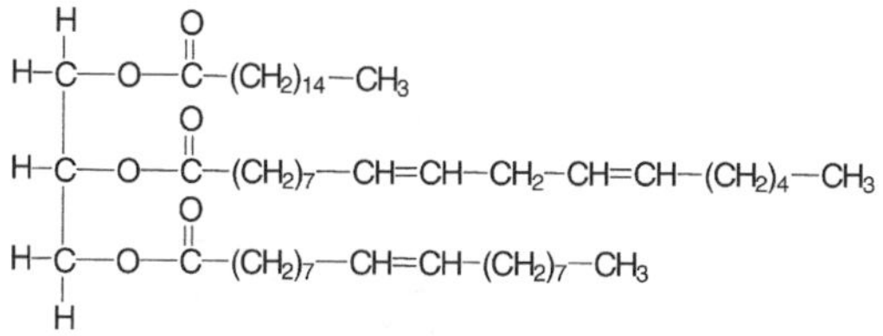

(4.11)

If we reconsider our simple fat with only three fatty acids, there should be 27 possible triglycerides. However, as the 1- and 1′ positions of the glycerol molecule are usually treated as indistinguishable, the total comes down to 18. The range of possibilities is illustrated in Table 4.6. It is important to note that the isomeric pairs of triglycerides, such as POP and PPO, are clearly differentiated. Such an abundance of data for a single fat forces us to simplify triglyceride composition data when we are comparing different fats. The most striking feature of Figure 4.17 is that fats of fairly similar fatty acid composition, such as lard and cocoa butter (see Table 4.3), can have very different triglyceride compositions. In lard there is a definite tendency for unsaturated fatty acids to occupy the outer positions of the glycerol molecule whereas in cocoa butter the reverse is true. This difference is in fact a general one between animal and plant fats, as is borne out by the further triglyceride compositions given in Figure 4.17.

If the fatty-acid residues at the two outer positions are different, the four substituents groups attached to the central carbon atom of the glycerol moiety are different and the triglyceride molecule is asymmetric and optical activity would be expected. It has been possible to show that a

Table 4.6 *The triglycerides of lard. Simple three letter codes identify the triglycerides; thus PMS has palmitic acid at position 1, myristic acid at position 2 and stearic at position 3 (1′). O = oleic acid; L = linoleic acid. As with the fatty acid composition of a particular fat there is considerable variation triglyceride composition from sample to sample although the broad features remain consistent. Based on data quoted by F. D. Gunstone* et al. *in The Lipid Handbook (see Further Reading).*

No double bonds:

PMS	SMS	PPP	PPS	SPS	PSS	SSS	Others			
0.4	0.4	0.5	2.0	2.0	0.4	0.4	0.5			

One double bond:

POP	POS	SOS	PMO	SMO	MPO	PPO	SPO	PSO	SSO	Others
0.6	1.9	1.5	0.4	0.7	0.8	7.9	12.8	0.9	1.6	0.6

Two double bonds:

POO	SOO	OMO	OPO	OSO	PPL	SPL	Others			
5.2	6.1	1.6	18.4	1.2	1.8	2.1	1.5			

Three double bonds:

OOO	SLO	OML	OPL	OSL	Others					
11.7	0.6	0.6	7.2	1.2	0.5					

Four or more double bonds:

OLL	OLO	LPL	Others							
1.4	1.5	0.5	0.5							

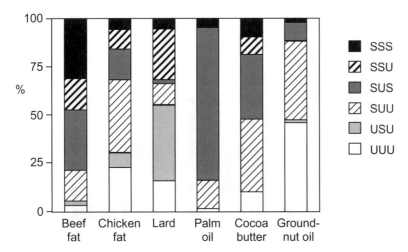

Figure 4.17 *The simplified triglyceride compositions of some animal and plant fats. All saturated fatty acids are shown as S, all unsaturated fatty acids as U.*

particular species of triglyceride in a natural fat does occur as a single enantiomer, as opposed to a racemic mixture. This indicates that the 1- and 1′-positions are differentiated during triglyceride biosynthesis. However, in many fats there is in fact very little difference between the fatty acid compositions of these two positions but in some, as implied by the data in Figure 4.18, there is a marked difference. Milk fats are exceptional in that their triglycerides fall into three broad classes. In the first, all three positions are occupied by long-chain fatty acids. The second class has long-chain acids at the 1- and 2-positions but short-chain at 1′. The third class has medium-chain fatty acids at the 1- and 2-positions and medium or short chain fatty acids at the 1′-position.

The determination of the triglyceride composition of a fat is not easy. The need to distinguish between isomeric pairs of glycerides such as 'SUS' and 'SSU', crucial to the difference between cocoa butter and lard, for example, makes special demands on chromatographic procedures. Thin-layer chromatography on silica gel impregnated with silver nitrate has been very successful as it will resolve these isomeric pairs. The silver ions are complexed by the double bonds of unsaturated triglycerides, whose mobility is thus reduced. The identity of a purified triglyceride can be established by means of pancreatic lipase. This enzyme specifically catalyses the hydrolysis of 1- and 1′-ester links of a triglyceride so that in the laboratory, as in the small intestine, the products from a triglyceride molecule will be the two fatty acids from the outer positions and one

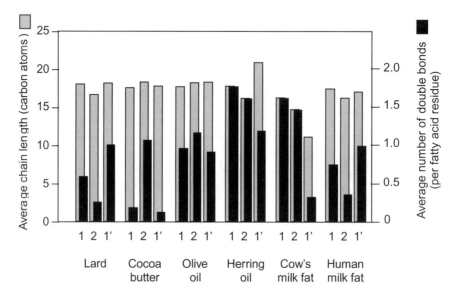

Figure 4.18 *Fatty acid distribution in a range of animal and plant fats. In order to simplify the presentation of what would otherwise be a massive amount of data the characteristics of the fatty acids esterified at each of the three positions of the glycerides have been expressed as their average chain length and average number of double bonds per fatty acid residue.*

2-monoglyceride. The liberated fatty acids can be isolated from the reaction mixture and identified by gas chromatography.

Melting and Crystallisation

For the food chemist the melting and crystallisation characteristics of a fat are physical properties of prime importance. Although the melting points of pure triglycerides are a function of the chain lengths and unsaturation of the component fatty acids, much as one might expect, the melting behaviour of fats is rather complex. Since natural fats are mixtures, each component having its own melting point, a fat does not have a discrete melting point but rather a 'melting range'. At temperatures below this range all the component triglycerides will be below their individual melting points and the fat will be completely solid. At the bottom of the range the lowest melting types, those of lowest molecular weight or most unsaturated, will liquefy. Some of the remaining solid triglycerides will probably dissolve in this liquid fraction. As the temperature is raised, the proportion of liquid to solid rises and the fat becomes increasingly plastic until, at the temperature arbitrarily defined as the melting point, there is little or no solid fat left.

This melting behaviour is illustrated in Figure 4.19 for cocoa butter and lard. The key difference between the shapes of the melting curves for these two fats is the steepness of the cocoa butter curve in the 32–36 °C region. This almost unique property of cocoa butter is due to the fact that about 80% of its triglycerides are of the SUS type, namely POP ~15%, POSt ~40%, and StOSt ~25%.* All these have very similar melting points. By contrast we have seen that lard contains a large number of different types, with no similarities between the major ones.

A second complication is that triglycerides are polymorphic, *i.e.* their molecules can pack into crystals in several different arrangements, each with its characteristic melting point and other properties. There are three basic types of crystal arrangement (polymorphic form), known as α, β', and β, in order of increasing stability. A particular fat will sometimes show a small number of variants within the β', and β classes so that some fats (*e.g.* cocoa butter) may demonstrate as many as six different polymorphic types, each with its own distinctive melting point. The melting points of the principal forms of a number of simple triglycerides are given in Table 4.7. When a melted triglyceride is cooled rapidly, it solidifies in the lowest-melting, unstable, α-form. If this is slowly heated,

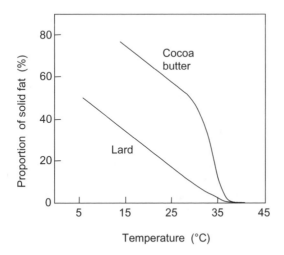

Figure 4.19 *Typical melting curves for samples cocoa butter and lard. The curves are based on data obtained by the use of the low-resolution NMR technique, the theory of which lies well beyond the scope of this book.*

* P = palmitic, O = oleic, St = stearic.

Box 4.3   Triglyceride crystals

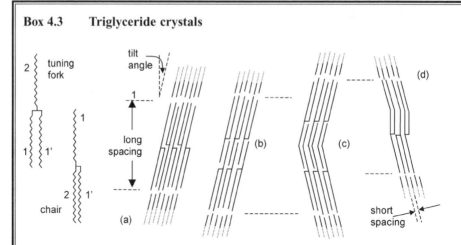

Figure 4.20   *The organisation of triglyceride molecules in fat crystals. In the stable crystal types shown here (β' and β, as discussed in the text) the triglyceride molecules stack alongside each other so that the terminal methyl groups form horizontal planes, the methyl terraces (heavy dashed lines). The dimensions of the long and short spacings and the tilt angle, established by X-ray crystallography, characterise the different forms. The long spacing reflects the different overlap patterns, double chain length (DCL) in (a) and (d), triple chain length (TCL) in (b) and (c).*

The fatty acid chains of an individual triglyceride molecule are organised into either a 'chair' or a 'tuning fork' arrangement as shown in Figure 4.20. Which is preferred depends on the symmetry of the triglyceride. The tuning fork arrangement is favoured in symmetrical triglycerides since it brings the two identical fatty acids at the 1 and 1' positions alongside each other. Similarly the chair arrangement is favoured in asymmetrical triglycerides. Neighbouring triglyceride molecules in a fat crystal are stacked in layers with the parallel hydrocarbon chains packed closely together and running more or less perpendicularly to the parallel planes of the glycerol groups and planes of terminal methyl groups. These planes are sometimes referred to as the 'methyl terraces'. Neighbouring triglyceride molecules overlap in different patterns to place the methyl terraces either two or three chain lengths apart. As with the choice between the tuning fork or chair arrangement the overlap pattern is dictated by the need to accommodate fatty acids of different chain length and the

it will melt and then resolidify in the β'-form. Repetition of this procedure will bring about a transition to the final, stable β-form. The β-form is also obtained by recrystallisation from solvent. Although each form, and many of their variations, has been fairly well characterised in terms of X-ray

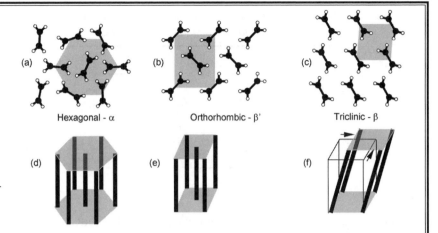

Hexagonal - α   Orthorhombic - β'   Triclinic - β

Figure 4.21 *Crystal packing in triglycerides. In* (a)–(c) *the zig-zag hydrocarbon chains of the fatty acid residues are viewed end on. The grey areas emphasise the geometrical relationships of the chains. In hexagonal and orthorhombic arrangements* (d) *and* (e) *the chains are shown perpendicular to the planes formed by their CH₃-ends, the so-called methyl terraces, while in the triclinic arrangement* (f) *the chains are tilted in both vertical planes. It is generally assumed that in triglycerides the chains are also tilted in the orthorhombic arrangement.*

distortions introduced by double bonds, as illustrated in Figure 4.20 (b, c and d).

In terms of melting point and crystal form the most important differences are between the three crystal types. In the crystalline state the hydrocarbon chains are arranged in extended zig-zags with all the carbon atoms in the same, vertical or near vertical, plane. Figure 4.21 (a, b and c) shows the chains viewed end on, *i.e.* along their long axes.

In the unstable α-forms, Figure 4.21 (a and d), the chains are arranged in a hexagonal pattern, but the planes of the hydrocarbon zig-zag are orientated randomly. The crystal structures of the β' and β-forms are formally described as orthorhombic and triclinic respectively (Figure 4.21). The tilting of the chains in the triclinic form, as shown in Figure 4.21f, allows the most compact packing of the chains. Although not allowed for in the formal definition of an orthorhombic crystal these fatty acid chains are also assumed to tilt.

diffraction pattern and infrared spectrum, the actual arrangements of the triglyceride molecules in the crystals remain to be described except for a handful of purified triglycerides.

It is the occurrence of large numbers of different species of triglyceride

Table 4.7 *The melting points /°C of the poly-morphic forms of simple triglycerides.*

	α	β′	β
Tricaprin	−15	–	32
Trilaurin	14	43	44
Trimyristin	32	44	56
Tripalmitin	44	56	66
Tristearin	54	64	73
Triolein	−32	−13	4

in a natural fat that makes the situation so complicated, but as a rule one polymorphic type predominates in a particular fat. Important examples of β-types are cocoa butter, coconut oil, corn oil, groundnut oil, olive oil, palm kernel oil, safflower oil, and lard; β′-types include cottonseed oil, palm oil, rapeseed oil, beef fat (tallow), herring oil, whale oil, and cow's milk fat.

The usefulness of a particular fat to a particular food application is crucially dependent on its melting and crystallisation characteristics. Fats to be spread on bread or blended with flour, *etc.*, in cake or pastry require the plasticity that is associated with a wide melting range. The term 'shortening' is nowadays applied to all manufactured fats and oils except those such as the margarines and 'low-fat' spreads that have a significant content of non-fatty material. The literal use of the term refers to the tendency of shortenings to reduce the cohesion of the wheat gluten strands in baked goods (see Chapter 5) and thereby 'shorten' or soften them.

The different polymorphic forms differ in the actual form of their crystals. The β′-tending fats crystallise in small needle-like crystals. At the correct mixing temperatures these fats consist of these crystals embedded in liquid fat matrix, giving a soft plastic consistency ideal for incorporating air bubbles and suspending flour and sugar particles as in cake batters. The β-tending fats form large crystals, which form clumps that give grainy textures. Though difficult to aerate, they are valuable in pastry making. If shortenings are blended from mixtures of β and β′ fats, the latter type dictates the crystallisation pattern.

The fatty acid composition, and distribution between the different positions of the triglyceride molecules, *i.e.* 1, 2 and 1′, all help to determine a fat's stable crystal form. In general where most of a fat's triglycerides have a broadly similar arrangement of fatty acids then it will readily form β crystals. However, fats with a more diverse mixture of triglyceride types or those dominated by asymmetrical triglycerides (*i.e.*

SSU or UUS) tend only to form β' crystals. Processes such as hydrogenation or interesterification will have the possibility for changing the physical characteristics of a fat by several different mechanisms. For example lard, whose 2 position is typically has around 60% palmitic acid normally crystallises in the β form. Interesterification redistributes the palmitic acid around all the positions of the triglyceride leading to a more diverse range of triglyceride types and a tendency to crystallise in the β' form.

Cocoa Butter and Chocolate

Cocoa butter is another good example of the importance of melting properties to a foodstuff – in this case chocolate. The reason for the most notable feature of chocolate, its sharp melting point (it melts in your mouth but not in your hand!), has already been considered, but we also expect chocolate to have a very smooth texture and a glossy surface. Cocoa butter can occur in six different polymorphic states with melting points ranging from 17.3 °C to 36.4 °C. Only the fifth of these (a β-3 type, melting point 33.8 °C) has the desired properties, and the special skill of the chocolate-maker lies in ensuring that the fat is in this particular state in the finished product. This is achieved by tempering. The liquid chocolate is cooled to initiate crystallisation and reheated to just below the melting point of the desired polymorphic type so as to melt out any of the undesirable types. The chocolate is then stirred* at this temperature for some time in order to obtain a high proportion of the fat as very small crystals of the desired type when it is finally solidified in the mould or when it is coating biscuits or confectionery.

Chocolate that has been incorrectly tempered or subjected to repeated fluctuations in temperature, as in, for example, a shop window, develops bloom. This is a grey film which resembles mould growth but is actually caused by the transition of some of the fat to a more stable polymorphic form which crystallises out on the surface. The migration of triglycerides from the nut centres of chocolates or the crumb of chocolate-coated biscuits can cause similar problems. Milk fat is an effective bloom inhibitor and is often included in small amounts in plain chocolates, besides accounting for about one-quarter of the total fat in a typical milk chocolate. Cocoa butter is obviously a very difficult fat to handle successfully in a domestic kitchen, and chocolate substitutes are marketed for home cake-decorating *etc.* These contain fats such as hardened (*i.e.* partially hydrogenated) palm kernel oil. Although they have an inferior,

---

* This stirring stage is known as 'conching', from the conch shell shape of the vessel in which it is carried out.

somewhat greasy texture, these artificially modified fats only have a single polymorphic state and therefore do not present the cook with the problems of tempering.

A fat such as hardened palm kernel oil is referred to as a cocoa butter replacer (CBR). Obviously such a fat cannot be used in combination with cocoa butter as their melting properties are different. A novel enzymic interesterification process (*see* below) for producing cocoa butter extenders (CBEs), which can be blended with cocoa butter but are rather cheaper, is now in commercial use.

Interesterification

The previous sections have shown that while we do have some capability for changing the properties of a fat by manipulating its fatty acid composition quite often the properties we may wish to alter depend more on the distribution of the fatty acids amongst the triglycerides rather than the overall fatty composition. To achieve this latter goal we need to resort to interesterification. Interesterification is the process of rearranging the fatty acids within the triglycerides of a fat or oil. The ester links between the fatty acids and glycerol are broken and then reformed but the fatty acids do not necessarily return to either their original position or even their parent glycerol molecule. When the reaction is carried out with a chemical catalyst the fatty acids end up randomly distributed around all the glycerol hydroxyl groups in the fat (Figure 4.22).

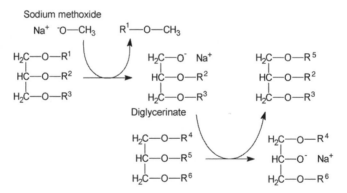

Figure 4.22   *Triglyceride interesterification catalysed by sodium methoxide. The methoxide ion extracts a fatty acid from a triglyceride molecule to generate a diglycerinate. This then extracts a fatty acid from another triglyceride molecule, a process than can be repeated indefinitely. As fatty acids at all three positions on the triglyceride react the continuation of sequence eventually leads to a completely random fatty acid distribution.*

After 'clean-up' (as with hydrogenation) and careful removal of all traces of water the fat is exposed to the catalyst, usually sodium methoxide, at temperatures around 50 °C for some 30 minutes. At the end of the reaction any traces of residual catalyst are readily washed out of the fat with water. Besides the manipulation of lard crystallisation referred to earlier this type of interesterification has recently been brought to bear on the *trans* acid problem in hydrogenated fats (*see* page 90). Some of the vegetable oil to be converted into margarine or shortening is fully hydrogenated so that it only contains saturated fatty acids such as stearic and palmitic, and of course no residual unsaturated *trans* acids. This is then mixed in appropriate proportions with totally unhydrogenated oil and the mixture subjected to interesterification. This can introduce just the right proportion of saturated acids into the oil to give an acceptable spreading fat without massive elimination of desirable polyunsaturates or the introduction of *trans* acids.

The specificity of enzyme catalysed reactions offers interesting variations on chemically catalysed interesterification. Many of the lipases secreted by fungi (for the purpose of utilising fatty materials) share the specificity of pancreatic lipase (*see* page 103) in that they only hydrolyse triglycerides at the 1 and 1′ positions. In the enzyme-catalysed interesterification reaction a fraction of palm oil rich in POP triglycerides is mixed with stearic acid. The mixture, diluted with hexane to ensure that both fat and fatty acid are dissolved and the resulting solution is not too viscous, is then pumped slowly through a bed of polymer beads in a reactor vessel at around 60 °C. The beads are coated with a fungal lipase, usually from *Mucor meihi*. The enzyme attacks the palmitic acid rich 1 and 1′ positions but in the near total absence of water the equilibrium position of the reaction is such that the enzyme catalyses not only the removal of fatty acids from the 1 and 1′ positions but also the reverse reaction, which will favour stearic acid insertion. The end result is the formation of a low cost fat with almost identical proportions of POP, POS and SOS to natural, and expensive, cocoa butter. As it also shares cocoa butter's polymorphic behaviour it can be used as a cocoa butter extender (the commercial term) to enhance the profitability of chocolate manufacture (the consumer's term).

Another interesting application of this technique is the preparation of fats with special nutritive characteristics. Human digestion of fats depends upon pancreatic lipase, with the fatty acids liberated from the 1 and 1′ positions together with 2-monoglycerides being absorbed by the mucosa of the small intestine. It is observed that long chain (*i.e.* C_{16} and above) saturated fatty acids are less well absorbed than their unsaturated counterparts, especially when they are positioned on the outer positions of dietary triglycerides. This is because when the lipase liberates them they readily form insoluble salts with any calcium ions about at the time. The calcium salts of the unsaturated fatty acids are much more soluble. As we have seen animal fats, including human milk fat, tend to have their saturated fatty acids in the 2 position, but plant fats are the reverse. This reduces the digestibility of plant fats when they are used in formula feeds for infants. The solution has been to use an enzymic interesterification with tripalmitin and oleic acid as the feedstocks to yield a product with oleic acid in the 1 and 1′ positions and palmitic acid in the 2 position. This is expected to find ready acceptance in infant formula feeds which mimic human milk much more closely than cow's milk. Cow's milk has more or less the correct distribution of fatty acids in its triglycerides but compared with human milk is low in linoleic acid, the essential fatty acid.

POLAR LIPIDS

The membranes of all living systems, not only the plasma membranes that surround cells but also those which enclose or comprise organelles such as mitochondria, vacuoles, or the endoplasmic reticulum, are composed of an essentially common structure of specialised proteins and lipids. A detailed account of the structure of membranes is outside the scope of this book, but the nature and properties of many of the lipid components are very important to food chemistry. Besides occurring in membranes, many polar lipids occur in small amounts in crude fats and oils and are also classically associated with egg yolk, where they are major constituents.

The defining feature of polar lipids is that they are *amphiphilic*. This is to say that while their molecules share with the triglycerides (and waxes) an affinity for non-polar environments, *i.e.* they are *lipophilic**, their molecules also include structural elements that have an affinity for aqueous, polar environments, *i.e.* they are *hydrophilic*. These two contrasting elements are apparent in the structure of phosphatidyl choline (Figure 4.23), otherwise known as lecithin.

Most of the important naturally occurring polar lipids are, like phosphatidyl choline, glycerophospholipids, having two long-chain fatty-acid residues, esterified with a glycerol molecule that also carries a phosphate group. The phosphate group is in turn esterified with one of a number of organic bases, amino acids, or alcohols, which add to the hydrophilic character of the molecule. The structures of a number of other

---

* Alternatively *hydrophobic*.

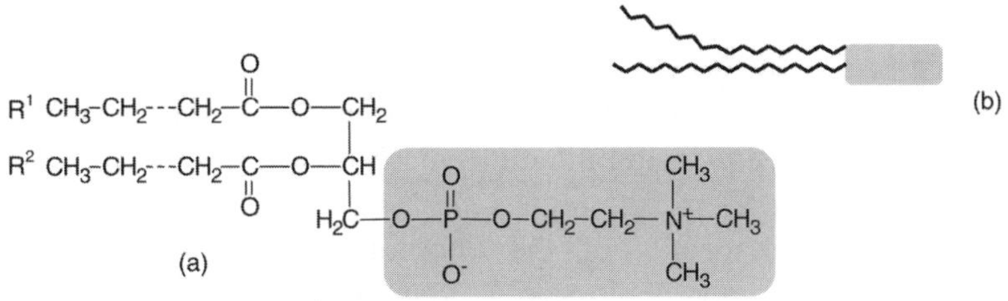

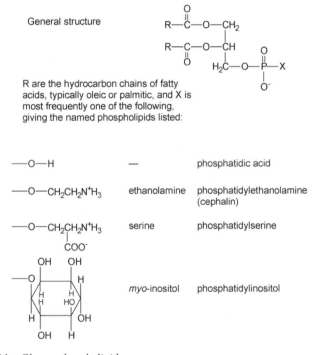

Figure 4.23 *Phosphatidyl choline. (a) The molecule has two long chain fatty acids (R¹ and R²), typically oleic or palmitic acid, esterified to the central glycerol. This also carries choline, attached through a phosphate group. The polar, hydrophilic, part of the molecule is shaded. (b) The small diagram shows how the molecule as a whole has a large hydrophilic head carrying a long hydrophobic tail.*

Figure 4.24 *Glycerophospholipids.*

glycerophospholipids are shown in Figure 4.24. Although a great many other phospholipid types have been described, from the point of view of the food chemist their properties are not significantly different from those shown here. One other polar lipid that should be mentioned is cholesterol

(4.12). Although it is not immediately obvious from the usual presentation of its chemical structure, as shown here, the hydrocarbon part of the cholesterol molecule has similar dimensions to the paired fatty acid chains of the glycerophopholipids and the molecule can occupy a similar position to the other polar lipids in membranes *etc*. Esters of cholesterol with fatty acids occur as components of the lipoproteins in the blood, but in membranes, and egg yolk, cholesterol is found unesterified.

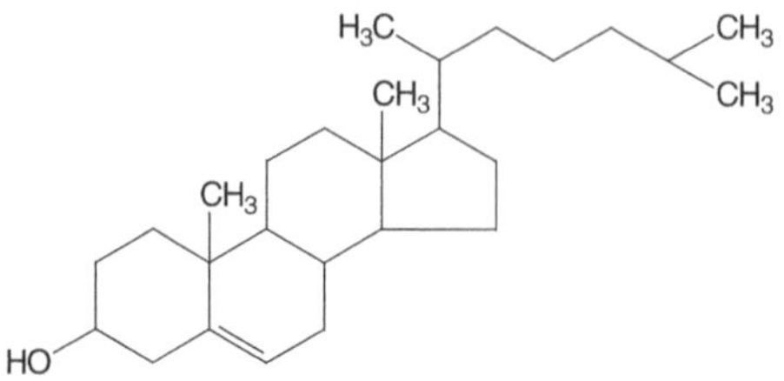

(4.12)

What is supplied commercially as lecithin actually contains a mixture of glycerophopholipids separated from crude soya bean oil. Typically it contains around 15% phosphatidyl choline, 12% phosphatidyl ethanolamine, and 10% phosphatidic acid, the remainder being mostly soya bean oil triglycerides. The importance of polar lipids in food systems is traditionally based on their ability to stabilise emulsions. Emulsions are colloidal systems of two immiscible liquids, one dispersed in the other continuous phase. Foods can provide examples of both 'oil-in-water' types, *e.g.* milk and mayonnaise, and 'water-in-oil' types, *e.g.* butter. In many cases the structure is complicated by the presence of suspended solid particles or air forming a foam.

Emulsions are prepared by vigorous mixing of the two immiscible liquids so that droplets of the disperse phase are formed. However, a simple emulsion usually breaks down rapidly as the dispersed droplets coalesce to form a layer which either floats to the surface or settles to the bottom of the vessel. Emulsion stability is enhanced by the presence of substances whose molecules have both polar and non-polar regions – such as phospholipids. Figure 4.25 shows how such molecules are able to orientate themselves at the interfaces between the two phases and thus form a barrier to the coalescence of the droplets.

Emulsifiers, and their non-food relatives, the soaps and detergents, are often described as surfactants (*surf*ace *act*ive ag*ents*). This is because one manifestation of their amphiphilic character is an ability to reduce the

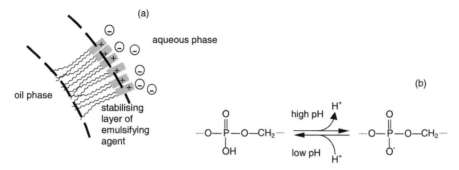

Figure 4.25   (a) *Part of the surface of an oil droplet in an oil-in-water emulsion stabilised by a phospholipid such as lecithin. The phospholipid molecules orientate themselves at the oil–water interface so that their polar groups interact with the aqueous continuous phase. If, as would be the case in the majority of food situations, the prevailing pH is low, the ionisable groups of the phospholipid will carry a net positive charge (b) which will in turn attract negatively charged ions present in the aqueous phase. The mutual repulsion of these layers of negative charges keeps the droplets apart, preventing them from coalescing.*

surface tension of water (*see* Chapter 12) in which they are dissolved. Not all surfactants have large lipophilic and hydrophilic groups. Polar organic solvents that are miscible with both water and non-polar solvents, such as ethanol (C_2H_5OH) and acetone (($CH_3)_2CO$) are both regarded as surfactants in that they lower the surface tension of water. The hydrophilic groups in many phospholipids include both phosphate groups (negatively charged at the pH values normally found in food) and quaternary nitrogen atoms (positively charged) and are therefore classified as *zwitterionic*, (*see* Chapter 5). Many other surfactants are acidic in character (*i.e.* anions) or have hydrophilic groups that do not ionise, the so-called non-ionic, surfactants. There are no cationic surfactants, where the hydrophilic group is positively charged at ordinary pH values, that are permitted food additives.

For fat to be digested in the small intestine, it is emulsified with bile salts, derivatives of cholic acid secreted by the liver. Sodium taurocholate (4.13) is a typical example. It may not be obvious how taurocholate can operate as an emulsifier since its formula appears to show no clearly marked polar region. However, a more three-dimensional portrayal of its structure, (4.13a) reveals that is able to present a lipophilic face to a fat globule surface and a hydrophilic face to the surrounding water.

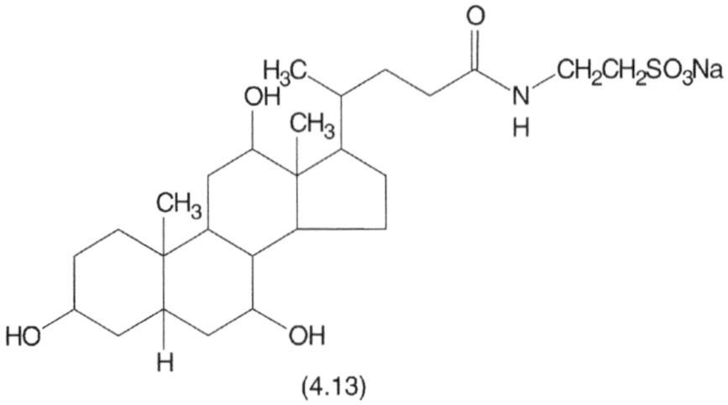

(4.13)

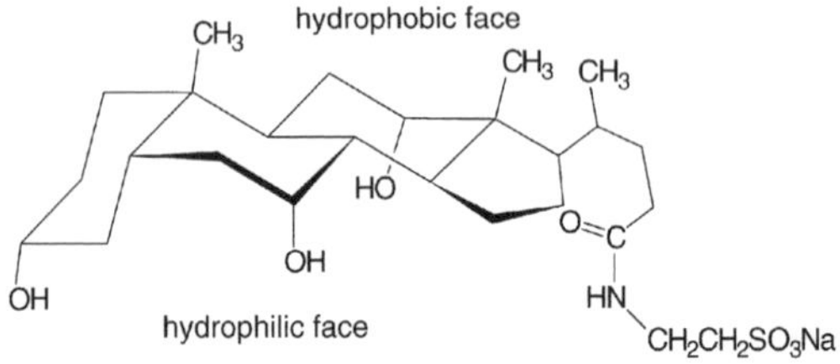

(4.13a)

The best known source of emulsifying properties in food preparations is egg yolk. Approximately 33% of the yolk of a hen's egg is lipid (protein is about 16%), of which about 67% is triglyceride, 28% is phospholipid, and the remainder is mostly cholesterol. Lecithins are the predominant phospholipids with small amounts of cephalins, lysophosphatidyl cholines (*i.e.* lecithins with only one fatty residue, on the 1-position), and sphingomyelins (4.14).

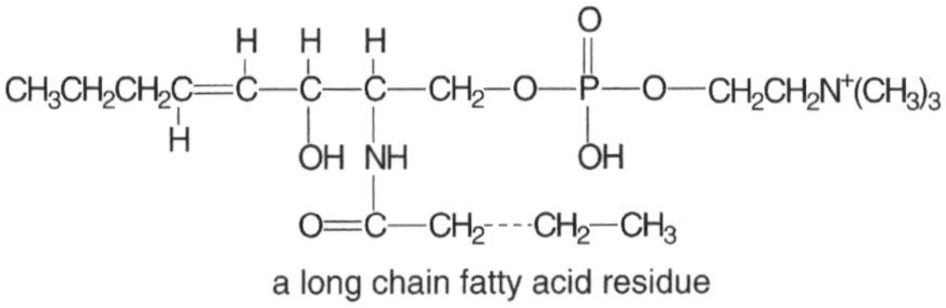

a long chain fatty acid residue

(4.14)

Yolk actually consists of a suspension of lipid/protein particles in a protein/water matrix. Specific associations of lipids and proteins such as these are known as lipoproteins and are employed in animal systems whenever lipid material has to be transported in an aqueous environment such as the blood. The best-known applications of egg yolk as an emulsifier are in mayonnaise and sauces such as hollandaise, but its role in cake-making is just the same. The lecithin recovered from soya bean oil during its preliminary purification is increasingly valued as an alternative natural emulsifying agent, particularly for use in chocolate and other confectionery, and also where a specifically vegetarian product is demanded.

In 1965 Griffin introduced the HLB (Hydrophile-Lipophile Balance) system, a scale of 1 to 20, for quantifying the emulsifying capability of surfactants. Although originally proposed for use with cosmetics, HLB values were widely applied to food emulsifiers. The HLB for an emulsifier was originally defined by the equation:

$$HLB = 20[1 - (S/A)]$$

where S is the *saponification number* of the ester and A is the *acid number* of the resulting acid. These are, to say the least, outdated chemical parameters and it is common nowadays to regard S as the overall molecular weight of the emulsifier and A as the molecular weight of the hydrophilic part of the molecule. These definitions gives no recognition to the differences in affinity for water between different hydrophilic groups; contrast the two hydroxyl groups of a monoglyceride with the phosphate and quarternary nitrogen of lecithin. Many quoted HLB values have been derived by empirical methods based on giving values to particular chemical groups or laboratory based experimental comparisons. The HLB values of a selection of commercially important emulsifiers is given in Table 4.8. Surfactants with low HLB values between 4 and 6 are the most

Table 4.8 *HLB values of a range of permitted emulsifiers. The structures of the semi-synthetic emulsifiers asterisked are shown in Figure 4.26. The fact that the value for sodium stearoyl-2-lactylate is greater than 20 is a typical case of the rather more empirical approaches to HLB value calculation often adopted nowadays.*

Emulsifier	E number	Type	HLB value
Oleic acid	470a	Anionic	1.0
Glycerol dioleate	471	Nonionic	1.8
Sorbitan tristearate (Span 65)*	492	Nonionic	2.1
Glycerol monooleate	471	Nonionic	3.4
Sorbitan monooleate (Span 80)	494	Nonionic	4.3
Soy lecithin (crude)	322	Zwitterionic	8.0
Diacetylated tartaric acid esters of monoglycerides*	472e	Anionic	8.0
Sucrose monolaurate*	473	Nonionic	15.0
Polyoxyethylene sorbitan monopalmitate* (Tween 40)*	434	Nonionic	15.6
Sodium stearoyl-2-lactylate*	481	Anionic	21.0

lipophilic and best suited to water-in-oil emulsions whereas for oil-in-water emulsions are HLB values in the 8-18 range are required. High HLB values indicate a high degree of hydrophilicity and solubility in water.

Alongside the uncertain origins of their values the principal reason for the limited application of HLB values in food is that food systems are far too complicated for such a simplistic approach. In particular almost all foods which contain emulsified fat also contain protein, and protein molecules are extremely effective emulsifying agents. The normal folding of the polypeptide chains of proteins (*see* Chapter 5) ensures that outer surface of a protein molecule is dominated by hydrophilic amino acid side chains. In contrast the interior of the molecule is dominated by hydrophobic amino acids. A major contributor to the maintenance of the correct folding of the protein is the drive to separate these hydrophobic side chains from the aqueous environment surrounding the protein. When a protein molecule encounters an oil droplet it naturally unfolds to expose its hydrophobic elements to the oil surface and at the same time present a hydrophilic surface to the surrounding water – Figure 4.25 on a larger scale. Loops consisting of hydrophilic sections of the polypeptide chains will protrude some distance above the surface of the oil droplet. The adsorption of the protein onto the oil surface is effectively an irreversible denaturation.

The major role of food emulsifiers is usually assumed to be in the foods that are most obviously emulsions, like mayonnaise, where an oil-in-water emulsion (actually olive oil and vinegar) is stabilised by egg yolk. However, food emulsions are much more widespread. For example, the fat in a cake batters is dispersed by the egg yolk phospholipids. Soya lecithin is often incorporated into commercial cake recipes to reduce the level of egg (much more expensive) required. The proteins from the meat are responsible for emulsifying the fat in sausages, especially the featureless pink objects revered in Britain. One reason why so much fat pours out of sausages and burgers when they are grilled is that the thermal denaturation of the meat proteins (in particular the so-called sarcoplasmic proteins, *see* Chapter 5) destroys their fat binding capabilities. The term 'emulsifying salts' refers to substances often included alongside added emulsifiers in meat products and processed cheese such as the sodium or potassium salts of phosphate, citrate, or tartrate which promote emulsification by solubilising proteins, which are then able to act as the true emulsifying agents. The solubilised proteins also bind any added water, such as that used to dissolve the emulsifying salts, until such time as it is released when cooking denatures the proteins. Stabilisers, which are also not emulsifiers in their own right, are also found in many food products that contain emulsions, such as ice cream and instant desserts. Most stabilisers are polysaccharides such as locust bean gum, xanthan gum, carageenans and alginates. They enhance emulsion stability by increasing the viscosity of the aqueous phase.

A small proportion of emulsifier, usually lecithin, is an essential ingredient in chocolate. Although chocolate contains very little water it does contain large quantities of sugar, milled to very small (<30 μm) particles. The sugar particles obviously have a very hydrophilic surface and would tend to clump together in liquid chocolate making the liquid very viscous. The non-fatty particulate material from the bean, mostly polysaccharides and protein, would join in this behaviour. The emulsifier molecules bind to the surface of the particles and give them a lipophilic surface that keeps them apart and the viscosity of the liquid chocolate low. Melted chocolate needs to be free flowing if it is to be used to coat biscuits *etc.*, or flow around elaborate moulds for Easter eggs.

Milk Fat, Cream and Butter

Dairy products provide examples of both oil-in-water and water-in-oil emulsions. The fat in milk occurs in very stable globules mostly between 4 and 10 m in diameter. There are $1.5–3.0 \times 10^{12}$ globules in one litre of milk. Globules of such small size should, according to Stoke's Law, take some 50 hours to float to the top of a pint bottle of milk and establish the familiar layer of cream, but we know that in fact half an hour will suffice. This implies that much larger particles are involved, with effective diameters up to 800 μm. This clustering of globules is clearly no ordinary coalescence, since 'evaporated milk', which is sterilised after canning by holding at temperatures above 100 °C for several minutes, shows no tendency to 'cream'. The explanation for this curious behaviour lies in the nature of the 'milk fat globule membrane' which surrounds each globule. This consists largely of specific lipoproteins and polar lipids arranged around the globule in a similar arrangement to that shown in Figure 4.25. The lipids of this membrane, including those associated with proteins to form lipoproteins, are typical of those found in the plasma membrane of animal cells, including cholesterol and its esters with long-chain fatty acids, phosphatidyl ethanolamine, choline, serine and inositol, and sphingomyelin. The membrane is most likely acquired as the milk fat globule is secreted through the plasma membrane of the mammary gland cells into the milk duct. Clustering is caused by crosslinking through one specific type of protein that occurs in the aqueous phase of the milk in very small amounts – macroglobulin. Heating above 100 °C for a few minutes (but not pasteurisation) denatures this protein so that creaming is prevented.

Creaming is also prevented in homogenised milk where the fat globules are reduced to about 1 m in diameter by passage through very small holes at high pressures and velocities around 250 m s^{-1}. The vast increase in total surface area of the fat globules (from around 75 m^2 per litre of milk to around 400 m^2) results in extra proteins being adsorbed from the aqueous phase of the milk. These proteins prevent coalescence of the globules but do not interact strongly with the macroglobulin, so that creaming no longer occurs.

Butter-making results in the formation of a water-in-oil emulsion. Cream with a fat content of 30–35% is inoculated with a culture of bacteria and incubated for a few hours. The bacteria produce the characteristic flavour compounds such as diacetyl (butanedione). The cream is then mechanically agitated (*i.e.* churned) sufficiently to disrupt the fat globule membranes and cause coalescence.

During mixing, a proportion of the aqueous phase, known as buttermilk, is trapped as small droplets. These are prevented from coalescing by the rigidity of the fat and by the layer of proteins and polar lipids which forms at the fat/water interfaces. Butter normally contains about 20% water, to which salt is added - nowadays as a flavouring, but originally to deter the growth of unwanted micro-organisms. Margarine has buttermilk and salt added to it during the final blending stages, and, since it lacks its own emulsifying agents, soya lecithin and other emulsifiers have also to be added.

Synthetic Emulsifiers

Figure 4.26 shows a range of semi-synthetic emulsifiers and their structures. They are 'semi-synthetic' in the sense that they are based on naturally occurring fats or oils subjected to a variety of chemical processes. The processes used do not result in the neat formation of a fairly pure, in the sense of consisting of a single chemical species, product. Instead what is given a single name is actually a mixture. For example, polyglyceryl polyricinoleate (PGPR) is often used in place of lecithin in chocolate. As shown in outline in Figure 4.27, PGPR is synthesised from glycerol and ricinoleic acid (from castor bean oil) but the final product will contain individual molecules differing in both the number of glycerol residues and the number of ricinoleic acid residues.

In terms of bulk usage monoglycerides are by far the most important as well as being the starting point for the industrial synthesis of many other types. They are manufactured by interesterification of triglycerides with excess glycerol in the presence of sodium hydroxide. The fatty acids become randomly distributed about all the glycerol molecules present resulting in a mixture of 'monoglycerides' that also contain some glycerol,

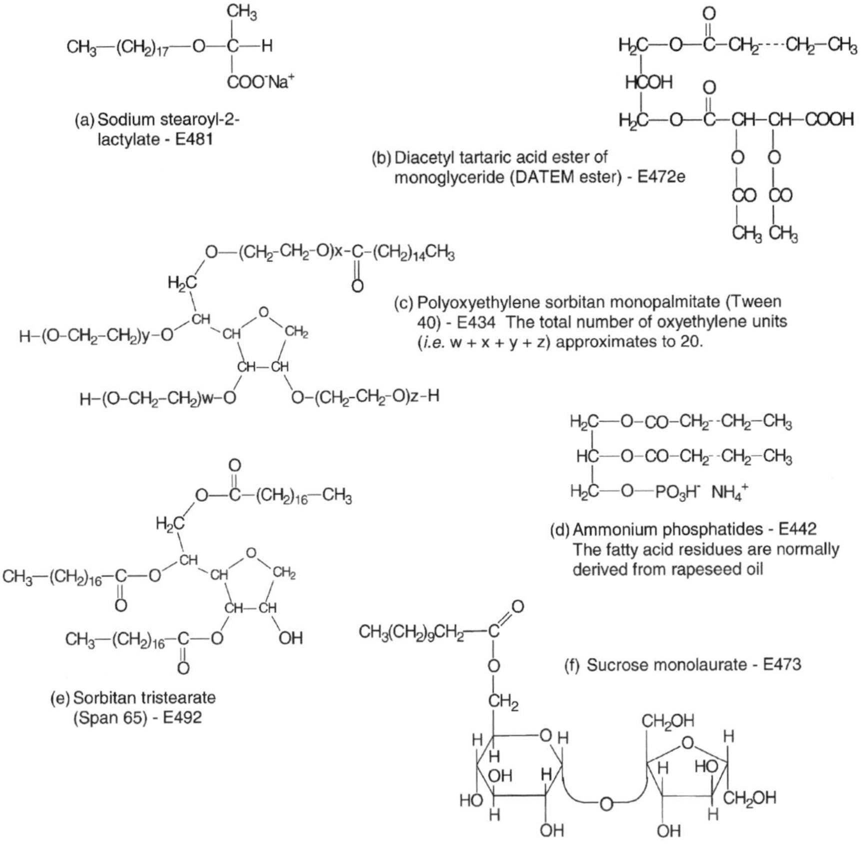

Figure 4.26 *The structures of a number of semi-synthetic emulsifiers permitted for use in food. As discussed in the text many of these formulae should be regarded as representative of a class of closely related substances, with variations in the fatty acid complement etc. This is particularly so in the sorbitan compounds (c) and (e) where six-membered anhydride rings also occur and the locations and total numbers of fatty acid residues varies. 'Tween' and 'Span' are widely used quasi-commercial designations for the two types of sorbitan based emulsifiers, applied regardless of the actual manufacturer.*

and di-and triglycerides. If required, fairly pure monoglycerides can be obtained from the mixture by fractional distillation under vacuum.

Baked goods, including bread, burger buns, cakes and biscuits account for a high proportion of total semi-synthetic emulsifier use. Monoglycerides of saturated fatty acids are useful for slowing down the rate of staling. The straight hydrocarbon chains are able to occupy the central core of the amylose and amylopectin helices, slowing down retrogradation (*see* page 50). Diacetyl tartaric acid esters of monoglycerides (DATEMS) are

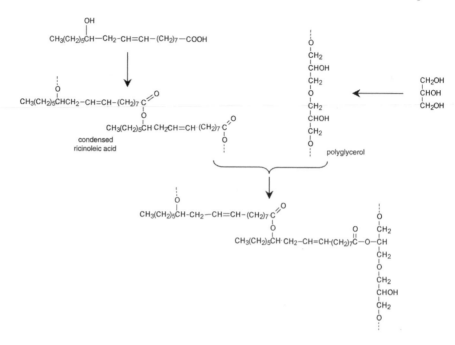

Figure 4.27 *An outline of the synthesis of polyglyceryl polyricinoleate (PGPR). The polyglycerol moiety normally has 2–4 glycerol units, the polyricinoleate moiety around 5 fatty acid units. There may be several polyricinoleate chains attached to the polyglycerol moiety.*

included at levels of around 0.5% (of the flour weight) in many commercial bread recipes, where they not only delay staling but also increase the loaf volume and improve the crumb texture, apparently by binding to gluten proteins.

Phytosteroids

Cholesterol and bile salts were mentioned earlier in this section and thereby provide some slight justification for including a brief consideration of a another group of steroids here, although they can hardly be classified as polar. Although the their presence in plant tissues has been recognised for a long time their recent inclusion in some margarines and low fat spreads, whose consumption can have beneficial effects on serum cholesterol levels, has brought them to prominence.

The three sterols shown in Figure 4.28 are abundant in plants, where, together with smaller amounts of the phytostanols, they play the same role in cell membranes as cholesterol does in animal cells. Western diets

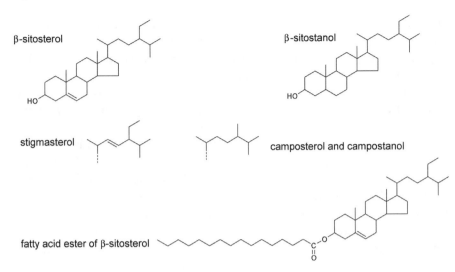

Figure 4.28 *Structures of the common phytosterols and phytostanols. The phytostanols differ from the phytosterols in being fully saturated; otherwise the differences between them are entirely within the side chains. The fatty acid (normally stearic) esters are the forms added to margarines.*

normally deliver around 200–400 mg daily of these plant sterols (with vegetarians towards the top of this range), about the same as the amount of cholesterol, plus about one tenth this amount of phytostanols. Although one of them, β-sitosterol, has been used for the reduction of serum cholesterol levels in hypercholesteraemic patients for many years it is only recently that the possibilities for their wider application as 'nutraceutical' food additives has arisen. It was discovered that attaching fatty acid chain greatly increased their solubility in oils and fats and made them effective at very much lower doses.

The usual sources of phytosterols are vegetable oils, such as soya and corn, where they are extracted along with lecithin and tocopherols during the refinement of the oil prior to hydrogenation. One important, if unlikely, source is 'tall oil'. Tall oil is extracted from pinewood during the manufacture of wood pulp for paper making.

Daily consumption of 2 g or more of phytosterols has been shown, in studies of patients with elevated serum cholesterol, to reduce levels of serum cholesterol particularly the low density lipoprotein (LDL) form, by 8–13%. It is an elevated level of this form of LDL-cholesterol, rather than the total cholesterol figure, that is a major risk factor in heart disease. Bearing in mind that in Western Europe the typical daily consumption of spreading fats, butter, margarine *etc.*, is around 20 g levels up to 20% of phytosterols (as esters) are being used to supplement margarines and low fat spreads. It has been calculated that consumption of 3 g per day of phytosterols could lower the risk of heart disease by as much as 15–40%, depending on other risk factors. The US Food and Drug Administration (FDA) have approved phytosterols addition to this level. It is not illegal to market them in Europe but the EC legislative position is presently confused.

Phytosterols work by inhibiting the uptake of cholesterol from the small intestine but the exact mechanism remains unresolved. Only a small proportion of the phytosterol is absorbed and even that which is absorbed is rapidly eliminated *via* the bile. There appear to be no significant safety issues in the consumption at even the highest levels suggested here. However, the widespread adoption of margarines and low fat spreads supplemented by phytosterols faces a number of obstacles, in particular the cost. At the present time supplemented products (in Britain at least) cost around five times as much as their unsupplemented counterparts! The question of their suitability for healthy young children does not seem to have been addressed as extensively as it has for hypercholesteraemic adults. In real families maintaining two distinct piles of sandwiches may not be quite as simple as one would wish and a more targeted pharmaceutical approach to the delivery of phytosterols may be more appropriate.

FURTHER READING

P. F. Fox and P. L. H. McSweeney, 'Dairy Chemistry and Biochemistry', Blackie, London, 1998.

'Food Lipids. Chemistry, Nutrition, and Biotechnology', eds. C. C. Akoh and D. B. Min, Dekker, New York, 1998.

H. Lawson, 'Food Oils and Fats. Technology, Utilization, and Nutrition', Chapman and Hall, New York, 1995.

'Egg Science and Technology', eds. W. J. Stadelman and O. J. Cotterill, 4th edn., Food Products Press, New York, 1995.

M. I. Gurr and J. L. Harwood, 'Lipid Biochemistry: an Introduction', 4th edn., Chapman and Hall, London, 1991.

'Rancidity in Foods', eds. J. C. Allen and R. J. Hamilton, Blackie, Edinburgh, 1994.

E. Dickinson, 'An Introduction to Food Colloids', Oxford University Press, Oxford, 1992.

'Diet and Heart Disease', ed. M. Ashwell, British Nutrition Foundation, London, 1993.

'Free Radicals and Heart Disease', ed. O. I. Arnoma and B. Halliwell, Taylor and Francis, London, 1991.

'The Lipid Handbook', ed. F. D. Gunstone, J. L. Harwood and F. B. Padley, Chapman and Hall, London, 1986.

'Fatty Acids in Foods and Their Health Implications', ed. C. K. Chow, Dekker, New York, 1992.

S. T. Beckett, 'The Science of Chocolate', The Royal Society of Chemistry, Cambridge, 2000.

'Industrial Chocolate Manufacture and Use', 2nd edn., ed. S. T. Beckett, Blackie, Glasgow, 1994.

Chapter 5

Proteins

The proteins are the third class of macrocomponents of living systems, and therefore of foodstuffs, that we are to consider.* Proteins are polymers with molecular weights ranging from around 10000 to several million and are usually described as having a highly complex structure. In fact there is a great deal about the structure of proteins that is quite straightforward. The monomeric units of which they are composed, the amino acids, are linked by a single type of bond, the peptide bond, and the range of different amino acids is both strictly limited in number and essentially common to all proteins. Furthermore the 'polypeptide chain' of proteins is never branched. The special character of proteins lies in the subtlety and diversity of the variations, of both structure and function, that Nature works on this simple theme. The properties and functions of a particular type of protein depend entirely on the particular sequence of its amino acids, unique to that protein. Unlike the polysaccharides there cannot be anything vague about the exact length of the chain. If even one amino acid in the sequence is wrong then it is quite likely that the protein will lose its biological activity. It is the sequences of the amino acids in proteins that are defined by the sequences of bases in the DNA that makes up our genes.

A typical protein's amino acid sequence, in this particular case the α_s-casein of cow's milk (*see* page 142) is shown in Figure 5.1. The sequence may look random but is in fact very tightly controlled and reproduced exactly in every molecule of this protein that cows produce.

As we will see later on this chapter the proportions of the different amino acids in the proteins we consume are very important but the total quantity is at least as important as the 'quality'. Table 5.1 lists the protein contents of a wide variety of different foodstuffs.

---

* In spite of their fundamental role in living systems the nucleic acids, RNA and DNA, are of almost no significance as components of our diet.

```
NH₂- R  P  K  H  P  I  K  H  Q  G  L  P  Q  E  V  L  N  E  N  L  L  R  F  F  →
   -  V  A  P  F  P  Q  V  F  G  K  E  K  V  N  E  L  S  K  D  I  G  S  E  S  →
   -  T  E  D  Q  A  M  E  D  I  K  E  M  E  A  E  S  I  S  S  S  E  E  I  V  →
   -  P  N  S  V  E  Q  K  H  I  Q  K  E  D  V  P  S  E  R  Y  L  G  Y  L  E  →
   -  Q  L  L  R  L  K  K  Y  K  V  P  Q  L  E  I  V  P  N  S  A  E  E  R  L  →
   -  H  S  M  K  Q  G  I  H  A  Q  Q  K  E  P  M  I  G  V  N  Q  E  L  A  Y  →
   -  F  Y  P  E  L  F  R  Q  F  Y  Q  L  D  A  Y  P  S  G  A  W  Y  Y  V  P  →
   -  L  G  T  Q  Y  T  D  A  P  S  F  S  D  I  P  N  P  I  G  S  E  N  S  E  →
   -  K  T  T  M  P  L  W  -COOH
```

Figure 5.1 *The amino acid sequence of bovine αₛ-casein. Each amino acid is represented by a standard code letter (see Figure 5.2). The start and finish of the sequence are referred to as the amino (NH₂) and carboxyl (COOH) ends respectively. The basis of these terms will become apparent in Figure 5.3.*

Table 5.1 *The total protein contents of a variety of foods and beverages. These figures are taken from 'McCance and Widdowson' (see Appendix II) and in all cases refer to the edible portion. They should be regarded as typical values for the particular food rather than absolute figures to which all samples of the food comply.*

Food	Total protein (%)	Food	Total protein (%)
White bread	8.4	Canned baked beans	5.2
Wholemeal bread	9.2	Lentils (dried)	24.3
Rice	2.6	Frozen peas (boiled)	6.0
Pasta	3.6	Beansprouts (raw)	2.9
Cornflakes	7.9	Tofu (steamed)	8.1
Cow's milk (whole)	3.2	Runner beans (boiled)	1.2
Human milk	1.3	Cabbage (raw)	1.7
Soya milk	2.9	Mushroom (raw)	1.8
Cheese (Cheddar)	25.5	Sweetcorn (canned)	2.9
Cheese (Brie)	19.3	Eating apples (raw)	0.4
Cheese (Parmesan)	39.4	Bananas	1.2
Yoghurt (plain)	5.7	Raisins	2.1
Ice cream (dairy)	3.6	Almonds	21.1
Whole egg	12.5	Peanuts (dry roasted)	25.5
Beef (lean, raw)	20.3	Jam	0.6
Lamb (lean, raw)	20.8	Plain chocolate	4.7
Chicken (lean, raw)	20.5	Milk chocolate	8.4
Beefburgers (raw)	15.2	Potato crisps	5.6
Pork sausages (raw)	10.6	Beer (bitter)	0.3
Cod fillet (raw)	17.4	Stout	0.3
Tuna (canned)	27.5	Lager	0.2
New potatoes	1.7		

AMINO ACIDS

All but two of the amino acids that occur in proteins have the same
general formula (5.1). (The structural formulae shown in Figure 5.2 show
how proline and hydroxyproline do not quite fit this formula.) The central

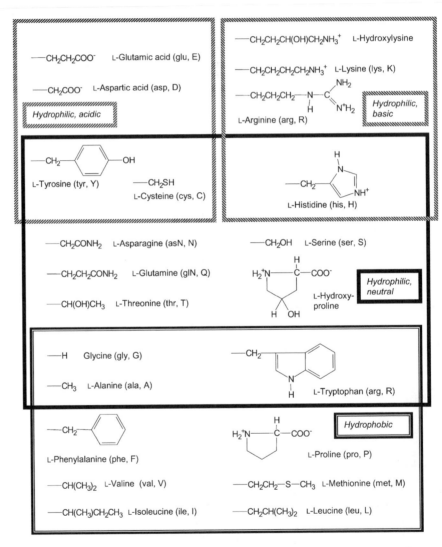

Figure 5.2 *The side chains of the amino acids commonly found in proteins.* L-*Proline and*
L-*hydroxyproline are not strictly amino acids but 'imino' acids as their*
nitrogen atom is attached to two carbon atoms, the second one being, in
effect, the far end of the side chain curled round and attached to form a ring.
Apart from L-*hydroxylysine and* L-*proline the names of all these amino acids*
have two abbreviated forms, shown here in brackets. The single letter 'codes'
are used when writing long amino acid sequences, as in entire proteins.

carbon atom carrying the side chain (R) that characterises the particular amino acid, as well as the carboxyl and amino groups, is known as the '*α*-carbon'*. Except in the case of glycine, where R = H, each of the four groups attached to the *α*-carbon atom is different, *i.e.* the *α*-carbon is asymmetrically substituted. This means that amino acids, like sugars (*see* page 10), are optically active. All the amino acids found in proteins are members of the L-series, *i.e.* they have the optical configuration shown in (5.2). This formula also shows the amino and carboxyl groups in their ionized states, giving a zwitterion, the form which prevails at neutral pH values. D-Amino acids do occur in nature but not in proteins; they are found in bacteria as components of the cell wall and in certain antibiotics. Free amino acids are not very important to food chemists although they do contribute to the flavour of some foods, as mentioned in Chapters 2 and 7.

$$\begin{array}{ccc}
\text{R} & & \text{NH}_3{}^+ \\
| & & | \\
\text{H}_2\text{N}-\text{C}-\text{H} & \text{or} & \text{H}-\text{C}-\text{COO}^- \\
| & & | \\
\text{COOH} & & \text{R} \\
(5.1) & & (5.2)
\end{array}$$

The structures of the side chains of the 20 amino acids found in most proteins are shown in Figure 5.2. Two other amino acids, hydroxylysine and hydroxyproline, are also included. They occur in a number of the structural proteins of animals, notably collagen (*see* page 159). There are various approaches to the classification of the amino acids, but the most useful is to consider them in terms of the properties of their side chains rather than their chemical structures. The ionisation of some side chains is obviously dependent on the pH of the protein's environment or the special conditions that may prevail at the active site of an enzyme.

PROTEIN STRUCTURE

The amino acids of a protein are linked together by so-called 'peptide bonds', formed by the loss of a molecule of water when the amino group of one reacts with the carboxyl of another. The peptide bond has an amide structure, as shown in Figure 5.3.

Of course the actual reaction involved in protein synthesis in living systems is much more complicated than the simple arrow in Figure 5.3

---

* Chemists use the Greek letters *α*, *β*, *γ*, *etc.* to identify the carbon atoms in sequence along a chain staring from the atom *next* to a major functional group such as, in this case, the carboxyl group. The amino acids that occur in proteins are strictly '*α*-amino acids' since their amino group is carried on the *α*-carbon.

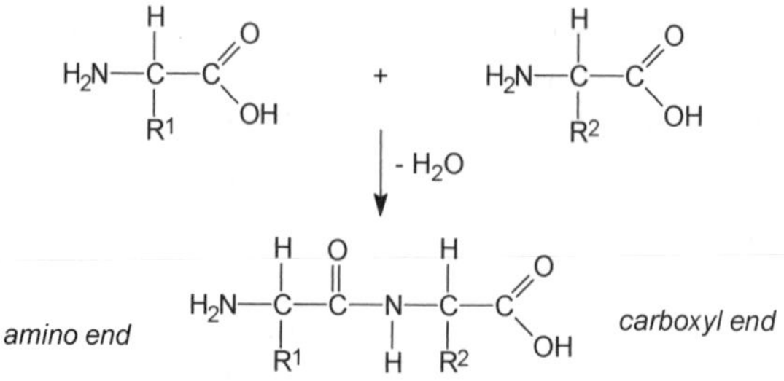

Figure 5.3 *The formation of a peptide bond. The 'amino' and 'carboxyl' ends of the chain are frequently referred to as the 'N' and 'C' termini respectively.*

might imply – not least because the enzymes concerned, part of the cell structures known as ribosomes, not only have to forge the links but also make sure the amino acids are combined in the correct sequence.

The electrons of the carbonyl group are delocalised to give the C–N bond considerable double-bond character. As a result there is no free rotation about the C–N bond and, as shown in Figure 5.4, all six atoms lie fixed in a single 'amide plane'. Figure 5.4 shows how a length of polypeptide chain can be visualised as a series of amide planes linked at the α-carbon atoms of successive amino acid residues. The polypeptide chain in Figure 5.4 has been drawn with scant regard for the tetrahedral distribution of the bonds about the α-carbon atoms. In reality the chain takes up more compact folding arrangements that can be defined by the angles of rotation about the C_α–N and C_α–C bonds. When a particular pair of such angles is repeated at successive α-carbon atoms, the polypeptide chain takes on a visibly regular shape which will usually be a helix but at the extreme will be a zig-zag arrangement not unlike that in Figure 5.4. The folding of the polypeptide chain in a protein is determined by its amino acid sequence in ways that molecular biologists are now beginning to understand to the extent that sophisticated computer programs can predict, with considerable accuracy, the most likely configuration that a given sequence of amino acids would take up. Within the molecule the unique patterns of folding of the peptide chain are maintained by bonds of

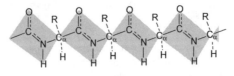

Figure 5.4 *The 'amide planes' of the polypeptide chain.*

various types. Covalent links, so-called sulfur bridges, are found linking cysteine* residues (5.3).

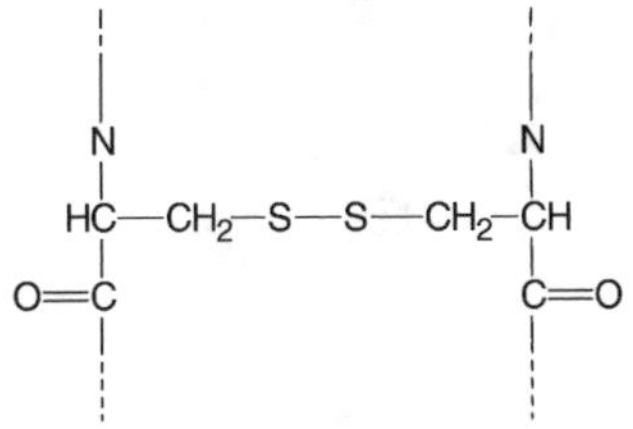

$$O=C \qquad \qquad C=O$$

$$HC-CH_2-S-S-CH_2-CH$$

(5.3)

Hydrogen bonding is especially important in maintaining ordered spatial relationships along the polypeptide chain. When these repeatedly join the carboxyl oxygen of one amino acid to the amino hydrogen of the amino acid next but three along the chain, the well-known α-helix results, as shown in Figure 5.5.

The side chains of the amino acids in helical structures such as these point outwards (*i.e.* away from the axis of the helix) and are themselves able to form bonds that stabilise the overall folding of the polypeptide chain in globular proteins.

Cysteine–cysteine bridges have already been mentioned, but inspection of Figure 5.2 will reveal ample scope for hydrogen bonding between side-chains, particularly involving the amide groups of glutamine and aspar-agine. Hydrophobic side chains tend to be orientated so that their interaction with the aqueous environment of the protein is minimised. Thus such residues are usually found directed towards the centre of the folded molecule, where they can be amongst other non-polar residues. The tendency of these amino acids to act in this way is a major factor in the maintenance of the correct folding of the polypeptide chain and is usually referred to as 'hydrophobic bonding', a less than ideal term (*see* Chapter 12).

Whatever biological role a particular protein has, it is always dependent on the correct folding of the backbone to maintain the correct spatial relationships of its amino acid side chains. Not surprisingly, extremes of pH and also high temperatures disrupt the forces maintaining the correct folding, which leads to the 'denaturation' of the protein. In a few cases highly purified proteins can be persuaded to return to their correct arrangement after denaturation, but in food situations this is highly unlikely to occur. It is much more probable that the unfolded proteins will

---

* The dimer of two molecules linked through their sulfhydryl groups is known as cystine.

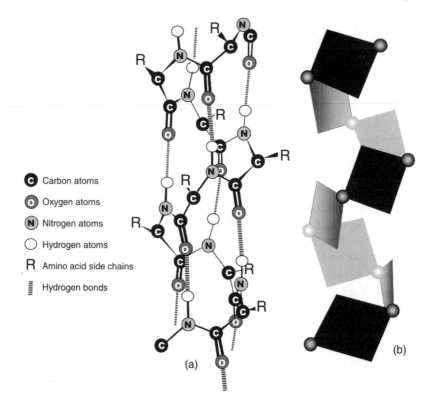

C Carbon atoms

O Oxygen atoms

N Nitrogen atoms

○ Hydrogen atoms

R Amino acid side chains

▤ Hydrogen bonds

Figure 5.5 *The α-helix. In* (a) *only the atoms of the backbone and those involved in hydrogen bonds are shown. In* (b) *the corresponding sequence of amide planes are shown linked at the α-carbon atoms. There are 3.6 amino acid residues per turn of the helix. Other helical patterns do occur, with other patterns of hydrogen bonding, but these do not give such a compact structure without straining the hydrogen bonds away from their preferred linear arrangement of the four atoms N–H···O=C.*

form new interactions with their neighbours that lead to precipitation, solidification, or gel formation. For example, the white of eggs is almost entirely water (∼88%) and protein (∼12%). When egg white is heated, denaturation leads to the formation of a solid gel network in which the water is trapped. Similarly, when liver is cooked, the proteins contained in the liver cells are denatured; if cooking is continued for too long, an unpalatable hard texture results from the rigid network of polypeptide chains.

The denaturation of proteins in foodstuffs is not necessarily undesirable. Vegetables are blanched in steam or boiling water before freezing to inactivate certain enzymes, particularly lipoxygenase (*see* page 252), whose activity can lead to the generation of off-flavours.

Proteins are found in numerous roles in living systems. Enzymes, the catalysts upon which the chemical reactions that comprise all life processes depend, are proteins. The carrier molecules, such as haemoglobin, which carries oxygen in the blood, and the permeases, which control the transport of substances across cell membranes, often against concentration gradients, are also proteins. Another group of proteins are the immunoglobulins, which form the antibodies that provide an animal's defence against invading micro-organisms. These three classes of proteins, with enzymes by far the most numerous, are all characterised by their physiological function. In structural terms they share a common globular arrangement of their polypeptide chains, as illustrated by the diagram of myoglobin in Figure 5.6.

Like many enzymes and other carrier molecules, myoglobin has a prosthetic group, *i.e.* a non-protein component, which participates in the catalytic or carrier function. In this case the iron in the centre of the porphyrin system forms a coordinate link with the oxygen molecule. It

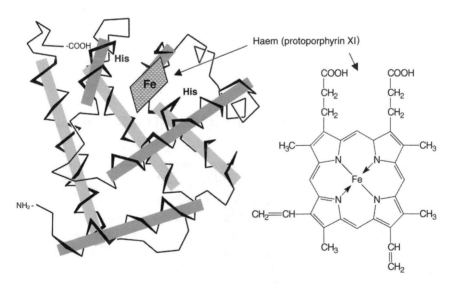

Figure 5.6 *Myoglobin. Only the backbone ($-C_\alpha-N-C-C_\alpha-N-C-C_\alpha-$) of the polypeptide chain is shown, with the 'angles' between successive 'bonds' showing the positions of the α-carbon atoms. The shaded 'rods' mark the axes of the regions of the polypeptide chain arranged in the α-helix configuration. The central shaded area shows the position of the haem prosthetic group, surrounded by amino acid side chains forming a hydrophobic pocket. The two histidine residues that are linked to the haem iron atom are marked 'His'.*

should be pointed out that the proportion of the polypeptide chain of myoglobin that is in the α-helix configuration (about 75%) is particularly high; most proteins have no more than 10% α-helix.

If a protein has angles of rotation about the C_α–N and C_α–C bonds repeated along the entire length of its polypeptide chain, this will obviously rule out any sort of compact, globular folding arrangement. Instead we can expect extended, highly ordered molecules. Proteins with such a configuration have structural roles in animal tissues, and one in particular, collagen, plays a crucial role in texture of meat. We will return to meat and muscle proteins later in this chapter.

The third group of proteins are those with a nutritional function either in the transmission of nutrients from mother to offspring (the casein of milk) or in the storage of nutrients to be utilised by an embryo (seed proteins of plants and egg proteins of birds). In these cases the physical characteristics of the protein will be of secondary importance to its overall chemical composition. Gluten, for example, the principal protein of wheat, is rich in glutamine, so that it has a higher nitrogen content compared with other proteins. Amongst their other distinctive characteristics, seed proteins and caseins have in common a tendency to form more or less ill-defined aggregates. Biologically, such behaviour ensures that large amounts of nutrient can be concentrated without the problems of osmotic pressure usually associated with high solute concentrations. For the analyst this behaviour has hindered efforts to elucidate the structure of these proteins, and, as we shall see, it is only recently that the structure of gluten has begun to be properly understood.

ESSENTIAL AMINO ACIDS AND PROTEIN QUALITY

The protein in our diet provides the amino acids from which the body synthesises its own proteins, the major constituent of our tissues. The proportions of the different amino acids supplied by a range of foodstuffs is presented in Table 5.2 The action of hydrolytic enzymes, first in the stomach and then in the small intestine, breaks down food proteins to their component amino acids. On absorption into the bloodstream they then become part of the body's amino acid pool. Breakdown of the body's own tissue proteins (an essential part of the process of renewal of ageing or redundant cells) also contributes to this pool. The amino acid pool is drawn upon not only for protein synthesis but also to provide the raw materials for the synthesis of purines, pyrimidines, porphyrins, and other substances.

The balancing of the pattern of amino acid supply against the needs for synthesis is a major function of the liver. Many so-called non-essential

Table 5.2 *The amino acid composition of dietary proteins. The data (calculated
from the data in 'First Supplement to McCance and Widdowson's The
Composition of Foods', eds. A.A. Paul, D.A.T. Southgate and J. Russell,
HMSO, London, 1980) shows the proportions of each amino acid in the
total protein content of the food.*

	Fresh peas	Wheat flour	Chicken breast	Beef steak	Whole egg	Cow's milk	Human milk	Cod fillet
g per 100 g of total protein								
Isoleucine	4.7	3.9	4.8	4.9	5.6	4.9	5.3	5.2
Leucine	7.5	7.0	7.8	7.6	8.3	9.1	9.9	8.3
Lysine	8.0	1.9	9.3	8.7	6.3	7.4	7.1	9.6
Methionine	1.0	1.6	2.5	2.6	3.2	2.6	1.5	2.8
Cysteine	1.2	2.6	1.3	1.2	1.8	0.8	2.0	1.1
Phenylalanine	5.0	4.8	4.7	4.3	5.1	4.9	3.8	4.0
Tyrosine	3.0	2.6	3.6	3.7	4.0	4.1	3.0	3.4
Threonine	4.3	2.7	4.3	4.5	5.1	4.4	4.5	4.7
Tryptophan	1.0	1.1	1.1	1.2	1.8	1.3	2.3	1.1
Valine	5.0	4.4	5.0	5.1	7.6	6.6	6.8	5.6
Arginine	10.0	3.6	6.5	6.4	6.1	3.6	3.8	6.2
Histidine	2.4	2.1	3.1	3.5	2.4	2.7	2.5	2.8
Alanine	4.5	3.1	6.0	6.1	5.4	3.6	4.2	6.7
Aspartic acid*	11.9	4.4	9.4	9.1	10.7	7.7	9.1	10.2
Glutamic acid[†]	17.3	32.9	17.1	16.5	12.0	20.6	17.4	14.8
Glycine	4.3	3.2	5.1	5.6	3.0	2.0	2.5	4.6
Proline[‡]	4.1	1.3	4.3	4.9	3.8	8.5	9.9	4.0
Serine	4.7	5.6	4.1	4.3	7.9	5.2	4.3	4.8

*Includes asparagine. [†]Includes glutamine. [‡]Includes hydroxyproline. Many of the more common
analytical procedures fail to separate these pairs of amino acids.

amino acids can be synthesised by mammals, provided that adequate
supplies of amino nitrogen and carbohydrate are available. However, there
are the other, essential amino acids (identified in Figure 5.7), which cannot
be synthesised by mammals and must be supplied in the diet.* Ideally the
protein in the diet should provide the amino acids in the same relative
proportions as the body's requirements, but of course such an ideal is
never attained. Excess supplies of particular amino acids are broken down,
with the carbon skeletons oxidised to provide energy or converted into fat
for storage. The nitrogen is either converted into urea for excretion *via* the
kidneys or utilised in the synthesis of any non-essential amino acids that
may be in short supply. Difficulties arise when it is one or more of the

---

* The rate at which humans can synthesise histidine, though adequate for adults, is insufficient to meet
the demands of rapidly growing children.

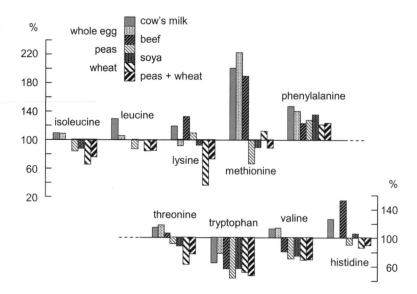

Figure 5.7 *The essential amino acids in some important food proteins. The values shown for each amino acid indicate the proportions (expressed as a percentage) relative to the proportion in human milk. Thus the provision of methionine by the animal proteins is seen to compare very favourably with this standard whereas the provision of lysine by wheat protein is inadequate. The total amino acid composition of human milk and some other important foodstuffs are shown in Table 5.2.*

essential amino acids that is in short supply. Absence of even one particular amino acid will result in cessation of all protein synthesis, since almost all proteins have at least one residue of all the amino acids shown in Figure 5.1 (except hydroxyproline and hydroxylysine).

The diet of a typical European normally contains more than ample supplies of both total protein and the individual essential amino acids, but for many of the people of Africa and Asia the supply of protein in the diet is inadequate in terms of both total quantity and supply of particular essential amino acids. Nutritionists have established that the proportions of the different amino acids required by the human infant correspond closely to the amino acid composition of human milk, so that this is now accepted as the standard against which the nutritional value of other foodstuffs is judged. Figure 5.7 shows the relative proportions (molar rather than weight) of the essential amino acids of a number of key foodstuffs in comparison with human milk.

It is immediately apparent from these data that the universality of the roles of the different amino acid side chains in the maintenance of protein structure leads to broadly similar amino acid compositions in foodstuffs of quite diverse origins. The animal protein foods (eggs, milk, and meat) do not differ significantly from human milk with regard to any amino acid, but the plant protein foods do present some problems. The low proportion of lysine in wheat protein means that wheat is only about 50% as efficient as human milk as a source of protein. An amount of wheat protein that provides just adequate levels of lysine provides wasteful levels of the other amino acids. The legumes such as soya and peas provide ample proportions of lysine but are deficient in methionine, so that they are also inefficient as sources of protein. A diet containing a mixture of both cereal and legume proteins will obviously be much more efficient than either alone. The efficiency of an individual protein, foodstuff, or whole diet can be described numerically by calculation from data on the relative proportions of its amino acids. The proportions of each essential amino acid are compared with those in human milk protein and the lowest of these (as a percentage) is the 'Chemical score'. These values (a selection are shown in Table 5.3) are a reasonable guide, but only a guide, to the actual performance of proteins in human nutrition. Experiments that measure efficiency in terms of the proportion of the dietary protein's nitrogen that the body utilises show that we never attain theoretical efficiencies. The inevitable inefficiencies of our digestive system and the effects of cooking are usually blamed for this and the wide variations in published data, most of which has been obtained in the context of the nutrition of farm animals rather than humans. The protein content of typical 'well balanced' European diets gives a Chemical Score of around 70. Adults in the UK typically consume between 60 and 90 g of protein per day so that any reasonably balanced diet (whether or not vegetarian) will provide at least 3 g of even the least abundant amino acids.

Table 5.3 *The nutritional efficiencies of proteins by 'Chemical Score' and experiment. The wide range of potential effects of cooking means that a single experimental value for meat has no been recorded.*

Protein source	Chemical score	Approximate experimental value
Human milk	100	94
Whole egg	100	87
Cow's milk	95	81
Peanuts	65	47
Beef	57	–
Wheat	53	49

Although plant proteins are used less efficiently than animal proteins one should not overlook the fact the much greater efficiency of agricultural processes in providing plant protein sources. This is a consideration that should apply to the supply of food in the affluent as well as the undernourished countries of the world.

The problems posed by the low levels of lysine and methionine in what would otherwise be ideal proteins are exacerbated by the tendency of these two amino acids to undergo reactions during the storage or processing of foods that destroy their nutritional activity. The Maillard reaction between the lysine amino groups and reducing sugars is the most important and was considered in Chapter 2, but this is not the only route by which the lysine in a protein may be lost. During processing at high temperatures, especially under alkaline conditions, several amino acids undergo changes in their side chains that result in nutritional losses. Cysteine and serine are both liable to be converted into dehydroalanine, which can form links with lysine side chains that result in irreversible links between different sections of the polypeptide chain. Affluent readers should also note that it is the hydrogen sulfide from the cysteine in egg proteins that tarnishes their silver egg spoons. Besides reducing the number of lysine residues available, cross-link formation of this type between neighbouring polypeptide chains will tend to prevent the assimilation of much of the remainder of the protein molecule, since unfolding and access for the proteolytic enzymes of the intestine will be prevented. The reactions involved in these processes, and losses of some other amino acids, are shown in Figure 5.8.

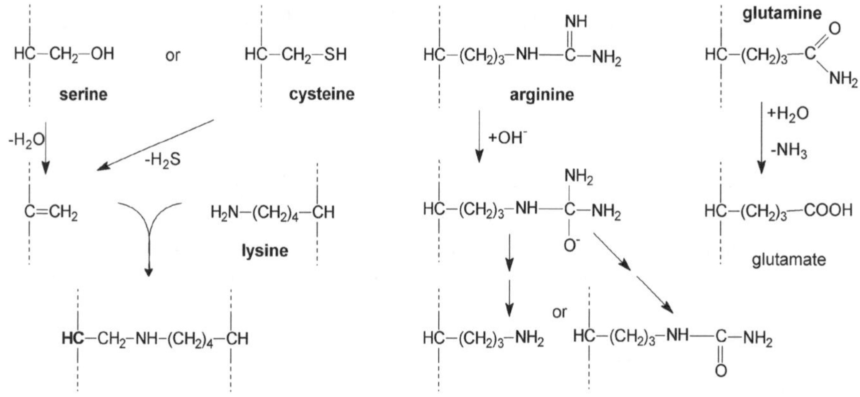

Figure 5.8   *Pathways of amino acid breakdown that occur when proteins are subjected to heat at alkaline pH values. Asparagine is converted into aspartic acid in the same reaction as shown for glutamine. The loss of water from serine is to some extent a reversible reaction but the product is just as likely to have the D configuration and therefore loses any nutritional value.*

ANALYSIS

Measuring the amount of protein in a food material is far from straightforward and the value of the data obtained for a particular application depends on the method used. For this reason the analysis of protein is given more prominence in this book than that of other food components.

The quality of a food protein in nutritional terms can really only be determined in feeding trials, but sufficient is now known about protein digestion and the effects of processing techniques for fairly accurate predictions to be made. These predictions require knowledge of both the total protein content of the food material and the protein's amino acid composition. Over the years many colour reactions specific to particular amino acids have been described. Unfortunately, few of them are quantitative and they cannot give the comprehensive data that are required. For these reasons chromatographic techniques are the analysts' automatic choice for the determination of amino acid composition. Well established systems are available for one- and two-dimensional paper chromatography. The amino acid spots can be detected by means of the well-known reaction with ninhydrin, triketohydrindene hydrate, to produce a blue colour (5.4). Proline and hydroxyproline produce a yellow colour.

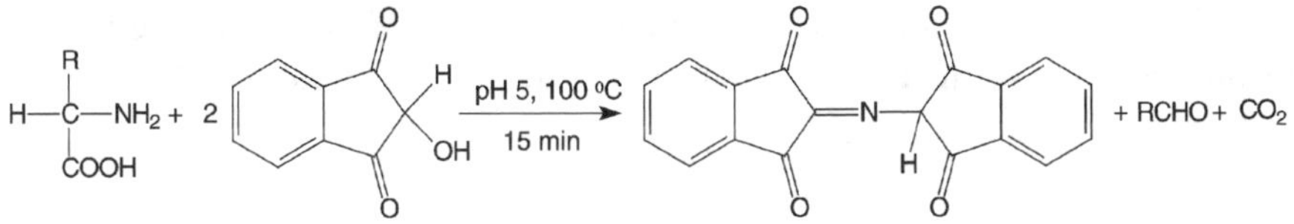

(5.4)

For routine work most laboratories use ion exchange column chromatography. After the protein hydrolysate has been applied to the ion exchange resin in the column, it is eluted with a series of solutions of increasing pH and ionic concentration that elute the amino acids in a highly reproducible sequence. Although a simple demonstration of the ion exchange separation of amino acids can be performed with ordinary laboratory apparatus, most laboratories use commercially manufactured 'amino acid analysers' in which all the operations of sample application, elution, and the detection and estimation of the amino acids by the ninhydrin reaction as they emerge from the column are automated. The acid hydrolysis that is required before these methods are applied inevitably converts the amino acids glutamine and asparagine into their corresponding acids, so that usually the content of these four amino acids is reported as glutamic acid plus glutamine and aspartic acid plus asparagine. Acid hydrolysis also destroys tryptophan, and it is therefore necessary to determine the tryptophan content of an alkaline hydrolysate by specific chemical tests.

The determination of the total protein content of a foodstuff or raw material is by no means as simple as one might imagine. Almost any chemical test, if it is to be unique to proteins, will depend on the presence of a particular amino acid side chain. Therefore analytical results will be subject to the variations in the proportions of the 'target' amino acid in the proteins of the test material and will generally need to be referred to some arbitrarily selected 'standard protein'. An example of this approach is the *Lowry method*, more popular with biochemists than food chemists but only applicable to proteins in solutions at concentrations of up to 300 $\mu g\,cm^{-3}$. In alkaline solution, in the presence of copper ions, tyrosine residues reduce phosphomolybdotungstate to a blue compound that can be determined spectrophotometrically.

Another colorimetric/spectrophotometric procedure, also limited to proteins in solution, which is fairly uniform in its response to different proteins is the *Biuret method*. Again in alkaline conditions, which ensure that the polypeptide chain is unfolded and fully accessible, Cu^{2+} ions are complexed by the nitrogen atoms of the polypeptide chain to give a characteristic violet colour (5.5). Potassium tartrate is included in the reaction mixture to ensure that excess Cu^{2+} ions are not precipitated as $Cu(OH)_2$. Although much less sensitive than the Lowry method, giving a linear response up to around 10 mg cm^{-3}, it has the great merit of being very easy to use.

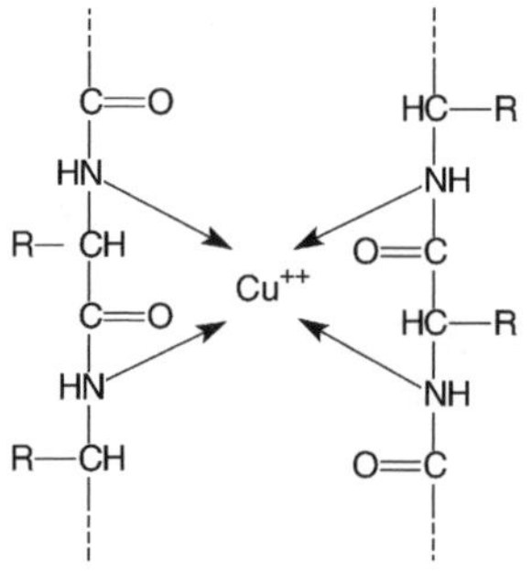

(5.5)

By far the most commonly adopted method for protein determination is the *Kjeldahl procedure*. This is based on the assumption that the proportion of non-protein nitrogen in a food material is too small to be significant and that a determination of total nitrogen (excluding nitrate and nitrite, which are not measured anyway) will therefore be an accurate reflection of the total protein content. The proportion of nitrogen in most proteins is roughly 16% (by weight) so that a factor of 6.25 is widely used to convert nitrogen content to protein content. Of course, variations in the amino acid composition of different foods lead to the need for slightly different factors being used in accurate work with certain foodstuffs. For example, cereal proteins have an unusually high proportion of glutamine, resulting in a higher than usual nitrogen content and the need for a lower conversion factor, 5.70. About one-third of the amino acids of gelatine are glycine, which also raises the nitrogen content, and a factor of 5.55 is required. Meat requires the standard 6.25, but the higher factors of 6.38 and 6.68 are required for milk and egg, respectively.

In the Kjeldahl procedure the entire food sample is digested at high temperatures with boiling concentrated sulfuric acid under reflux and with a heavy metal salt (*e.g.* copper sulfate) present as a catalyst. A quantity of sodium sulfate is also added to raise the boiling point. Under these conditions organic material is oxidised, with any organic nitrogen being retained in solution as ammonium ions. On completion of the digest an aliquot is withdrawn and transferred to an apparatus known as a Markham still. Here it is first made alkaline and then the liberated ammonia is steam-distilled off into an aliquot of boric acid. The quantity of ammonia can then be obtained by titration. Modern laboratories use a semi-automated system for these procedures.

FOOD PROTEIN SYSTEMS

Milk

The milk of the domestic cow, *Bos taurus*, is an important protein source for humans, particularly children. Milk is an aqueous solution of proteins, lactose, minerals, and certain vitamins that carries emulsified fat globules. It also carries particles known as *casein micelles* that consist of a number of related protein species, collectively known as the casein proteins, together with phosphate, calcium and a trace of citrate. If the fat is removed from milk, we call the product skimmed or skim milk. If the casein micelles are also removed (they are precipitated out at pH values around 4.6), the residual solution is known as serum (by analogy with blood) or more commonly, whey. The whey obtained in cheese-making has a slightly different composition, as some of the casein is solubilised and some of the lactose will have been converted into lactic acid by the action of bacteria.

The proportions of the different casein and whey proteins vary quite considerably, as does the total protein content of the milk, but the values shown in Table 5.4 are typical.

Table 5.4 *Typical values for the amounts of the major proteins in skimmed milk. The α_s-casein fraction consists of two proteins α_{s1} and α_{s2} that occur in the approximate ratio of three to one.*

	Grams per litre	*Percentage of total protein*
Casein proteins (total)	~27	~80
α_s-Casein	10–16	31–45
β-Casein	8–12	24–34
κ-Casein	10–15	3–5
γ-Casein	4–6	1.0–1.5
Whey proteins (total)	~6	~20
Lactalbumin	1.0–1.5	3.3–5.0
Lactoglobulin	2–4	6.6–13.3
Immunoglobulins	0.7–0.8	2.3–2.7
Others	1–2	3–7

The Greek letters used in the names of the different proteins were originally assigned on the basis of the protein's mobility on electrophoresis. The 's' of α_{s1}-casein refers to its sensitivity to precipitation by calcium ions. Although the prefixes α_{s1}, β, *etc.* define particular protein species, it is now recognised that there are a number of different versions of each that differ slightly in their amino acid sequences. For example, α_{s1}C-casein differs from α_{s1}B-casein (the most common variant) in having a glycine residue instead of a glutamic acid residue at position 192 in the polypeptide chain. One actually can identify the breed of cow from which a sample of milk was obtained by detailed examination of the proportions of these variants.

The casein micelles of milk are roughly spherical particles with diameters of up to 600 nm. However, approximately half of the total casein is found in micelles with diameters between 130 and 250 nm, with the remainder fairly equally divided between micelles above and below this range. Milk has about 10^{15} micelles per litre. A typical micelle contains 10^4–10^5 casein molecules (the molecular weights of α_s-, β-, and κ-casein are 2.35, 2.40, and 1.90×10^4, respectively). γ-Casein is believed to be an artefact of the laboratory procedures used to separate and purify casein molecules – a fragment of β-casein resulting from a limited breakdown caused by proteolytic enzymes in milk.

For many years chemists have sought to describe the structural arrangement of the casein molecules within the micelle. Although many details are still to be resolved the casein micelle structure that is most widely accepted nowadays is based on the concept of submicelles, introduced by Slattery and Evard in 1975. The submicelles are more or less spherical aggregates of 25–30 molecules of α_S-, β-, and κ-caseins. The average proportions of the different casein molecules in the sub-micelles will obviously reflect those of the milk as a whole but it is likely that the proportions will vary widely between individual submicelles. In particular it is assumed that some will be very much richer in κ-casein than others.

The association of the casein molecules to form submicelles depends on the unusual characters of all three types of casein. The polypeptide chains of all three have a predominance of polar amino acids towards their N-terminal ends. In particular phosphoserine residues (5.6) occur in the hydrophilic regions of α_S- and β-caseins and are responsible for binding calcium ions and cross-linking between submicelles *via* chains of so-called 'colloidal' calcium phosphate clusters. In contrast κ-casein has no phosphate residues. Instead one or more of the threonine residues in the polar C-terminal region carry a trisaccharide unit, α-N-acetylneuraminyl-(2→6)-α-galactosyl-(1→6)-N-acetylgalactosamine (5.7) which ensures hydrophilic character for the C-terminal region.

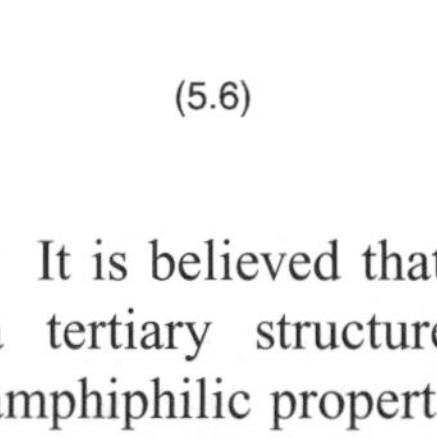

(5.6) 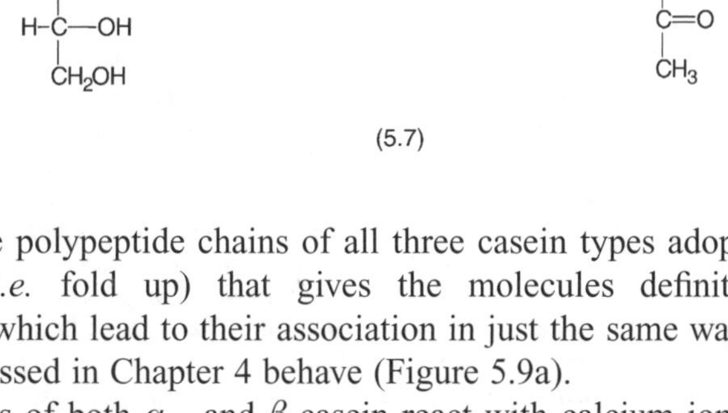   (5.7)

It is believed that the polypeptide chains of all three casein types adopt a tertiary structure (*i.e.* fold up) that gives the molecules definite amphiphilic properties which lead to their association in just the same way as the polar lipids discussed in Chapter 4 behave (Figure 5.9a).

The phosphate groups of both α_S- and β-casein react with calcium ions to link the submicelles together, either directly or more likely through chains of colloidal calcium phosphate (CCP) particles. Occasional magnesium and citrate ions are also involved. All the available phosphoserine residues carry calcium ions, as will the available carboxyl groups of amino acids such as glutamic acid (Figure 5.9(b)). The key to the submicelle concept is the suggestion that the κ-casein molecules aggregate together in

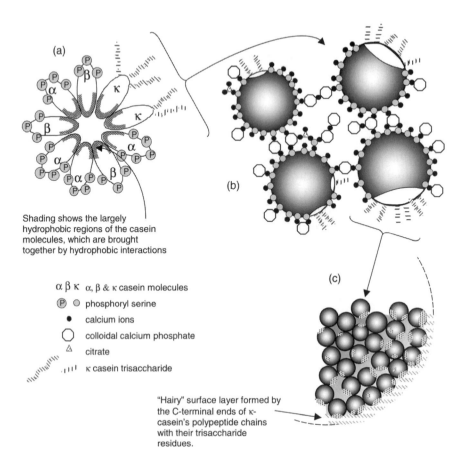

Shading shows the largely
hydrophobic regions of the casein
molecules, which are brought
together by hydrophobic interactions

α β κ   α, β & κ casein molecules

Ⓟ   ○   phosphoryl serine

●   calcium ions

○   colloidal calcium phosphate

△   citrate

,ıııı   κ casein trisaccharide

"Hairy" surface layer formed by
the C-terminal ends of κ-
casein's polypeptide chains
with their trisaccharide
residues.

Figure 5.9   *The casein micelle, based on the original concept of Slattery and Evard and
subsequently refined by Schmidt and Holt. (a) A cross section of a typical
submicelle. Although the three types of casein molecules are shown here with
a 'rugby ball' shape evidence is accumulating to show that their actual shapes
are more complex and provide for more precise interlocking than can be
shown here. (b) Cross links between submicelles consist of chains of colloidal
calcium phosphate particles attached to the phosphoserine residues of α- and
β-casein. Cross links cannot be formed by regions dominated by the κ-casein
trisaccharide. (c) Formation of the full sized micelle. A submicelle containing
little or no κ-casein will form links with other micelles in all directions and
will therefore tend to be found in the centre of a micelle. However, a
submicelle with a high, and localised, proportion of κ-casein will tend to form
links in a limited range of directions and will tend to be found towards the
outside of the micelle. As more and more submicelles become attached the
surface of the micelle inevitably gets flatter. This gives fewer and fewer
opportunities for binding additional submicelles until a limiting size, defined
by surface curvature rather than diameter, is reached.*

the submicelle so that their highly hydrophilic, but non-calcium-binding, ends form an area, resembling a polar ice cap, where no cross-linking between submicelles can occur. Even if the κ-casein molecules are not as highly localised as Figure 5.9(b) implies there will still be a tendency of submicelles with little or no κ-casein to predominate at the core of the micelle. The increase in size as more submicelles are added (Figure 5.9(c)) results in an inevitable tendency for the non-linking, κ-casein, areas to come to dominate the surface and ultimately prevent the indefinite increase in the size of the micelle. Electron microscopy of casein micelles has confirmed that that have the raspberry like appearance that one would expect of an aggregate of roughly spherical particles.

There are two important features of this structure. Firstly it provides for a self-limiting size for the micelle with the κ-casein dominated surface that has been recognised for many years. It is extremely difficult to visualise any biological process that could equip a micelle consisting of α_S- and β-casein with a κ-casein coat once it reached a certain size! From a physiological point of view some sort of protein aggregate is essential since at this concentration a simple protein solution would be too viscous for ordinary secretion. The other is that binding CCP to protein ensures that large amounts of both calcium and phosphate can be supplied to the infant without risk of impossibly large crystals of insoluble calcium phosphate being formed. The CCP particles (typically $Ca_9(PO_4)_6$) are amorphous.

The most abundant serum proteins, α-lactalbumin and β-lactoglobulin, have attracted considerable attention from protein chemists. They have been interested in their physical properties and structure, but these studies have revealed little of the reason for their presence in milk beyond their contribution to the total protein content. The immunoglobulins, in contrast, have proved to be of great interest to food scientists. The role of the macroglobulins (one class of immunoglobulins) in the creaming phenomenon was discussed in Chapter 4, but the biological role of immunoglobulins is much more important. Immunoglobulins are popularly known as antibodies. They are proteins synthesised in various parts of the body (including the mammary gland) in response to the invasion of the tissues by foreign matter, particularly bacteria, viruses, and toxins. Their reaction with these antigens is remarkably specific and facilitates the neutralisation or destruction of the invader by other parts of the body's defence mechanisms. It is now well established that exposure of a mother to many common pathogenic bacteria, especially those causing intestinal diseases such as diarrhoea which are very dangerous to the new-born, leads to the appearance of appropriate antibodies in the mother's milk. These are undoubtedly effective in protecting the new-born infant from many dangerous diseases. However closely the manufacturers of infant milk products based on cow's milk match the nutrient content of human milk, it is becoming increasingly clear that they will not be able to match the other, unique, advantages of human milk for feeding human babies.

Cheese

The precipitation of casein to form a curd is the fundamental process involved in cheese-making. In the case of yoghurt and some cottage cheeses the precipitation is caused entirely by low pH. The growth of lactobacilli in the milk is accompanied by their fermentation of the lactose to L-lactic acid (5.8). The lactobacilli used are sometimes those naturally present in the milk, but usually a starter culture of a strain having particularly desirable characteristics is added to the milk. There is not a total precipitation of the casein at the pH of these types of fermented milk product. Associations between the casein micelles give the gel-like texture that characterises yoghurt.

$$
\begin{array}{c}
COOH \\
| \\
HO\!-\!\!\!-\!\!\!\!-\!H \\
| \\
CH_3
\end{array}
$$
(5.8)

In the production of hard cheeses such as Cheddar, Stilton and Gruyére, microbial action brings the pH of the milk down to around 5.5. At this point rennet is added to bring about extensive precipitation and curd formation. Rennet is a preparation of the enzyme chymosin (in the past known as rennin) obtained from the lining of the abomasum (the fourth stomach) of calves. Chymosin specifically catalyses the hydrolysis of one particular peptide bond in κ-casein, as shown in Figure 5.10. The κ-casein is split into two fragments. One, the para-κ-casein, remains as part of the micelle; it includes the hydrophobic section of the molecules. The other, the κ-casein glycopeptide, is lost into the whey. The macropeptide fragment carries the carbohydrate units. The loss of their carbohydrate coats means that the formation of strong cross links between micelles is no longer blocked and a curd rapidly develops.

The curd is held for several hours, during which time the acidity increases and the curd acquires the proper degree of firmness. When this is achieved, the curd is chopped into small pieces to allow the whey to run out as the curd is stirred and finally pressed. Salt is added before pressing to hasten the removal of the last of the whey and to depress the growth of unwanted micro-organisms. The final stage of cheese manufacture, ripen-

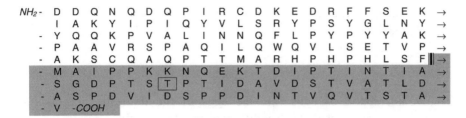

Figure 5.10 *The amino acid sequence of κ-casein using the single letter code for amino acids. The bar marks the peptide bond between phenylalanine-105 and methionine-106 that is hydrolysed by the enzyme chymosin during cheese manufacture, liberating the fragment (shaded) known as the κ-casein glycopeptide into the whey. The boxed threonine residue (133) is the normal site of attachment for the trisaccharide. Trisaccharide units are also found occasionally at a number of nearby threonine and serine residues.*

ing, requires low-temperature storage for a considerable period. Rennet enzymes (notably pepsin) and other proteases from the micro-organisms cause a limited degree of protein breakdown, which is important for both the texture and flavour (many small peptides have distinctive bitter or meaty tastes) of the finished cheese. Lipases from the milk liberate short-chain fatty acids from the milk fat triglycerides, which also contribute to the flavour.

In recent years the increasing yields of milk per cow have meant that fewer calves are born in relation to the amount of milk being converted into cheese. As a result rennin supplies have not kept pace with demand. This has forced cheese makers to examine other enzymes as substitutes or in mixtures with calf rennin. Proteases from fungi and porcine pepsin have been used but they suffer from the disadvantage of causing excessive breakdown of other proteins[*]. The modern solution brings genetic engineering to bear on the problem. The DNA coding for calf rennin has been successfully inserted into the chromosomes of the yeast *Kluyeromyces lactis*, the fungus *Aspergillus niger*, and the bacteria *Escherichia coli*. These micro-organisms can then be persuaded to produce large quantities of the enzyme, identical in every way to that produced by the calf.

The special relationship between rennin action and the structure of casein obviously did not evolve to enable us to make cheese, but the details of casein digestion in the calf or in man have not yet been

---

[*] Cheese manufactured using fungal rennet is usually marketed as 'vegetarian'. Bearing in mind that cows have to give birth to a calf before they can produce any milk, and that only a small proportion of calves go on to enjoy a fulfilling career of their own in milk production, it seems rather eccentric to focus on the source of chymosin as the qualification for vegetarian status.

investigated in sufficient detail to provide an explanation for the origin of the relationship.

Egg

The yolk fraction of eggs (*i.e.* the eggs of the domestic hen, *Gallus domesticus*) was considered briefly on page 116 and this section will be devoted to the 'white' or albumen fraction. Egg white consists almost entirely of protein, 11–13%, and water. Many of the applications of eggs in cooking can be explained, as we shall see, in terms of the properties of this protein.

In the raw egg the white is not homogeneous but occurs in four layers, two thick and two thin. The inner thick layer nearest the yolk is continuous with the chalazas, the stringy structures that suspend the yolk in the centre of the egg. The protein composition of the different layers is identical except that thick, *i.e.* more viscous, layers have a higher proportion of the protein ovomucin. For all culinary purposes the different layers of the white are thoroughly mixed and the white can therefore be regarded as a simple aqueous solution of all but one of the globular proteins listed in Table 5.5. The exception is the ovomucin which apparently forms fibres which make a major contribution to the viscosity of the white.

The predominant protein, ovalbumin, is a glycophosphoprotein, *i.e.* a protein with both phosphate and carbohydrate groups attached. Up to two phosphates are attached to serine residues (as in α- and β-caseins) and an asparagine residue carries the carbohydrate unit. This consists of two or more *N*-acetylglucosamine (3.1) units carrying several mannose units. Ovalbumin is readily denatured, the process we observe when egg white sets on heating. The polypeptide chain of ovalbumin includes six cysteine residues. Two of these form a sulfhydryl bridge (5.3) which helps to

Table 5.5 *The major proteins of hen's egg white. Con-albumin is also known as ovotransferrin.*

	Approximate percentage of total proteins
Ovalbumin	54
Conalbumin	12
Ovomucoid	11
Ovomucin	1.5–1.3
Lysozyme	3.4
Ovoglobulins	8

maintain the three-dimensional structure of the protein. When the temperature is raised the polypeptide chain unfolds and new sulfhydryl bridges become possible, not only within the molecule but also linking neighbouring molecules together, forming the rigid gel of cooked egg white.

Ovalbumin is also particularly susceptible to denaturation at air-water and at-water interfaces. It will be recalled (from page 131) that the polypeptide chains of globular proteins tend to fold up so that amino acids with hydrophobic side chains are located in the centre of the molecule away from the surrounding water. If the water on one side of a protein molecule is displaced by air or fat the advantage of keeping the hydrophobic amino acids tucked away in the centre of the molecule disappears and the molecule unfolds, *i.e.* becomes denatured. As before, the denatured molecules readily interact together but instead of forming a solid gel they form a film around air bubbles or fat droplets, stabilising respectively foams (as in meringue) or emulsions (as in cake batters). Subsequent heating causes further denaturation to reinforce these films.

Beating egg white to a stable foam is a difficult process to control. If the beating goes on for too long too much of the ovalbumin becomes denatured, leaving too little undenatured protein to bind the water. Beating must stop as soon as peak stiffness is achieved. Further beating to incorporate more air will only weaken the foam.

Conalbumin, another glycoprotein, is a metal binding protein analogous to the lactoferrin in milk (*see* page 382). It behaves similarly to ovalbumin but has been implicated in one rather odd aspect of cookery technique. Glass bowls are preferred to plastic bowls for beating egg white as the surface of even well-cleaned plastic bowls is likely to be contaminated with traces of fat which interferes with foam formation. However, copper bowls are claimed to be the very best. It has been suggested that the conalbumin binds with Cu^{2+} ions from the bowl and that the resulting copper–protein complex is more resistant to surface denaturation, thereby providing some protection against overbeating.

Ovomucoid, also a glycoprotein, gives the albumen its high viscosity. Although some viscosity is important for foam stability too much makes whipping difficult. This is why it is important that eggs for meringue or cake making should not be used straight from the refrigerator when their low temperature would cause too high a viscosity.

Foam stability is enhanced when the pH is close to the isoelectric points of the proteins. Many of a protein's amino acid side chains include ionisable groups, notably aspartic and glutamic acids, lysine, arginine and histidine. The proportion of a particular amino acid side chain that is ionised is clearly a function of the hydrogen ion (H^+) concentration, *i.e.* pH:

$$-COO^- + H^+ \longleftrightarrow -COOH \qquad \text{aspartic \& glutamic acids}$$

$$-NH_2 + H^+ \longleftrightarrow -NH_3 \qquad \text{lysine}$$

$$=NH_2 + H^+ \longleftrightarrow =NH_3 \qquad \text{arginine}$$

$$>\!\!N + H^+ \longleftrightarrow >\!\!N^+H \qquad \text{histidine}$$

The isoelectric point of a protein is that pH at which the numbers of positively charged groups just equals the number negatively charged groups, *i.e.* the protein molecule carries no *net* charge. The protein molecules do not repel each other so much when they are not charged so that they can interact together more strongly. Cooks use cream of tartar (potassium hydrogen tartrate) or a trace of vinegar to achieve this acidification.

Meat

When we consume the proteins of milk and eggs we are consuming proteins that evolved purely as a food; when we consume meat or study the processes involved in the conversion of muscle tissue into the slice of roast beef on our plate, we must remember that the biologist's view of its function, as well as the view of the animal itself, will be quite different from that of the food chemist or nutritionist. The structure of muscle tissue* is exceedingly complex, and readers with biological or biochemical interests are strongly advised to supplement this rather rudimentary account with the details to be found in text-books of cell biology, mammalian physiology, or biochemistry.

A typical joint of meat from the butcher's shop is cut from a number of muscles, each having its own independent attachment to the skeleton, its own blood supply, and its own nerves. Each muscle is surrounded by a layer of connective tissues consisting almost entirely of the protein collagen. This layer contains and supports the contractile tissues of the muscle and at its extremities provides the connections to the skeleton. The contractile units of the muscle are the muscle fibres. These are exceptionally elongated cells, $10-100$ μm in diameter but often as long as 30 cm. Examined in cross-section, the muscle fibres can be seen to be organised into bundles separated by connective tissue which is continuous with that surrounding the muscle and providing the connections to the skeleton. Blood vessels, adipose (*i.e.* fatty) tissue, and nerves are also embedded in the connective tissue.

---

* Although the term 'meat' usually encompasses fish and poultry as well as other mammalian tissues such as liver, kidney and intestine (the offals), this account is largely restricted to the properties of the skeletal muscle tissue of mammals such as pigs, sheep and cattle.

The muscle fibres have most of the features found in more typical animal cells. Surrounding the fibre is the cell membrane, known as the sarcolemma (the prefix 'sarco-' is derived from the Greek for 'flesh'). Along the length of the cell, just inside the sarcolemma, are numerous nuclei. Muscle fibres have two elaborate intracellular membrane systems, the transverse tubules, or T-tubules, which are invaginations of the sarcolemma, and the sarcoplasmic reticulum, which is the counterpart of the endoplasmic reticulum of other cells. Both participate in the transmission of the signal from the motor nerve endings on the surface of the fibre to the contractile elements of the fibre – the myofibrils. The T-tubules and sarcoplasmic reticulum wind around each myofibril so that the contraction events in each myofibril are accurately synchronised. Innumerable mitochondria are also found lying between the myofibrils to ensure that the supply of adenosine triphosphate, ATP, the chemical fuel for the contraction process, is maintained.

The myofibrils are themselves composed of bundles of protein filaments arranged as shown in Figure 5.11(a). The thin filaments are composed mostly of the protein *actin*. together with smaller amounts of *tropomyosin* and *troponin*, as shown in Figure 5.11(b). The thick filaments are aggregates of the very large (molecular weight 5×10^5), elongated protein myosin arranged as shown in Figure 5.11(c). The sequence of events in muscle contraction is as follows:

(i) The nerve impulse is received at the membrane surface of the fibre, the sarcolemma, transmitted throughout the cell by the T-tubules and provokes the release of Ca^{2+} ions from the reservoir of vesicles formed by the membranes of the sarcoplasmic reticulum.

(ii) The Ca^{2+} ions bind to the troponin of the thin filaments. This causes a change in the shape of the troponin molecules which move the adjoining tropomyosin protein complex. The change in the position of the tropomyosin exposes the 'active site' of the actin molecules.

(iii) The 'activated' actin molecules are now able to react repeatedly with the myosin of the thick filament and ATP, the energy source for contraction as shown in Figure 5.12.

(iv) The heads of the myosin molecules are at different angles to the thick filaments at different stages in the cycle, so that with each turn of the cycle, and with the hydrolysis of one ATP molecule, the myosin head engages with a different actin molecule, the two filaments move with respect to each other, and contraction occurs.

(v) When the nerve impulse ceases, calcium is pumped back into the

(a) The arrangement of thick and thin filaments

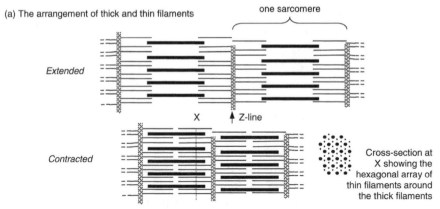

Extended

X   Z-line

Contracted

Cross-section at
X showing the
hexagonal array of
thin filaments around
the thick filaments

(b) The thin filament, length ~1μm, diameter ~7nm

The twisted double chain of
actin molecules has two
corresponding strands of
tropomyosin molecules and
complexes of troponin
molecules at regular intervals.

(C) The thick filament, length ~1.5μm, diameter ~16nm

The myosin molecules (length
~160nm) are packed together
so that that heads project in six
rows to interact with the actin
molecules of the surrounding
thin filaments. Going from one
end of the filament to the other
the myosin molecules change
direction, resulting in a central
region lacking heads.

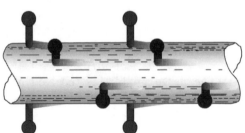

Figure 5.11   *The myofilaments of skeletal muscle.*

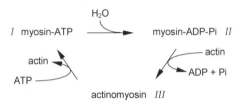

Figure 5.12   *The myosin/actinomyosin cycle. When the muscle is in the process of contracting the cycle functions continuously as shown. When the concentration of Ca^{2+} ions drops the cycle comes to rest at stage II as linkages between actin can no longer be formed. If the supply of ATP fails the cycle comes to rest at stage III.*

sarcoplasmic reticulum and the cycle comes to rest at stage II as the tropomyosin returns to its original position and the actin can no longer interact with a myosin head. The muscle is now relaxed and will be returned to its original extended state by the action of other muscles.

The energy, as ATP, for contraction and for the active uptake of calcium by the sarcoplasmic reticulum, is derived from two sources. When moderate levels of muscle activity are demanded, pyruvic acid (2-oxopropanoic acid) (5.9), derived from carbohydrate breakdown or from acetyl units, themselves derived from fatty-acid breakdown, is oxidised in the mitochondria of the muscle fibres. The energy made available by this oxidation is utilised in the mitochondria for the phosphorylation of adenosine diphosphate to the triphosphate. Three molecules of ATP are formed for every $\frac{1}{2}O_2$ reduced to H_2O. The oxygen is carried from the blood to the mitochondria as the oxygen adduct of myoglobin, the source of the red colour of fresh meat.

$$
\begin{array}{c}
\text{COOH} \\
| \\
\text{C}=\text{O} \\
| \\
\text{CH}_3
\end{array}
$$
(5.9)

When short bursts of extreme muscle activity are demanded, for example when one runs for a bus or when a bird like a pheasant briefly takes to the air, the limited rate at which the lungs can supply of oxygen to the muscles *via* the bloodstream is insufficient give an adequate supply of ATP. At these times the muscle obtains its energy by the less efficient but anaerobic and more rapid conversion of glycogen or glucose into lactic acid by the metabolic process of glycolysis. Only two molecules of ATP per molecule of glucose are obtained in glycolysis, compared with 36 when the glucose is oxidised completely to CO_2. In the muscles of the living animal anaerobic glycolysis cannot be maintained for many seconds before the accumulation of lactic acid becomes excessive and the consequent pain brings the exertion to a halt. As soon as demands on the muscles ease, the lactic acid is readily oxidised back to pyruvate; then either the pyruvate can be oxidised to CO_2 or some of it can be reconverted back into glucose in the liver.

On the death of the animal there is still a demand for ATP in the muscle from the pumping systems of the sarcoplasmic reticulum and other reactions, even though the muscles are no longer being called upon to contract by signals from the nervous system. As there is obviously no further supply of oxygen from the blood stream, the supply of ATP is maintained by glycolysis for some time until one of two possible circumstances brings it to a halt. If the animal was starved and otherwise stressed by bad treatment before slaughter, the muscle's reserves of glycogen will have been depleted and hence will be rapidly exhausted after slaughter, and glycolysis will quickly come to a halt. However, if the animal's muscles did have adequate glycogen reserves, glycolysis will still cease, but only after a few hours when the accumulation of lactic acid lowers the pH to around 5.0–5.5, sufficient to inhibit the activity of the glycolytic enzymes. When the ATP level in the muscle has consequently fallen, calcium will no longer be pumped out of the sarcoplasm into the vesicles of the sarcoplasmic reticulum. This calcium allows the myosin to interact with the actin, but the lack of ATP will prevent the operation of the full cycle, which would normally lead to contraction. The thick and thin filaments will therefore form a permanent link at stage III and the muscle will go rigid – the state of rigor mortis. It is clearly impossible to get meat cooked before it goes into rigor, and meat cooked during rigor is said to be exceedingly tough.

The answer to the problem of rigor mortis and, as we shall see, other aspects of meat quality, lies in butchers ensuring that the critical pH range of 5.0–5.5 is achieved. (Suffice to say, it is the application of sound pre- and post-slaughter practices rather than the possession of a pH meter that makes a good butcher.) The most obvious effect of the low pH is to deter the growth of the putrefactive and pathogenic micro-organisms that spread from the hide, intestines, *etc.* during the preparation of the carcass. During the conditioning (*i.e.* hanging) of the carcass the toughness due to rigor mortis disappears. A highly specific protease, active only at around pH 5 and also requiring the presence of Ca^{2+} ions, has been detected in muscle tissue. What use it is to the live animal is totally obscure, but in meat it catalyses the breakdown of the thin filaments at the point where they join the Z-line filaments. This creates planes of weakness across the muscle fibres which ensure the properly conditioned meat is suitably tender.

Muscle tissue contains a great deal of water, 55–80%. Some of this water is directly bound to the proteins of the muscle, both the myofilaments and the enzyme proteins of the sarcoplasm. The bulk of the water, however, occupies the spaces between the filaments. Protein denaturation on cooking releases some of this water to give meat its desirable moistness, and in the raw state a moist and shiny cut surface of a joint of meat is preferred. Although the sarcoplasmic proteins will bind less water at pH values near to their isoelectric points, which do happen to be mostly around pH 5, the principal cause of moisture release, drip, from well conditioned meat is that, as the pH falls, the gap between neighbouring filaments is reduced and some water is squeezed out. Although the butcher would obviously not wish too much moisture to be lost from a joint before it is weighed and sold, the increased mobility of the muscle water has another beneficial effect besides giving the meat an attractive moist appearance.

The further effect is, as we shall see, on the colour of the meat. It has already been mentioned that the red colour of myoglobin varies from muscle to muscle, from species to species, and with the age of the animal. Some typical data for cattle and pigs are given in Table 5.6 (based on results quoted by R. A. Lawrie, 1998 – see 'Further Reading'). As a general rule, muscle used for intermittent bursts of activity, which will be fuelled predominantly by anaerobic glycolysis, has only low levels of myoglobin and few mitochondria and will thus be pale in colour. The breast, *i.e.* the flight muscle, of poultry is an extreme example. In contrast, muscles used more or less continuously, such as those in the legs of poultry, will be fuelled by oxidative reactions and will consequently require high levels of myoglobin and be much darker.

The state of the iron in myoglobin also affects its colour. The structure of myoglobin was shown in Figure 5.6 but we need to concentrate on the iron atom. The state of the iron atom differs in the three important forms of myoglobin. In its normal state, referred to simply as 'myoglobin' and abbreviated to 'Mb', the iron is in its reduced, ferrous state, otherwise known as Fe^{2+} or Fe^{II}. When it binds an oxygen molecule (in the living animal as part of the system for transferring oxygen from the haemoglobin of the bloodstream to the muscle cell mitochondria) the result is known as oxymyoglobin (MbO_2). In spite of the oxygen's presence the iron is not oxidised and oxygen molecule can be released again when the local oxygen concentration is low. Oxymyoglobin is bright red in colour (its

Table 5.6 *The myoglobin content of different muscles.*

Animal	Age	Muscle	Myoglobin content g per 100 g fresh weight
Calf[a]	12 days	*Longissimus dorsi*[b]	0.07
Steer[c]	3 years	*Longissimus dorsi*[b]	0.46
Pig[d]	5 months[d]	*Longissimus dorsi*[b]	0.030
Pig[e]	7 months[e]	*Longissimus dorsi*[b]	0.044
Pig[f]	7 months[f]	*Rectus femoris*[g]	0.086

[a] Veal. [b] The principal muscle of 'rib of beef' or 'short back bacon'. [c] Beef. [d] Pork. [e] Bacon. [f] Pork sausage. [g] The principal muscle of 'leg of pork'.

Box 5.1   Myoglobin and free radicals

As shown in Figure 5.13, four of the six coordination positions of the iron are occupied by nitrogen atoms of the porphyrin ring. The fifth, perpendicular to the plane of the ring, is occupied by the nitrogen atom of a histidine in the polypeptide chain. The sixth position, opposite to the fifth, is where oxygen binds reversibly in its role as an oxygen carrier between the haemoglobin of the blood and cytochrome oxidase of the muscle cell mitochondria. In the live animal the iron this extremely rapid simple reaction:

$$Mb + O_2 \leftrightarrow MbO_2$$

is all there is to since the myoglobin only encounters the very low oxygen concentration in the muscle cell or the high concentration when it collects oxygen from the blood. However, in meat the essentially irreversible formation of metmyoglobin takes place at intermediate oxygen concentrations (Figure 5.14). The actual mechanism of the reaction leading to the formation of the superoxide anion is probably more complex than that shown in Figure 5.13.

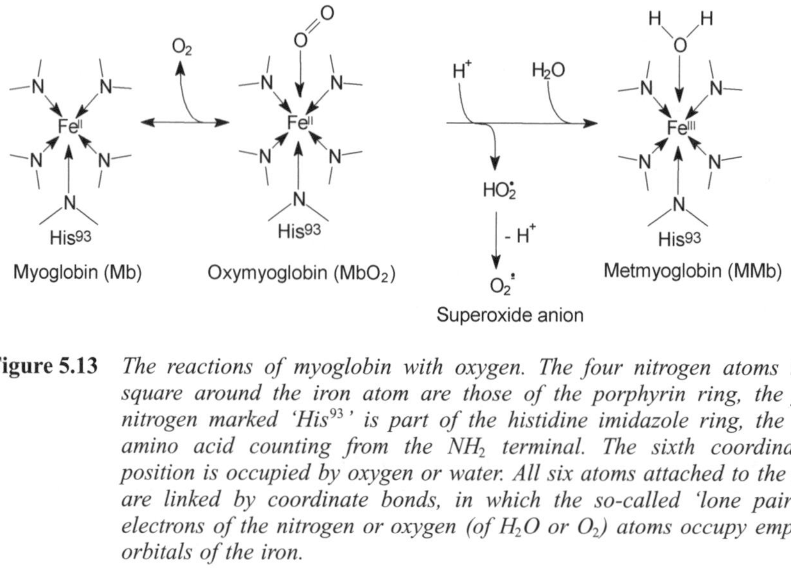

Figure 5.13   *The reactions of myoglobin with oxygen. The four nitrogen atoms in a square around the iron atom are those of the porphyrin ring, the fifth nitrogen marked 'His^{93}' is part of the histidine imidazole ring, the 93rd amino acid counting from the NH$_2$ terminal. The sixth coordination position is occupied by oxygen or water. All six atoms attached to the iron are linked by coordinate bonds, in which the so-called 'lone pair' of electrons of the nitrogen or oxygen (of H$_2$O or O$_2$) atoms occupy empty d orbitals of the iron.*

spectrum has a double peak, λ_{max} values 544 and 581 nm) whereas myoglobin is a dark purplish red (λ_{max} 558 nm)*.

---

* The difference between the colours of arterial and venous blood is based on a similar relationship between the two forms of haemoglobin.

Like other free radicals the superoxide anion is extremely unstable but in its short life it is still capable wreaking havoc. With the exception of the bacteria that are obligate anaerobes, *i.e.* cannot survive in the presence of oxygen, all organisms have a pair of enzymes, superoxide dismutase (SODM, always pronounced as with Gomorrah) and catalase, whose joint role is to destroy this dangerous by-product of oxygen metabolism:

$$2O_2^{\cdot -} \xrightarrow[\text{SODM}]{O_2} O_2^{2-} \xrightarrow[\text{catalase}]{H^+} OH^- + \tfrac{1}{2}[O_2]$$

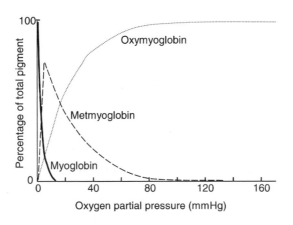

Figure 5.14 *The effect of oxygen concentration on the proportions of the different forms of myoglobin. In air at atmospheric pressure the concentration of oxygen is equivalent to a partial pressure of 170 mmHg.*
(From J. C. Forrest, *et al.* 'Principles of Meat Science', W. H. Freeman, San Francisco, 1975.)

The interior of the muscle tissue of a carcass will obviously be devoid of oxygen, and the freshly cut surface of a piece of raw meat is, as one would expect, the corresponding shade of dark red/purple. Soon, diffusion of oxygen into the surface begins to convert myoglobin (Mb) into oxymyoglobin (MbO$_2$) and the meat acquires an attractive bright-red appearance as a layer of MbO$_2$ develops. The importance of an adequate post-mortem pH fall is that it is the free moisture of the meat through which the oxygen diffuses and if the pH has remained high there is much less free moisture available. The respiratory systems of the muscle continue to consume any available oxygen until the meat is actually cooked, so that the MbO$_2$ layer never gets to be more than a few millimetres deep. At intermediate oxygen concentrations, as prevail at the interface between the MbO$_2$ and Mb layers, the iron of the myoglobin is oxidised by the oxygen it binds to convert it into the ferric state (Fe^{III} or Fe^{3+}). This has two consequences. The most obvious is that the metmyoglobin (MMb) which is produced is a most unattractive shade of brown (λ_{max} 507 nm). Meat kept too long at room temperature therefore develops a brown layer just below the surface, and as time proceeds this widens until the surface of the meat looks brown rather than red – a sure sign that the meat is no longer fresh. The other consequence is that the oxygen departs as a free radical referred to as superoxide. the protonated form of the superoxide anion. This free radical may well have time to initiate the chain reactions of fatty-acid autoxidation described in Chapter 4 that lead to the rancidity of the fatty parts of the meat. Thus there is a clear association between the appearance of the raw meat and its flavour when cooked.

When meat is cooked, the myoglobin is denatured along with most other proteins. The unfolding of the polypeptide chain displaces the histidine whose nitrogen atom was linked to the iron of the porphyrin ring. This changes the properties of the iron so that it is now readily oxidised to the Fe^{III} state by any oxygen present. The result is that, except for the anaerobic centre of a large joint, the meat turns brown. The pink colour at the centre of a roast joint is due to a haem derivative in which histidine nitrogen atoms (perhaps, but not necessarily, from the original myoglobin molecule) occupy both of the vacant co-ordination positions on opposite faces of the porphyrin ring.

Salting has always been an important method of meat preservation. Rubbing the carcass with dry salt, steeping in brine, or using modern injection methods all aim to raise the salt concentration in the tissue and sufficiently to inhibit microbial growth (*see* Chapter 9). In the case of bacon, low levels of nitrates are traditionally included with the salt. Under the influence of enzyme action in the meat the nitrate gives rise to nitrogen oxide (NO), which binds to the iron of myoglobin to form a red pigment, nitrosyl myoglobin (MbNO). This gives uncooked bacon and ham its characteristic red colour. When bacon or ham is cooked, denaturation of the MbNO leads to the formation of the bright pink pigment usually known as nitrosyl haemochromogen. This is a porphyrin derivative of uncertain structure but probably having two nitrogen oxide groups, one on each side of the porphyrin ring.

Although *post-mortem* changes in the myofibrillar structures are an important factor in the texture of meat, the consumer is much more aware of differences in texture that stem from the maturity of the animal and the part of the carcass the meat came from. These variations arise in the connective tissue of the muscle rather that the contractile tissues. The connective tissue of the muscle is composed almost entirely of fibres of the protein collagen. Collagen fibres consist of cross linked, longitudinally arranged tropocollagen molecules. Each tropocollagen molecule is 280–300 nm in length. The tropocollagen molecule itself consists of three very similar polypeptide chains. The conformation of each polypeptide chain is an extended left-handed helix (remember that the classic α-helix shown in Figure 5.5 is a compact right-handed helix), and the three chains wind together in a right-handed helix, Figure 5.15(a). The use of left- and right-handed helices at different levels of the organisation of the molecule mirrors that in rope and textile threads where it makes unravelling much less likely.

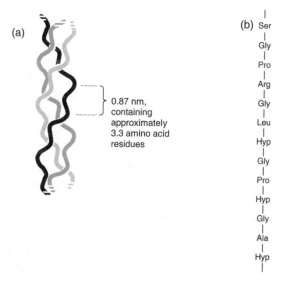

Figure 5.15 *The molecular structure of collagen. (a) A section of a tropocollagen molecule showing only the C–N–C–C–N–C– backbones of the three polypeptide chains. Each chain forms an extended left-handed helix. The three polypeptide chains twist around each other in a right-handed helix. (b) A section of the amino acid sequence of a tropocollagen polypeptide chain, 'Hyp' is hydroxyproline.*

The amino acid composition and sequence of tropocollagen are striking. Roughly one-third of all the amino acid residues are glycine; proline plus hydroxyproline account for 20–25%. The brief sequence given in Figure 5.15(b) shows that in fact glycine occurs at every third residue along the chain. Structural studies show that the glycine residues always occur at the points in the helical structure where the three chains approach closely. Only the minimal side chain of glycine – a solitary hydrogen atom – can be accommodated in the small space available. The two imino acids are important, since their particular form of the peptide bond gives the ideal bond angles for the tropocollagen helix. The structure is maintained by hydrogen bonding between the three chains and also by hydrogen bonds involving the hydroxyproline hydroxyl group. The hydroxylation of proline to hydroxyproline occurs after the protein has been synthesised. The observation that ascorbic acid, vitamin C, is the reducing agent in the reaction can be readily correlated with the clinical signs of the deficiency of the vitamin discussed in Chapter 8. However strong an individual tropocollagen molecule is, the strength of the fibre as a whole will depend on lateral cross links to ensure that neighbouring molecules do not slide over each other under tension. A number of types of cross links have been identified; two are shown in Figure 5.16. Cross links are most abundant in connective tissues where greatest strength is required, such as the Achilles tendon.

It is a common observation that meat from young animals is more tender than that from older ones. It has now been realised that this is due to an increase, as the animal ages, in the number of cross links rather than an increase in the proportion of total connective tissue. There are, of course, variations in the amount of collagen between different muscles of the same animal, which are very important to meat texture. The most tender cuts, and the most expensive, are those with the lowest proportion of connective tissue. For example, in beef the connective tissue content of the *Psoas major* muscle (fillet steak) is one-third of that in the *Triceps brachii* muscle (stewing beef).

When meat is heated, the hydrogen bonds that maintain collagen's structure are weakened. Very often the fibres shorten as the polypeptide chains adopt a more compact helical structure. If the heating is prolonged, as in a casserole, not only hydrogen bonds but also some of the more labile cross links will be broken. The result will be solubilisation of the collagen, some of which will be leached out and cause gelatinisation of the gravy as it cools and the hydrogen bonds are re-established.

Muscle tissue only accounts for a modest proportion of an animal's total collagen; most is located in the skin (or hide) and the bones.

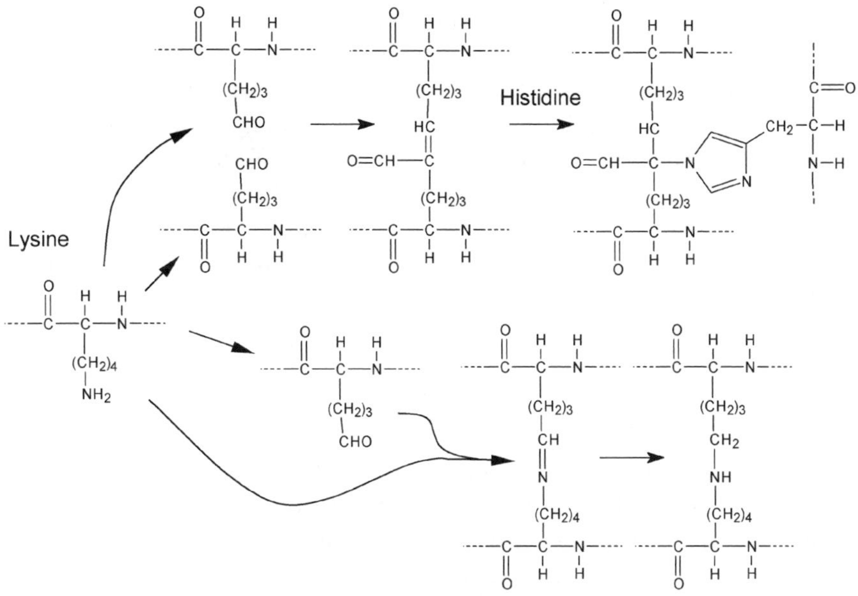

Figure 5.16 *The formation of tropocollagen cross links involving lysine side chains. The initial conversion of lysine to its aldehyde derivative is catalysed by the enzyme lysine oxidase but all subsequent reactions are spontaneous. The two linkages shown here are normally found linking neighbouring tropocollagen molecules. The attachment of the histidine side chains provides for a three way linkage. Other types of links are found supplementing the hydrogen bonds between chains within the same tropocollagen molecule.*

Although these tissues are of little importance as food they are the usual source of collagen for conversion into the important food ingredient gelatin. Gelatin is widely used as a gelling agent, particularly in confectionery, desserts (the 'jelly' of children's parties, 'jello' in the USA), and meat products.

When gelatin is manufactured from bones they are first crushed and then steeped in dilute hydrochloric acid for up to two weeks to solubilise and remove the calcium phosphate. The residue can then be processed along similar lines to cattle hides. This involves an initial treatment with lime (calcium hydroxide, $Ca(OH)_2$) for several weeks. The alkaline conditions break down the cross links between the collagen chains. When the lime is eventually washed out other proteins and fatty residues are also washed away. After neutralization the residue is then extracted with hot water at progressively higher temperatures. The gelatin, as it now is, dissolves, and after filtration the solution can be concentrated and finally dehydrated.

Rather like the polysaccharide gel formers described in Chapter 3 gelatin is insoluble in cold water but dissolves readily in hot water. When the solution is cooled the polymer molecules revert from their disorganised 'random coil' state to the helical arrangement similar to that in the original connective tissue. In this state segments of the polymer chains can associate together, forming junction zones based on hydrogen bonding, as occurs in polysaccharide gels.

Bread

Bread has been eaten in the temperate zones of the world for thousands of years. It is only in the last few that chemists and biochemists have been able to address themselves seriously to the fundamental problem of bread. That is: why, of all the cereals, wheat, rye, barley, oats, sorghum, maize, and rice, is it only wheat that will give bread with a leavened, open, crumb structure. The so-called rye bread sold in Britain actually contains 50% wheat flour to give it a structure acceptable to British tastes.

The grains (*i.e.* the seeds) of cereals have a great deal in common. They consist of three major structures: (i) the embryo or germ of the new plant, (ii) the endosperm, which is the store of nutrients for the germinating plant, and (iii) the protective layers of the seed coat, which are regarded as bran by the miller. The endosperm is about 80% of the bulk of grain. White flour is almost pure endosperm, and, since we are primarily interested in bread, what follows refers to the endosperm and endosperm constituents of wheat. The endosperm consists of tightly packed, thin-walled cells of variable size and shape. The cell walls are the origin of the small proportions of cellulose and hemicelluloses in white flour. The cells are packed with starch granules lying in a matrix of protein. The protein, some 7–15% of the flour, is of two types. One type (about 15% of the total) consists of the residues of the typical cytoplasmic proteins, mostly enzymes, which are soluble in water or dilute salt solutions. The remaining 85% are the storage proteins of the seed, insoluble in ordinary aqueous media and responsible for dough formation. These dough-forming proteins are collectively referred to as gluten. The gluten can be readily extracted from a flour by adding enough water to form a dough, leaving the dough to stand for half an hour or so, and then finally kneading the dough under a stream of cold water, which washes out all the soluble material and the starch granules. The resulting tough, viscoelastic, and sticky material contains about one-third protein and two-thirds water.

Flours from different wheat varieties vary in protein content. In general, flours that are good for bread-making (*i.e.* give a good loaf volume) are given by the spring-sown wheat varieties grown in North America. These varieties tend to have higher protein contents (12–14%). Good bread-making wheats are most often the type described as 'hard' by the miller, *i.e.* the endosperm is brittle and disintegrates readily on milling. The baker's description of a good bread-making flour as 'strong' refers to the characteristics of the dough: it is more elastic and more resistant to stretching than the dough of a 'weak' flour. Weak flours are essential for biscuits and shortcrust pastry. These flours are usually obtained from the 'soft' winter wheats (actually sown in the autumn!) that are grown in Europe*. Their protein content is usually less than 10%. In spite of the general correlations between protein content and milling/baking properties, there are enough exceptions to demonstrate that it is the properties of different wheat proteins rather than their abundance that is the major factor. The so-called 'patent' or 'household' flour that is sold for domestic use is blended from different types to give intermediate properties that provide a reasonable compromise for home baking.

The gluten proteins can be fractionated on the basis of their solubilities. The most soluble, the gliadins, can be extracted into 70% ethanol. The gliadins constitute between a third and a half of the total gluten. The remainder are the glutenins, which are extremely difficult to dissolve fully. A solution consisting of $0.1 \, \text{mol dm}^{-3}$ ethanoic acid, $3 \, \text{mol dm}^{-3}$ urea, and $0.01 \, \text{mol dm}^{-3}$ cetyltrimethylammonium bromide (cetrimide, a surfactant) will dissolve all but 5% of the glutenins. What is known of the molecular structure of the gluten proteins goes a long way towards explaining their remarkable dough-forming and solubility characteristics. Using elaborate electrophoretic techniques, it has been shown that a single variety of wheat may have over 40 different gliadin proteins. The glutenins are much more difficult to characterise. They are very large aggregates (molecular weights above 1 million and mostly around 2 million) of individual proteins known as glutenin subunits. These glutenin 'macropolymers' can be broken down into their subunits by treatment with reagents such as mercaptoethanol, which break the disulfide bridges between cysteine sulfhydryl groups:

---

* In recent years British bread-making grists, *i.e.* the blends of grain being milled for a particular batch of flour, have contained increasing proportions of European grown strong wheat varieties, newly developed by plant breeders. Gluten extracted in an industrial version of the process described here, dried, milled and referred to 'Active Gluten', is also added to flour to enhance its protein content and baking performance. The manufacture of Quorn™, the mycoprotein food ingredient, was originally conceived as an outlet for the considerable quantities of starch that would be a by-product of gluten manufacture.

Box 5.2   Wheat genes and chromosomes

The numbers of different, but closely related, gliadins and glutenin subunits that are found not only in the single species, *Triticum aestivum*, but also occur in a single cultivar (variety) of wheat, are a consequence of the special genetic character of modern bread wheat plant. The nuclei of the vast majority of animal and plant cells (excluding of course the germ cells required for sexual reproduction) contain one double set of each of their chromosomes, *i.e.* their cells are *diploid*. The cell nuclei of *T. aestivum* are exceptional in having three copies of each set (referred to as genomes AA, BB and DD respectively) of 6 paired chromosomes so that we refer to bread wheat as *hexaploid*. This is the result of nuclear fusion rather than conventional cross-breeding between wheat ancestors. Bread wheat's ancestors are believed to include relatives of a number of modern species:

T. tauchii	} goat-faced grasses	diploid	DD
T. speltoides		diploid	BB
T. monococcum	einkorn wheat	diploid	AA
T. turgidum	emmer wheat	tetraploid	AABB

Modern durum wheat (*T. durum*), used for pasta, is a tetraploid relative of emmer wheat. Archaeologists believe that that a hexaploid ancestor of modern bread wheat arose in the eastern Mediterranean area no longer ago than

---

$$\ldots -protein-S-S-protein- \ldots + 2HSCH_2CH_2OH \rightarrow$$

$$\ldots -protein-S-SCH_2CH_2OH + HOCH_2CH_2S-S-protein- \ldots$$

Some features of the gliadins and glutenin subunits are shown in Table 5.8.

The lack of solubility of these proteins is readily explained by their amino acid compositions. They have a low proportion of charged amino acids compared with more typical proteins (*e.g.* glutamic acid, $\sim$2% in gliadins, 13% in α-casein; histidine *plus* lysine *plus* arginine, $\sim$3% in gliadins, 12% in α-casein). The contents of glutamine (36–45 mol%) and proline (15-30 mol%) are exceptionally high compared with other proteins. These two amino acids contain an above average proportion of nitrogen and are also particularly easily utilised by the germinating seed as sources

7500 BC. *T. aestivum* had a much larger grain and higher protein content than its ancestors. The impact of this event on early agriculture, and the history of mankind, was considerable. Possibly through the effects of human intervention as it was cultivated the species lost other characteristics of a wild grass so that it became easier to reap and thresh but at the same time became dependant on human cultivation for seed distribution and competition with weeds.

The genome and chromosome locations of the genes for most gliadins and HMW glutenin subunits are shown in Table 5.7.

Table 5.7 *The loci of the gliadin and glutenin genes. Subunit 2* is similar, but not identical, to subunit 2.*

		Genome		
	Chromosome	*AA*	*BB*	*DD*
HMW glutenin subunits	1	1 & 2*	6, 7, 8, 9, 13, 14, 15, 16, 17, 19, 21 & 22	2, 3, 4, 5, 10, 11, & 12
Gliadins	1	ω	Most γ	A few β
Gliadins	6	α	Most β	A few γ

Table 5.8 *Essential features of the gluten proteins.*

		% Total gluten protein	*Molecular weight (× 10⁻³)*	*Distinctive features of amino acid composition*
Gliadins	Sulfur rich (α, β & γ)	51–64	32–42	Glutamine 32–42%, proline 15–24%, Cysteine ~2%, phenylalanine 7–9%
	Sulfur poor (ω)	7–13	55–79	Glutamine 42–53%, proline 20–31%, Cysteine 0%
Glutenin	Low MW	19–25	36–44	Glutamine 33–39%, glycine 13–18%,
subunits	High HMW	7–13	90–124	Cysteine 0.5–1%

Box 5.3   More details of gluten proteins

The different classes of gluten proteins have different consensus sequences. For example the consensus sequence P-Q-Q-P-F-P-Q-Q- is found in ω-gliadins, and several, including G-Q-Q-, P-G-Q-G-Q-Q- and G-Y-Y-P-T-S-L-Q-Q- are found in HMW glutenin subunits. In most departures from the consensus involving a change of amino acid (as oppose to an insertion or deletion) the replacement amino acid has similar physical properties so that changes in the overall structure of the protein are minimal. In the example of α-Gliadin$_A$, two short sections of which are shown in Figure 5.17, the full DNA sequence has been established and we know that of the seven changes in these two sections (Figure 5.17) five (at positions 8, 14, 56, 58 and 66) are the result of single base changes in the codon for the amino acid.

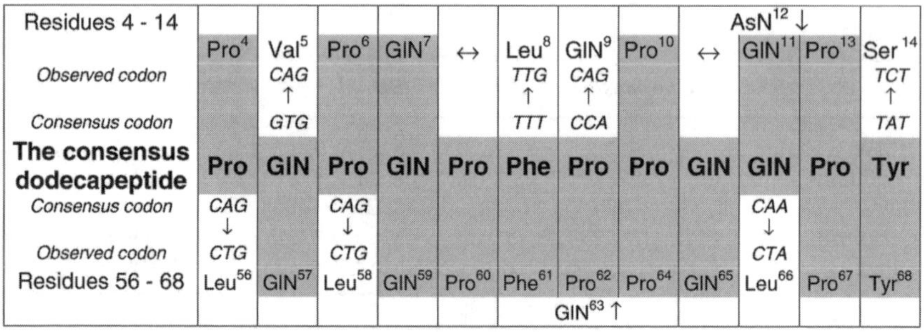

Residues 4 - 14									AsN^{12} ↓			
	Pro^4	Val^5	Pro^6	GIN^7	↔	Leu^8	GIN^9	Pro^{10}	↔	GIN^{11}	Pro^{13}	Ser^{14}
Observed codon		*CAG* ↑				*TTG* ↑	*CAG* ↑					*TCT* ↑
Consensus codon		*GTG*				*TTT*	*CCA*					*TAT*
The consensus dodecapeptide	**Pro**	**GIN**	**Pro**	**GIN**	**Pro**	**Phe**	**Pro**	**Pro**	**GIN**	**GIN**	**Pro**	**Tyr**
Consensus codon	*CAG* ↓		*CAG* ↓							*CAA* ↓		
Observed codon	*CTG*		*CTG*							*CTA*		
Residues 56 - 68	Leu^{56}	GIN^{57}	Leu^{58}	GIN^{59}	Pro^{60}	Phe^{61}	Pro^{62}	Pro^{64}	GIN^{65}	Leu^{66}	Pro^{67}	Tyr^{68}
						GIN^{63} ↑						

Figure 5.17   *Masked repetitions in the amino acid sequence of α-Gliadin$_A$. Two typical sections of the sequence are compared with the postulated consensus sequence. Shading shows where the residues are identical. In these sections there are two insertions, asparagine at position 12 and glutamine at position 63, and two deletions, between positions 7 and 8 and 62 and 63. The DNA codons for the positions involved in mutations are also shown, italicised.*

of both nitrogen and carbon. By comparison the metabolic pathways for the breakdown of the other nitrogen rich amino acids, arginine, lysine, and histidine are much more complex. Although glutamine may be regarded as a hydrophilic amino acid its abundance leads to the formation of large numbers of hydrogen bonds between polypeptide chains (5.10) many with an intermediate water molecule (5.10a).

The polypeptide chains of ordinary, globular, proteins are often found to be organised into *domains*. These are usually regions that fold up together to form compact units in which some part of the biological activity is localised. Although very is little is known for certain about the three-dimensional structure of gluten proteins it is common to divide their sequences up into three (or more) domains. The central domain (normally II) is dominated by the β-spiral structure whereas in the domains at the extremities (normally I and III) the polypeptide chain is more globular in character, with a great deal of α-helix and other less well-defined structures (Figure 5.18).

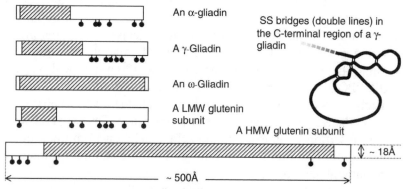

An α-gliadin

A γ-Gliadin

An ω-Gliadin

A LMW glutenin subunit

A HMW glutenin subunit

SS bridges (double lines) in the C-terminal region of a γ-gliadin

$\updownarrow \sim 18\text{Å}$

$\longleftarrow \qquad \sim 500\text{Å} \qquad \longrightarrow$

Domains I and III are unshaded, Domain II is shaded, cysteine residues are shown↓

Figure 5.18 *The domain structure of gluten proteins. The C-terminal regions of α- and γ-gliadins are believed to have intra-molecular SS bridges between some of their cysteine residues that result in these molecules being more globular in shape. The central domains are dominated by repetitive sequences such as that shown in Figure 5.17 and these tend to result in the polypeptide chain taking up the β-spiral configuration. In the HMW subunits the central domain appears to consist entirely of a single β-spiral resulting in these molecules resembling a dumb-bell, with a central rod-like spiral region terminated by globular regions consisting largely of α-helix.*

(5.10)

(5.10a)

Complete amino acid sequences have now been obtained for a considerable number of gluten proteins and they reveal a number of fascinating features. The first is the extensive homology between the amino acid sequences of different proteins within the same class. For example the first 22 amino acids (counting as usual from the N-terminal end) of α-gliadins A, 2, 10, 11 and 12, , and γ-gliadin 1, are identical except at one position in β-gliadin 5. Similar homologies have been found within the ω-gliadin class and between many HMW glutenin subunits. These homologies indicate that in spite of the extensive changes in the wheat brought about by natural evolution and plant breeders over the centuries the common genetic ancestry is still apparent.

The amino acid sequences of all the different classes of gluten proteins also show a great deal of repetition. A typical example is α-gliadin A, whose first 95 amino acids are based on a 12-amino acid sequence (-PQPQPFPPQQPY-*) repeated 8 times. The repetition may not be immediately obvious as in each one of the repeats there are usually a small number of changes from the 'consensus' sequence. They may take the form of an insertion of an extra amino acid, a deletion from the sequence, or a substitution. Continuous sequences of 10–20 glutamine residues also occur in many gluten proteins.

Comparison of these known sequences with those in proteins whose secondary and tertiary structures are well established has suggested that gluten proteins are unusually rich in an arrangement known to protein chemists as a 'β-bend'. β-Bends are normally found in proteins where the polypeptide chain undergoes a sharp change in direction (such at the 'corners' between the α-helical sections of myoglobin (*see* Figure 5.6). The hydrogen bonds that in an α-helix link successive turns (*see* Figure 5.5) instead 'cut across the corner'. Scanning tunnelling electron microscopy (of a HMW gluten subunit) and physical measurements using the sophisticated techniques of Fourier transform infrared spectroscopy and far-UV circular dichroism[†] have confirmed this suggestion. The only other protein known to have an abundance of β-bends is the mammalian connective-tissue protein elastin. Ligaments and artery walls are very rich in elastin, and it is now believed that they owe their elasticity to this protein, with its large numbers of β-bends arranged in sequence into a 'β-spiral'. There is therefore a lot of support for the idea that the elastic component of dough's viscoelasticity can now be

---

* It may be appropriate to include here a reminder of the codes used for the amino acids mentioned in this section: P or pro = proline, Q or Gln = glutamine, F or phe = phenylalanine, Y or tyr = tyrosine.
† The theory of these techniques is way beyond the scope of this book.

accounted for in the amino acid sequences of its proteins. The β-spiral structure that has been proposed for one class of gliadin proteins is shown in Figure 5.19. The current view of the structure of glutenin that has emerged from these studies is a backbone of HMW subunits linked 'nose-to-tail' by disulphide bridges, carrying short, possibly branched, chains of LMW subunits.

Extensive comparisons of breadmaking performance of a wheat cultivar with its complement of gliadins and glutenin subunits have shown that the HMW glutenin subunits are major factors. Close genetic linkage (*i.e.* genes on the same chromosome lying very close to each other) means that subunits are considered in pairs. For example subunits 5 + 10 impart much better baking performance than 4 + 12. It is possible that the best subunit types are those that have multiple cysteine sulfdryl groups for forming branched gluten macropolymer chains promote strong doughs and high loaf volumes.

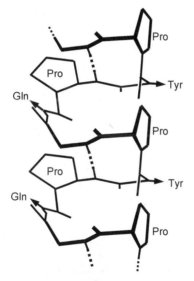

Figure 5.19 *The gluten β-spiral. The conformation of the polypeptide chain illustrated here was proposed for the repeating pentapeptide (-pro-tyr-pro-glN-glN-) of ω-gliadin but the β-spiral regions of other gluten protein are unlikely to be very different. The −N−C−C−N−C− backbone of the polypeptide chain is shown with proline rings, the N−H---O=C hydrogen bonds and arrowed the positions of the tyrosine and glutamine side chains. The spiral is stabilised by the hydrogen bonds forming across the β-spiral rather along it as in the α-helix. This allows a degree of stretch by rotation of the bonds around the glutamine residues. Interactions between the π-electron clouds of the stacked tyrosine side chains provides another stabilising force.*

When water is added to flour, a dough is formed as the gluten proteins hydrate. Some water is also, as mentioned in Chapter 3, taken up by the damaged starch granules. The viscoelastic properties of the dough depend upon the glutenin fraction, which is able to form an extended three-dimensional network. Links between the glutenin chains depend on different types of bonding. Hydrogen bonding between the abundant glutamine amide groups is probably most important, but ionic bonding and hydrophobic interactions also have a role. The gliadin molecules are assumed to have a modifying influence on the viscoelastic properties of the dough. It is now well established that the relative proportions of the high and low-molecular weight proteins, *i.e.* the glutenins and gliadins, are a major factor in the bread-making character of a flour. In general, a high proportion of glutenins results in doughs that require more mixing, are stronger, and give loaves of greater volume and therefore a more open, lighter, crumb. For biscuits (except crackers) and many types of cakes and pastry a good bread-making flour is most unsuitable. Made with a strong flour some types of biscuits would be hard rather then crisp and others tend to shrink erratically during baking. These are essentially problems for the commercial baker rather than the domestic cook. Sometimes L-cysteine is used as an additive in the flours for industrial scale pie making. In small, carefully controlled amounts, it weakens flour by breaking the sulfhydryl bridges between gluten proteins. If a large excess of L-cysteine is used the dough is converted into a ghastly slimy subtance.

Traditionally the bread-making properties of a flour were found to be improved by prolonged storage. Autoxidation of the polyunsaturated fatty acids of flour lipids (see page 90 *et seq.*) results in the formation of hydroperoxides, which are powerful oxidising agents. One consequence is a bleaching of the carotenoids in the flour, giving the bread a more attractive, whiter crumb. However, the most important of the beneficial effects of ageing are on the loaf volume and crumb texture. Over the first twelve months of storage of flour there is a steady increase in the loaf volume and the crumb becomes finer and softer. Longer storage results in worsening baking properties. For over 50 years it has been common practice to simulate the ageing process by the use of oxidising agents as flour treatments, at the mill or as additives at the bakery. The agents used have included chlorine dioxide (applied as a gaseous treatment by the miller), benzoyl peroxide [di(benzenecarbonyl) peroxide], ammonium and potassium persulfates, and potassium bromate and iodate. None of these are now permitted within the EC*.

---

* Prior to being banned in November 2000 chlorine was used in the manufacture of commercial cake flours used for 'high ratio' cakes. High ratio cakes (*e.g.* American style muffins) are made with very levels of water and sugar. The chlorine treatment damages the protein membrane around the starch

The elimination of these oxidising processes has been made possible by the introduction of ascorbic acid as an improving agent. This has largely come about through the adoption of the Chorleywood Bread Process for commercial breadmaking, to be discussed in detail at the end of this chapter. Its usefulness has undoubtedly been enhanced by its vitamin status, giving it a 'whiter than white' image with consumers, even though none of the vitamin activity survives the oven. The consequent use of unbleached flour has led to the appearance of bread with a distinctly yellow tinge to its crumb, but this is nowadays equated with naturalness rather than impurity, as once it would have been.

A satisfactory dough is one which will accommodate a large quantity of gas and retain it as the protein 'sets' during baking. The achievement of such a dough requires more than the simple mixing of the ingredients: mechanical work has also to be applied. In traditional bread-making the kneading of the dough provides some of this work, but the remainder is performed by the expanding bubbles of the carbon dioxide evolved by the yeast fermentation. At the molecular level these processes, collectively referred to as dough development, are not well understood, and the account of them that follows is unlikely to represent the final word on the subject.

It is generally believed that during development the giant glutenin molecules are stretched out into linear chains which interact to form elastic sheets around the gas bubbles. A number of chemical reactions are involved in this. The mechanical stresses are sufficient to break, temporarily, the hydrogen bonds that are important in binding together the different gluten proteins. During the early stages of mixing of the dough the polypeptide chains of both gliadins and glutenins will tend to become aligned alongside each other. This will give many more opportunities for hydrogen bond formation, and the resistance of the dough to mixing shows a sharp increase. Other reactions involve the sulphydryl groups of the proteins. Under mechanical stresses exchange reactions between neighbouring sulfhydryl groups will allow the glutenin subunits to take up more extended arrangements. Glutathiane (γ-glutamylcysteinylglycine, GSH) (5.11) has been identified as the low molecular weight sulfhydryl compound involved in these reactions (Figure 5.20). Adequate amounts (10–15 mg kg^{-1}) of glutathione are found in flour. It occurs naturally in flour in three forms: the free form (GSH), the oxidised dimer (GSSG) and bound to a protein molecule (PSSG).

---

granules to allow them to imbibe, and hold onto, the extra water. A similar effect is now achieved by subjecting the flour to an intense dry heat treatment. The oxidising agent azodicarbonamide ($H_2NCON=NCONH_2$) was permitted in Britain until a few years ago but remains in use in the USA, especially in the flour for burger buns.

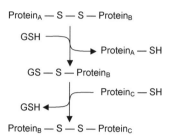

Figure 5.20 *Disulfide exchange reactions in dough. The cysteine residues most involved in reactions with GSH are those in the low molecular weight subunits. These subunits are suspected of being involved in forming links between the extended glutenin macropolymers.*

$$\text{HOOC}-\underset{\underset{\text{H}}{|}}{\overset{\overset{\text{NH}_2}{|}}{\text{C}}}-\text{CH}_2-\text{CH}_2-\underset{\underset{\text{O}}{\|}}{\text{C}}-\underset{\text{H}}{\overset{\text{H}}{\text{N}}}-\underset{\underset{\underset{\underset{\underset{\text{COOH}}{|}}{\text{CH}_2}}{|}}{\underset{\text{NH}}{|}}{\overset{\overset{\text{H}}{|}}{\text{C}}}-\text{CH}_2-\text{SH}$$

(5.11)

Traditional baking processes require prolonged periods (up to 3 hours) of fermentation before the dough is ready for the oven, and a great deal of research has taken place to find ways of accelerating the dough development process. After many attempts in both Britain and the USA it was at the laboratories of the Flour Milling and Baking Research Association (at Chorleywood, Hertfordshire) that the first really successful process was devised. Known internationally as the Chorleywood Bread Process (CBP), it depends upon the use of ascorbic acid (AA) as an improver coupled with high-speed mixing of the dough. When the flour, water, and other ingredients are mixed, the ascorbic acid is oxidised to dehydroascorbic acid (DHAA) (*see* Chapter 8) by ascorbic acid oxidase, an enzyme naturally present in flour:

$$AA + \tfrac{1}{2}O_2 \rightarrow DHAA + H_2O$$

The dehydroascorbic acid, an oxidising agent, effects its improving action by means of a second enzyme, also naturally present in flour, which catalyses the conversion of glutathione into its oxidised dimeric form:

$$DHAA + 2GSH \rightarrow AA + GSSG$$

which is of course inactive in the disulfide exchange reactions of dough development. The elimination of long fermentation periods have made the CBP very popular with bakers, even though it requires special mixers capable of 400 r.p.m. and monitoring the amount of work imparted to the dough – 11 watt hours (4×10^4 J) kg^{-1} of dough is required. The resulting bread need not be easily distinguishable from that produced by traditional bulk fermentation methods. However, it is often observed that the flavour of CBP bread is inferior: this is partly because the yeast has had less time in which to contribute ethanol and other metabolic by-products to the dough. The other cause is the tendency of bakers to under-bake their bread and thereby have less of the Maillard reaction taking place as a crust forms. For them this has two advantages, the absence of a crust makes the loaf less of a challenge to slicing machinery but even more important it means that more water is left behind in the loaf, allowing less flour to be used to produce a loaf of the same weight.

At the beginning of this section the question asked was 'what is so special about wheat?'. In as far as we know that the equivalent proteins of the other cereals appear to lack the special features of the HMW glutenin subunits that have preoccupied this section then we have an answer. However, issues such as the suggestion that dough viscosity and elasticity are attributable to, respectively, the gliadins and glutenins, have ceased to preoccupy the cereal chemists.

FURTHER READING

'The Chemistry of Muscle Based Foods', ed. D. A. Ledward, D. E. Johnstone, and M. C. Knight, The Royal Society of Chemistry, Cambridge, 1992.

'Egg Science and Technology', eds. W. J. Stadelman and O. J. Cotterill, 4th edn., Food Products Press, New York, 1995.

'Wheat Structure, Biochemistry and Functionality', ed. J. D. Schofield, The Royal Society of Chemistry, Cambridge, 2000.

R. A. Lawrie, 'Lawrie's Meat Science', 6th edn., Woodhead, Cambridge, 1998.

L. G. Phillips, D. M. Whitehead and J. Kinsella, 'Structure–Function Properties of Food Proteins, Academic Press, San Diego, 1994.

P. F. Fox and P. L. H. McSweeney, 'Dairy Chemistry and Biochemistry', Blackie, London, 1998.

Chapter 6

Colours

Colour is very important in our appreciation of food. Parents who may well deplore their children's selection of the most lurid items from the sweetshop will be found later in the greengrocery applying identical criteria to their own selections of fruit. Since we can no longer expect to sample the taste of food before we buy it, appearance is really the only guide we have to quality other than past experience of the particular shop or manufacturer. Ever since food preservation and processing began to move from the domestic kitchen to the factory, there has been a desire to maintain the colours of processed and preserved foods as close as possible to the colours of the original raw materials. Some foodstuffs acquire their recognised colours as an integral part of their processing, for example the brown of bread crust and other baked products, considered in Chapter 2, and the pink of cured meats, mentioned in Chapter 5. In this chapter we will be largely concerned with the most colourful foodstuffs, fruit and vegetables, and the efforts of chemists to mimic their colours. In recent years consumers have shown increasing concern over the use of chemically synthesised dyestuffs as food colorants, and some attention is also given here to the increasing use of naturally occurring pigments such as annatto and cochineal.

CHLOROPHYLLS

The chlorophylls are the green pigments of leafy vegetables. They also give the green colour to the skin of apples and other fruit, particularly when it is unripe. Chlorophylls are the functional pigments of photosynthesis in all green plants. They occur, alongside a range of carotenoid pigments, in the membranes of the chloroplasts, the organelles which carry out photosynthesis in plant cells. The pigments of the chloroplast are intimately associated with other lipophilic components of the membranes, such as phospholipids, as well as the membrane proteins. Algae and photosynthetic bacteria contain a number of different types of chlorophyll, but the higher plants that concern us contain only chlorophylls *a* and *b*, in the approximate ratio of 3 to 1.

The structures of chlorophylls *a* and *b* are shown in Figure 6.1. They are essentially porphyrins similar to those in haem pigments such as myoglobin except for the different ring substituents, the co-ordination of magnesium rather than iron, and the formation of a fifth ring by the linkage of position 6 to the γ-methine bridge. By analogy with pigments such as myoglobin we can guess that the porphyrin ring will associate readily with the hydrophobic regions of chloroplast membrane proteins. The extended phytol side chain facilitates close association with carotenoids and membrane lipids.

Absorption spectra of chlorophylls *a* and *b* are shown in Figure 6.2. The exact wavelengths of the absorption maxima and the extinction coefficients are not easily defined, as they vary depending on the identity of the organic solvent used and the presence of traces of moisture. These variations are sufficient to make the accurate determination of chlorophyll levels in leafy tissues by spectrophotometry rather difficult. Another difficulty is that in solution, especially in solvents that include methanol,

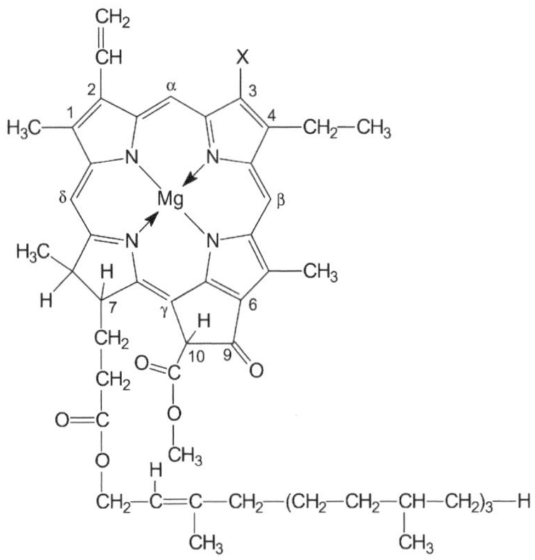

Figure 6.1 *Chlorophyll. In chlorophyll a X = CH₃ and CHO in chlorophyll b. The double bonds of the porphyrin ring system are shown positioned quite arbitrarily but in fact the π-electrons of systems such as this are completely delocalised. The C₂₀ isoprenoid alcohol esterified to the propionyl substituent at position 7 is known as phytol.*

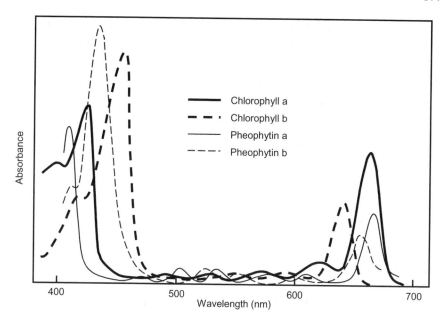

Figure 6.2 *Absorption spectra of chlorophylls and pheophytins.*

there is a tendency for oxidation or isomerisation to occur with replacement of the hydrogen at position 10 with OH or OCH_3. It is chlorophyll's stability, or lack of it, in vegetable tissues that is of interest to food chemists. Chlorophyll is lost naturally from leaves at the end of their active life on the plant. This breakdown accompanies a general breakdown of the chloroplast membranes, but the carotenoids are rather more stable so that autumn leaves and vegetables that are no longer fresh have a residual yellow colour. A number of different reactions have been implicated in chlorophyll breakdown including cleavage and opening of the porphyrin ring and removal of the phytol residue by the enzyme chlorophyllase, but there is very little known to support any of these hypotheses.

When green vegetables are heated, in ordinary cooking, when they are blanched prior to freezing, or during canning, there is evidence for the loss of the phytol side chain to give the corresponding chlorophyllide *a* and *b*, but the most important event is the loss of the magnesium. This occurs most readily in acid conditions, the Mg^{2+} ion being replaced by protons, to give pheophytins *a* and *b*. Pheophytins have a dirty brown colour (*see* Figure 6.2) and will be familiar as the dominant pigments in 'green' vegetables such as cabbage that have been overcooked. The acidity of the contents of plant cell vacuoles makes it difficult to avoid pheophytin formation, especially during the rigorous heating involved in canning peas. One approach is to keep the cooking water slightly alkaline by the addition of a small quantity of sodium bicarbonate. This is quite successful, but the alkaline conditions have an unhappy effect on the texture and flavour and losses of vitamin C are enhanced.

In the canning of peas chlorophyll loss is inevitable and artificial colour has to be resorted to. It is said that cupric sulfate was occasionally – and dangerously – used for this in the early years of this century; now, however, organic chemical dyes are universally used, a mixture of tartrazine and Green S (*see* Figure 6.15) being most popular. In the 18th and 19th centuries vegetables were preserved as pickles by cooking in vinegar. The pickled vegetables' bright green colour was owed to the Cu^{2+} ions leached from the copper pans (*see* Chapter 11). Since around 1860 legislation has ensured that our pickles are free of copper, and we have so grown to like the brown that nowadays caramel is added to enhance the contribution of the pheophytins. Growing doubts about the safety of synthetic dyestuffs, especially tartrazine, are encouraging food chemists to examine the possible use of chlorophyll derivatives as food colour. Chlorophyll itself is obviously unsuitable. Not only is it unstable but its insolubility in water makes it very difficult to apply. However, a derivative known as sodium copper chlorophyllin is now being used. This is the sodium salt of chlorophyllin (*i.e.* chlorophyll from which the phytol side has been removed) and has the magnesium replaced by copper. It has an acceptable blue–green colour, is moderately soluble in water, and, most important, does survive the heating conditions involved in canning. The amount of copper it contains is far too small to represent a toxicity hazard.

CAROTENOIDS

Carotenoid pigments are responsible for most of the yellow and orange colours of fruit and vegetables. They are found in the chloroplasts of green plant tissues alongside chlorophyll and in the chromoplasts of other tissues such as flower petals. Chemically they are classed as terpenoids, substances derived in nature from the metabolic intermediate mevalonic acid (6.1), which provides the basic structural unit, the isoprene unit (6.2). Terpenoids having one, two, three, or four isoprene units (hemi-, mono-, sesqui-, and di-terpenoids, respectively) are well known, but to food chemists the steroids which are triterpenoids (*i.e.* 30 carbon atoms) and the carotenoids, the only known tetra-terpenoid compounds, are the most important. Carotenoids occur in all photosynthetic plant tissues as components of chromoplasts, organelles which may be regarded as degenerate chloroplasts.

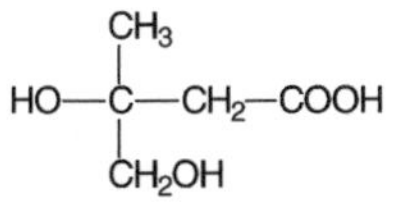

(6.1) 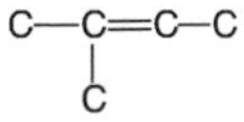   (6.2) 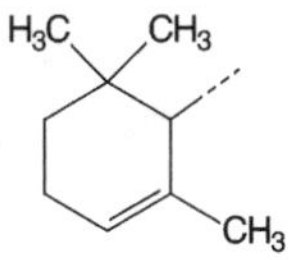   (6.3) 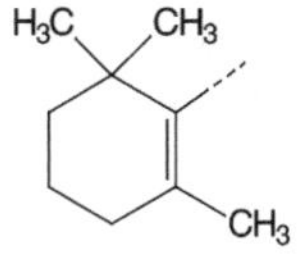   (6.4)

The carotenoids are divided into two principal groups, the carotenes, which are strictly hydrocarbons, and the xanthophylls, which contain oxygen. The structures of most of the carotenoids important in foodstuffs are shown in Figures 6.3 and 6.4. The simplest carotene is lycopene, and the numbering system shown illustrates clearly how the molecule is constructed from two diterpenoid subunits linked 'nose to nose'. Other carotenes are formed by cyclisation at the ends of the chain. Two possible ring structures result, the α-ionone (6.3) or the β-ionone (6.4). The acyclic end group occurring in lycopene and γ-carotene is sometimes referred to as the ψ-ionone structure.

The xanthophylls arise initially by hydroxylation of carotenes, and most plant tissues contain traces of cryptoxanthins, the monohydroxyl precursors of the dihydroxyl xanthophylls, such as zeaxanthin and lutein, which are shown in Figure 6.4. Subsequent oxidation reactions lead to the formation of epoxides, such as antheraxanthin. Neoxanthin is an example, rare in nature, of an allene, in which one carbon atom carries two double bonds. The apocarotenoids are a small group of xanthophylls in which

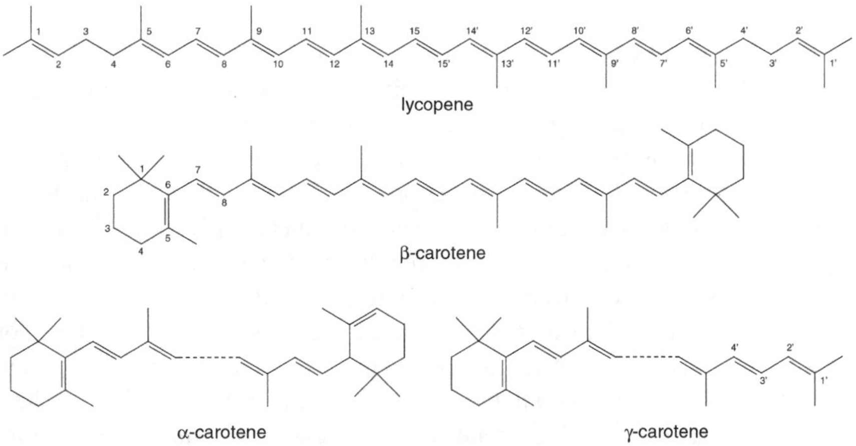

Figure 6.3 *Carotene structures. The central region common to all carotenoids has been omitted here in α- and γ-carotene and in Figure 6.4.*

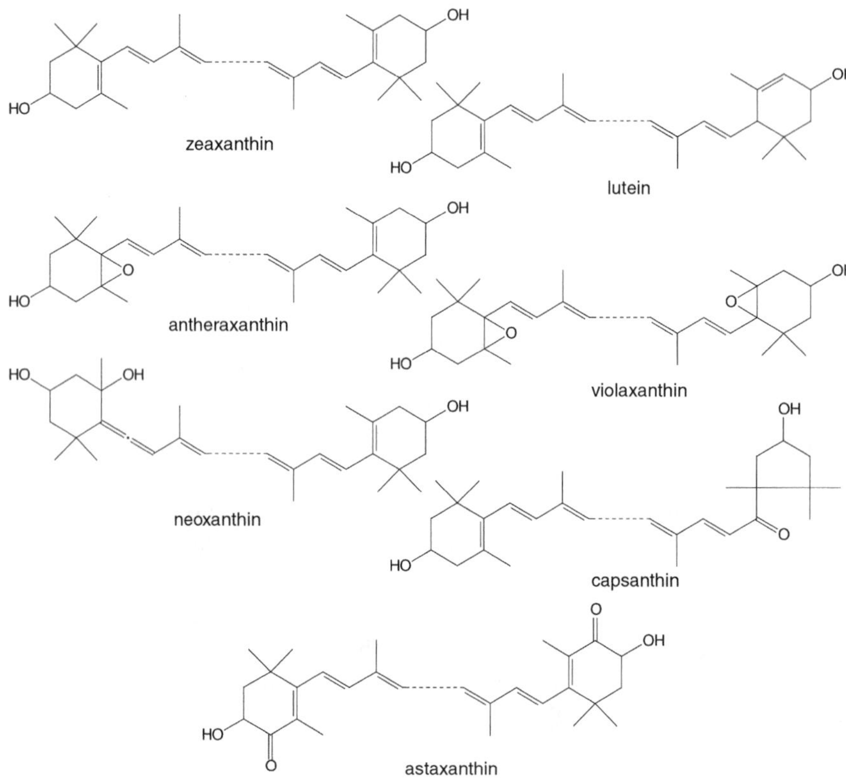

Figure 6.4 *Xanthophyll structures.*

fragments have been lost from one or both ends of the chain. Three examples are shown in Figure 6.5.

With the exception of the two acidic apocarotenoids, which will form water-soluble salts under alkaline conditions, carotenoids are only freely soluble in non-polar organic solvents. Their absorption spectra are generally rather similar, with the nature of the solvent causing nearly as much variation in λ_{max} and extinction coefficient values as is found between different carotenoids. The absorption spectra given in Figure 6.6 bear out the observation that xanthophylls are the dominant pigment in yellow tissues whereas carotenes tend to give an orange colour.

The distribution of carotenoids amongst the different plant groups or types of food materials shows no obvious pattern. Amongst the leafy green vegetables the carotenoid content follows the general pattern of all higher-plant chloroplasts, with β-carotene dominant and the xanthophylls lutein, violaxanthin, and neoxanthin all prominent. Zeaxanthin, α-carotene, cryptoxanthin (3-hydroxy-β-carotene), and antheraxanthin also occur in

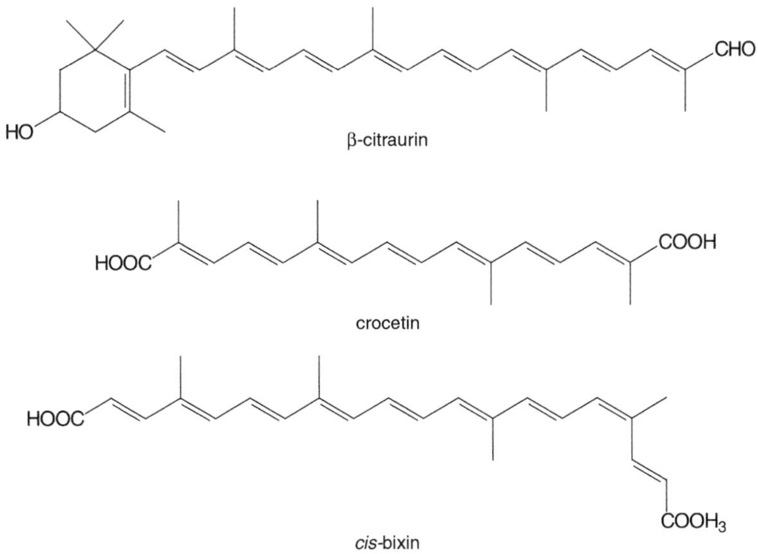

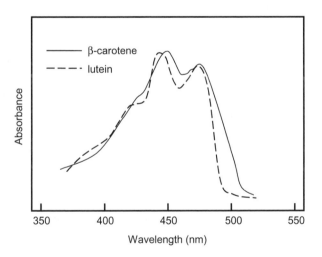

β-citraurin

crocetin

cis-bixin

Figure 6.5 *Apocarotenoid structures.*

Figure 6.6 *Absorption spectra of carotenoids.*

small amounts. The quantity of β-carotene in leaf tissue is usually between 200 and 700 μg g^{-1} (dry weight). Amongst fruit a wider range of carotenoids is found. Only rarely (*e.g.* the mango and the persimmon) are β-carotene and its immediate xanthophyll derivatives cryptoxanthin and zeaxanthin predominant. In the tomato, lycopene is the major carotenoid.

Orange juice contains varying proportions of cryptoxanthin, lutein, antheraxanthin, and violaxanthin together with traces of their carotene precursors. The apocarotenoid β-citraurin has also been reported. During juice processing the acidic conditions promote some spontaneous conversion of the 5,6- and 5',6'-epoxide groups into 5,8- and 5',8'-furanoid oxides (6.5).

(6.5)

Some carotenoids are restricted to just a few or even a single plant species; the capsanthin of red peppers is a good example. Of course the classic occurrence of carotenoids is in the carrot. Here β-carotene predominates together with a proportion of α-carotene. The carotenoids of the usual varieties of carrot comprise only 5–10% xanthophylls, located mostly in the yellow core. The total carotene content of carrots is 60–120 μg g^{-1} fresh weight, but some varieties are available with more than 300 μg g^{-1}.

Carotenoids are becoming increasingly popular as food colorants and many are now officially sanctioned (*see* Table 6.1). The two apocarotenoids, crocetin and bixin, have been used in a limited range of foods for many years. Crocetin is the major pigment of the spice saffron, where it occurs in the form of a glycoside, each of its carboxyl groups being esterified to molecules of the disaccharide gentiobiose. Saffron is extremely expensive, since it requires the some 150 000 stigmas from the flowers of the autumn flowering crocus, *Crocus sativus*, to produce 1 kg of the dried spice. *cis*-Bixin (*see* Figure 6.5) is the major constituent of a colouring matter known as annatto, which is the resinous material coating the seeds of *Bixa orellana*, a large shrub grown in a number of tropical countries. *cis*-Norbixin is a water soluble version, lacking the esterified methanol. Annatto is the traditional colouring for 'red' cheeses such as Leicester and is also used to colour margarine. Palm oil contains high levels of β-carotene (13–120 mg per 100 g) and is also used to colour margarine. Three carotenoids have been chemically synthesised on a commercial scale, β-carotene, β-apo-8'-carotenal (6.6) (E160e), and canthaxanthin (6.7) (E161g). These commercial preparations are being increasingly used in a wide range of products including margarine, cheese, ice-cream, and some baked goods such as cakes and biscuits.

Table 6.1 *Natural pigments commonly used as food colorants. As chemical treatments are required for the conversion of chlorophyll into copper chlorphyllin it is not usually claimed as 'natural' on food labels. Extracts containing crocin or crocetin are treated as spices rather than colorants and therefore not classified as food additives in the EC. It is debatable whether caramels (150a–d), discussed in Chapter 2, qualify as 'natural' pigments but they have been omitted from this list.*

Pigment	EC No.	Commercial source
Curcumin	E100	Tumeric *Curcuma longa*
Carminic acid	E120	Cochineal
Chlorophyll	E140	Dried ground grass, alfafa
Copper chlorophyllin	E141	Dried ground grass, alfafa
Mixed carotenes and β-carotene	E160a	Palm oil, carrots, *Dunaliella* sp. (algae)
Cis-bixin, *cis*-norbixin	E160b	Annatto
Capsanthin	E160c	Red peppers *Capsicum annum*
Lycopene	E160d	Tomato
Lutein	E161b	Marigold *Tagetes erecta*, alfafa
Crocin and crocetin		Saffron *Crocus sativus*
Betanin	E162	Beetroot
Anthocyanins	E163	Black grapes, red cabbage, elderberries *Sambucus nigra*

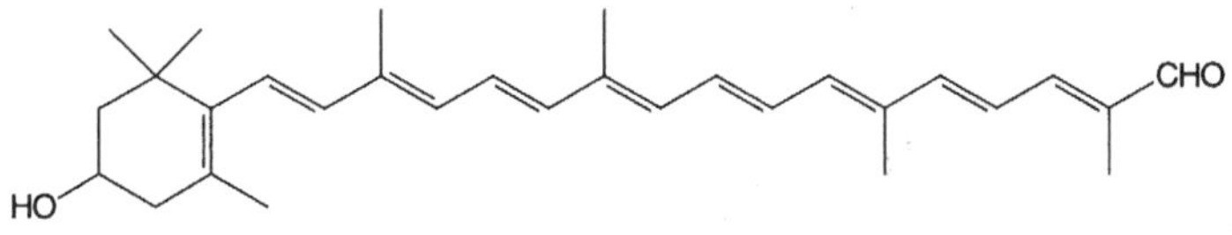

(6.6)

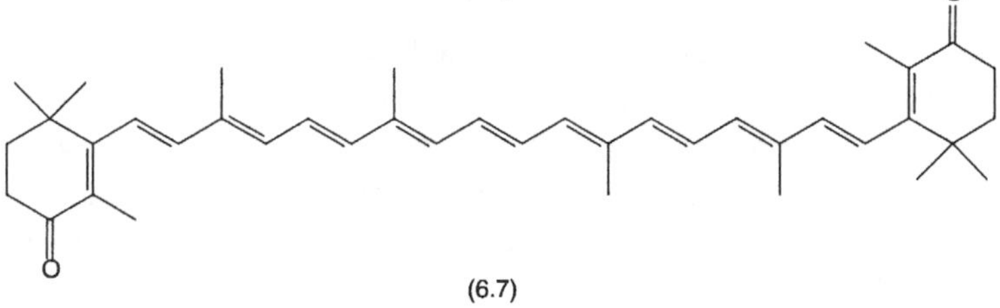

(6.7)

Although normally associated with plants, carotenoids do find their way into some animal tissues. Egg yolk owes its colour to the two xanthophylls lutein and zeaxanthin, with only a small proportion of β-carotene. These carotenoids, and the smaller amounts that give the depot fat of animals its yellowish shade, are derived from the vegetable material in the diet. The dark greenish-purple pigment of lobster carapace (and the shells of other shellfish) is a complex of protein with astaxanthin. When lobster is boiled, the protein is denatured and the colour reverts to the more typical reddish shade of a carotenoid. Astaxanthin is also the source of the pink colour of salmon flesh.

Carotenoids are generally quite stable in their natural environments, but when food is heated or when they are extracted into solutions in oils or organic solvents they are much more labile. On heating in the absence of air there is a tendency for some of the *trans* double bonds of carotenes to isomerise to *cis*. The structures of the resulting 'neocarotenes' have not been fully described, but as the number of *cis* double bonds increases there is a loss of colour intensity. The difference in colour between canned and fresh pineapple provides a demonstration of the effect. The isomerisation of 5,6-epoxides to 5,8-furanoid oxides has already been mentioned. In the presence of oxygen, particularly in dried foods such as dehydrated diced carrots, oxidation and bleaching occur rapidly. Hydroperoxides resulting from lipid breakdown are very effective in bleaching carotenoids. The final breakdown of β-carotene leaves a residue of β-ionone (6.8), a volatile compound which gives sun-dried (and partially bleached) hay its character-istic odour.

(6.8)

No specific physiological role has been identified for carotenoids as such in man but in recent years there has been a lot of speculation regarding the health benefits to be obtained from a diet rich in carotenoids. The role of the carotenoids with at least one β-ring (often referred to as provitamins A) as precursors of vitamin A is well established. Vitamin A activity is possessed by substances that are precursors of retinal, the non-protein component of the visual pigment of the retina. The enzyme catalysed formation of retinal from carotenoid precursors is shown in Figure 6.7. Subject to a sufficiency of vitamin A or its precursors in the diet the level of retinol in the tissues is physiologically regulated and high dietary levels (*i.e.* at the upper end of the normal nutritional range) do not raise the concentration of retinol in the tissues or bloodstream. Vitamin A is discussed further in Chapter 8.

Epidemiological studies, where the incidence of a particular disease in a population is studied in relation to the consumption of particular dietary

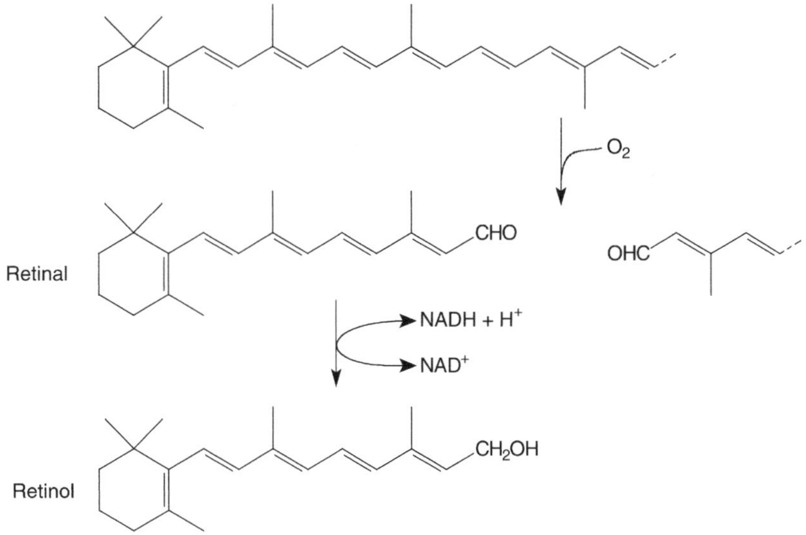

Figure 6.7 *Formation of retinol and retinal.*

components, have suggested that fruit and vegetable consumption is associated with a reduced incidence of many different types of cancer. This effect is most marked with cancers of the gastrointestinal tract (such as those of the mouth, oesophagus and stomach) and respiratory system (*e.g.* larynx and lung). Although attention was focussed on β-carotene it has been recognised that other plant constituents such as other carotenoids, ascorbic acid, flavonoids, fibre *etc.* could all be involved. Detailed dietary analysis has associated dietary α- and β-carotene and lutein with a reduced risk of lung cancer, and lycopene, but not the other carotenoids, with a reduced risk of prostate cancer. Unfortunately the obvious follow up to these observations, intervention trials in which β-carotene was supplied as a dietary supplement to smokers, produced the unexpected results of β-carotene supplements actually increasing the incidence of lung cancer. It seems likely therefore that other components of plant materials as well as particular carotenoids are involved and we need to concentrate on the quality of diets as a whole rather than their levels of individual components. Whether the quasi-pharmaceutical 'supplements' industry will embrace this concept remains to be seen.

The possibility that the antioxidant function of β-carotene and other carotenoids may have an antioxidant function, alongside vitamins C and E, has a bearing on the incidence of coronary heart disease is a separate issue, considered briefly in Chapter 4.

ANTHOCYANINS

The pink, red, mauve, violet, and blue colours of flowers, fruit, and vegetables are caused by the presence of anthocyanins. In common with other polyphenolic substances, anthocyanins occur in nature as glycosides, the aglycones being known as anthocyanidins. They are located in the vacuoles of plant cells, kept apart from the enzymes and other delicate structures in the cytoplasm. These are flavonoids*, *i.e.* substances based on the flavan nucleus (6.9).

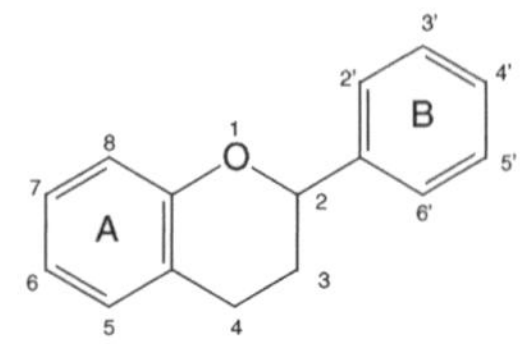

(6.9)

Six different anthocyanidins occur in nature, but the diversity of the patterns of glycosylation means that there are innumerable different anthocyanins. A single plant species will also contain considerable numbers of different anthocyanins. The structures of the six anthocyanidins are shown in Figure 6.8.

The form of the anthocyanidins shown in Figure 6.8 is correctly termed the flavylium cation, but, as we shall see, other forms occur, depending on the pH. Anthocyanins always have a sugar residue at position 3, and glucose often occurs additionally at position 5 and more rarely at positions 7, 3', and 4'. Besides glucose the monosaccharides most often found are galactose, rhamnose, and arabinose. Some rather unusual disaccharides also occur, such as rutinose [α-L-rhamnosyl-(1→6)-D-glucose] and

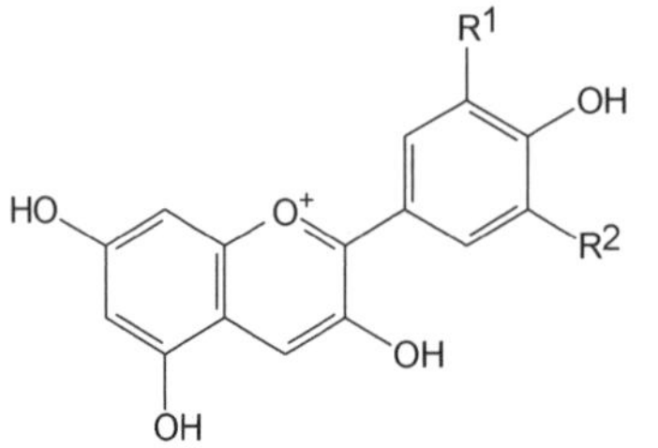

	R^1	R^2	λ_{max}*
pelargonidin	– H	– H	503
cyanidin	– OH	– H	517
peonidin	– OCH$_3$	– H	517
delphinidin	– OH	– OH	526
petunidin	– OCH$_3$	– OH	526
malvinidin	– OCH$_3$	– OCH$_3$	529

*The λ_{max} values shown are those of the corresponding 3-glucoside anthocyanins at pH 3.

Figure 6.8   *The structures of the anthocyanidins and the absorption maxima of anthocyanins.*

* Often spelt 'flavanoids'.

sophorose [β-D-glucosyl-(1→2)-glucose]. It is not uncommon for the 6-position of the sugar residue at position 3 to be esterified with phenolic compounds such as caffeic (6.10), *p*-coumaric (6.11), or ferulic (6.12) acids. The principal anthocyanins of a range of fruit are listed in Table 6.2. Cyanidin is by far the most widespread anthocyanidin, and it is clear that there is no link between the taxonomic classification of a plant and the identity of its anthocyanins.

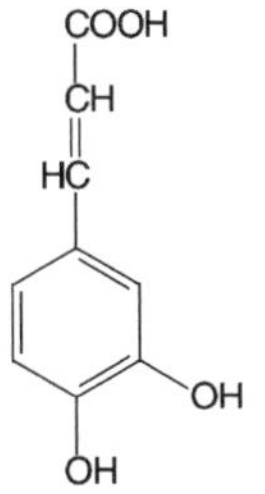

(6.10) 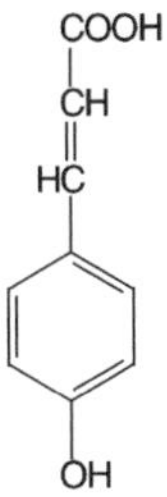 (6.11) 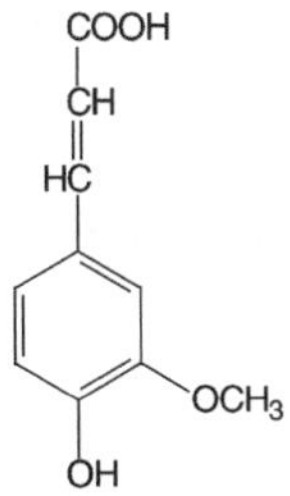 (6.12)

Grape anthocyanins are particularly interesting. Of the six anthocyanidins only pelargonidin is not found in grapes, and there is much more variety in the patterns of glycosylation and acetylation than in most plants. The classic European grape species is *Vitis vinifera*, which contains only 3-monoglucosides, whereas in the USA and other non-European countries species such as *V. riparia* and *V. rupestris* and their hybrids with *V. vinifera* are grown. All these species, but not *V. vinifera*, contain both 3-monoglucosides and 3,5-diglucosides. The diglucosides can be readily detected by a chemical test and also separate from the monoglucosides on

Table 6.2 *Fruit anthocyanins.*

Fruit	Principal anthocyanins
Blackcurrant *Ribes nigrum*	3-Glucosides and 3-rutinosides of delphinidin and cyanidin
Blackberry *Ribes fructicosus*	3-Glucosides and 3-rutinosides of cyanidin
Raspberry *Rubus ideaus*	3-Glucosides, 3-sophorosides and 3-rutinosides of pelargonidin and cyanidin
Elderberry *Sambucus nigra*	3,5-Diglucosides and 3-sambubiosides of cyanidin (sambubiose is β-D-xylosyl-(1→2)-β-D-D-glucose)
Strawberry *Fragaria* spp.	3-Glucosides of pelargonidin and cyanidin, 3-arabinoside of pelargonidin
European grape *Vitis vinifera*	3-Glucosides, 3-acetylglucosides and 3-coumarylglucosides of malvidin, peonidin, delphinidin, petunidin and cyanidin

chromatography. Detection of diglucosides in a red wine is therefore proof of its non-European origin.

Much less work has been done on the anthocyanins of vegetables. Cyanidin is the anthocyanidin of red cabbage, pelargonidin occurs in radishes and red-seeded varieties of beans, and delphinidin occurs in aubergines.

As shown in Figure 6.8 increasing substitution in the B-ring moves the absorbance maximum of anthocyanidins towards the red end of the spectrum but these changes are not sufficient to account for the great range of colours that anthocyanins can give to flower petals and fruit. The pH and the presence of other substances have a much greater influence on their colour than the nature of the ring substituents. Anthocyanins form hydrogen bonded complexes with other, colourless flavonoids that are common in plant tissues. Figure 6.8 shows the basic anthocyanidin structure in the flavylium cation form that predominates at low pH values; at pH 1 this is the only significant form. As the pH is raised, a proton is lost, a water molecule is acquired, and the carbinol pseudo-base is formed (*see* Figure 6.9). The pK_a value (*i.e.* the pH value at which the ionisation is half complete) for this transition is around 2.6. The flavylium cations are the red coloured forms of the anthocyanins and the carbinol pseudobases are colourless, so there is a gradual loss of colour intensity with rise in pH until strong blue colours are obtained at high pH values (Figure 6.10). Work with purified anthocyanins has clarified the pattern of the other ionisations including those at higher pH values. Above pH 3 the small residual proportion of the anthocyanins in the flavylium cation form begins to lose protons, forming first the quinoidal base, which is weakly purple in colour, and then the ionised quinoidal base, which is deep blue. At the same time the carbinol pseudobase gives rise to a small proportion of the colourless chalcone forms. After extended periods at elevated pH quite high proportions of chalcones are formed.

Although these equilibria have been described in some detail for purified anthocyanins, it must be remembered that in fruit juices and other natural plant materials the situation is neither as straightforward nor as well understood. Anthocyanins readily interact with other colourless flavonoids which abound in plant tissues. These associations, which depend on hydrogen bonding between hydroxyl groups, enhance the intensity of the colour, modify λ_{max} values, and tend to stabilise the blue quinoidal forms.

Anthocyanins also complex metal cations, but it is not known how important this is to the colour of fruit. Occasional unusual colour developments in canned fruit are undoubtedly due to interactions between the anthocyanins of the fruit and the metal, iron or tin, of the can when

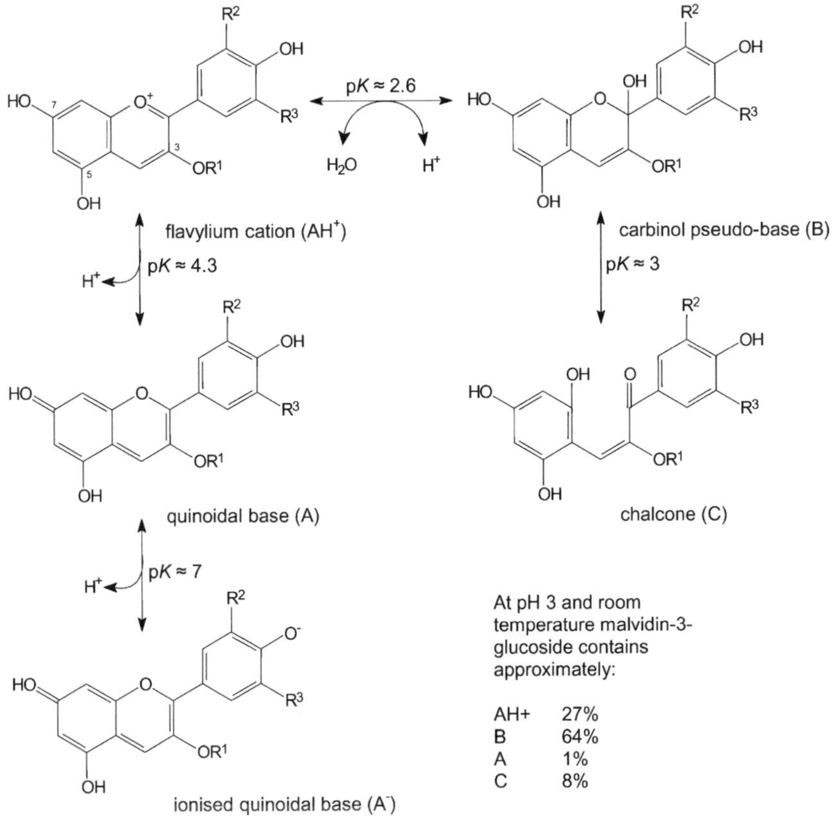

Figure 6.9 *Transitions in the anthocyanin structure. The presence of sugar residues at 5 or 7 will obviously influence the formation of quinoidal forms.*

the internal lacquering has failed. The unexpected appearance of colour in canned pears has been attributed to iron or tin complexes with leuco-canthocyanins, which are normally colourless. Another important reaction of anthocyanins is that with sulfur dioxide. Sulfur dioxide, usually as sulfite (SO_3^{2-}) or metabisulfite ($S_2O_5^{2-}$) (*see* Chapter 9), is routinely used as an antimicrobial preservative in wines and fruit juices. At high concentrations (1–1.5%) it causes a total irreversible bleaching[*] of anthocyanins, but at lower concentrations (500–2000 p.p.m.) it reacts with the flavylium cation to form a colourless addition compound, the chroman-

---

[*] Maraschino cherries are bleached with sulfur dioxide before being dyed with synthetic colours to garnish cocktails *etc.*

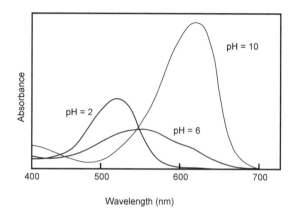

Figure 6.10   *The effect of pH on the absorption spectra of an anthocyanin.*
(Redrawn from R. Brouillard, in 'Anthocyanins as Food Colours' ed. P. Markakis, Academic Press, New York, 1982.)

4-sulfonic acid (6.13). Acidification or the addition of excess acetaldehyde (ethanal) will remove the sulfur dioxide. This is the basis of the insufficiently well known technique for dealing with red wine spills on white tablecloths – spill a few drops of white wine (which always contains sulfur dioxide) on the stain. If this requires the opening of another bottle then so be it.

The anthocyanins in red wine undergo a number of reactions with some of the other colourless flavonoids that were also originally present in the grape skins and seeds, such as catechins, (better known as a component of tea, *see* Figure 6.12). The formation of a link between the 4-position of the anthocyanin and the 8-position of the catechin leads to procyanidin formation – more strongly coloured compounds that are less affected by pH change or sulfur dioxide. By the end of the fermentation period as much as a quarter of the anthocyanins may have formed oligomers with other flavonoids. The brown of old wines is the result of extensive polymerisation of the anthocyanins and other flavonoids. This is accompanied by a mellowing of the flavour, as it is these, usually described as tannins, that give red wine its characteristic astringency.

In most food-processing operations the anthocyanins are quite stable, especially when the low pH of the fruit is maintained. Occasionally, however, the ascorbic acid naturally present can cause problems. In the presence of iron or copper ions and oxygen the oxidation of ascorbic acid to dehydroascorbic acid is accompanied by hydrogen peroxide formation. This will oxidise anthocyanins to colourless malvones (6.14), a reaction implicated in the loss of colour by canned strawberries.

Anthocyanins are becoming increasingly popular as food colours, replacing the synthetic dyes (discussed later in this chapter) in confectionary, soft drinks and similar products. They are of little value in dairy products as their colour at their less acid pH values is either weak or an unappealing blue or even purplish green, if other yellow pigments are present. Good sources of anthocyanins are not that common. One source that is being exploited is the residue, mostly skins, from pressing black grapes for wine. Red cabbage extracts are being marketed and elderberries are also likely to become important anthocyanin sources.

BETALAINES

The characteristic red–purple pigments of beetroot, *Beta vulgaris*, used to be described as 'nitrogenous anthocyanins'. However, it is now established that these pigments and those of the other members of the eight families usually grouped together as the order Centrospermae are quite distinct from the anthocyanins and never occur with anthocyanins in any member of these eight families. The betalaines are divided into two groups, the betacyanines, which are purplish red in colour (λ_{max} 534–555 nm), and the less common, yellow, betaxanthines ($\lambda_{max} \sim 480$ nm). The betalaines of *B. vulgaris* are the only ones of food significance, and this account will therefore be restricted to them. Their structures are shown in Figure 6.11.

Firstly, we have the betacyanines, which comprise about 90% of beetroot betalaines. They are all either the glycosides or the free aglycones of betanidine or its C-15 isomer. In beetroot, *iso* forms account for only 5% of the total betacyanines. About 95% of the betanidine and isobetanidine carry a glucose residue at C-5, and a very small proportion of these glucose residues are esterified with sulfate. Secondly, we have the betaxanthines, in beetroot represented by the vulgaxanthines, characterised by the lack of an aromatic ring system attached to N-1 and by the absence of sugar residues. A great diversity of betaxanthines are found in the

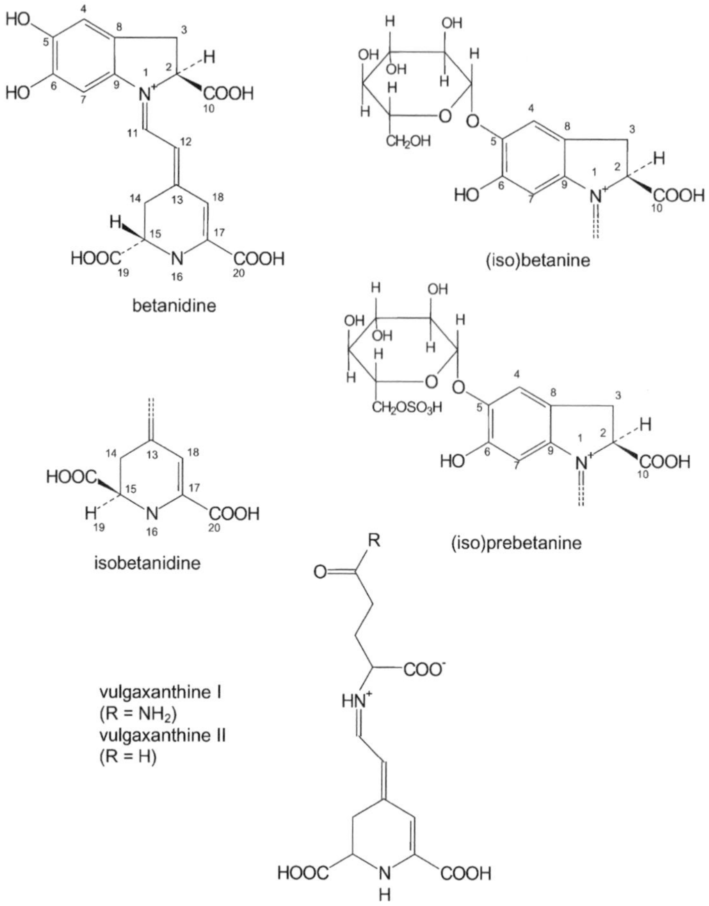

Figure 6.11 *Betalaine structures.*

Centrospermae, but only the two known as vulgaxanthines I and II are found in beetroot, in roughly equal proportions.

Although of a similar colour to the anthocyanins betacyanines differ in that their colour is hardly affected by pH changes in the range normally encountered in foodstuffs. Below pH 3.5 the absorption maximum of betanine solutions is 535 nm, between pH 3.5 and 7.0 it is 538 nm, and at pH 9.0 it rises to 544 nm. Betacyanins are fairly stable under food-processing conditions, although heating in the presence of air at neutral pH values causes breakdown to brown compounds. Concentrated beetroot extracts (containing around the 0.3% pigment, the rest being almost entirely sugar) are becoming popular as colorants for dairy and dessert products. Their instability to heat at neutral pH values means they are unsuitable for colouring cakes *etc.*

MELANINS

Although we do not necessarily regard them as desirable pigments in our fruits and vegetables, this is a convenient point to consider the brown, melanin-type pigments that arise when plant tissues are damaged. Everyone is familiar with the way in which pale-coloured fruit and vegetables such as apples, bananas, and potatoes quickly turn brown if air is allowed access to a cut surface. The browning occurs when polyphenolic substances, which are usually contained within the vacuole of the plant cells, are oxidised by the action of the enzyme phenolase, which occurs in the cytoplasm of plant cells. Tissue damage caused by slicing or peeling, fungal attack, or bruising will bring enzyme and substrates together. The enzyme phenolase (originally known as polyphenol oxidase) occurs in most plant tissues, and a related enzyme, tyrosinase, is found in mammalian skin. Phenolase is an unusual (but not unique) enzyme in that it catalyses two quite different types of reaction. The first of these, its cresolase activity, results in the oxidation of a monophenol to an *ortho*-diphenol (6.15).

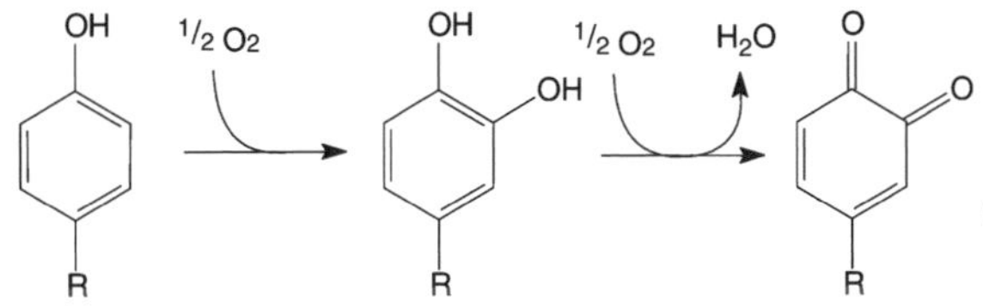

(6.15)

The second, catecholase, activity oxidises the *o*-diphenol to an *o*-quinone. *o*-Quinones are highly reactive and readily polymerise after their spontaneous conversion into hydroxyquinones (6.16).

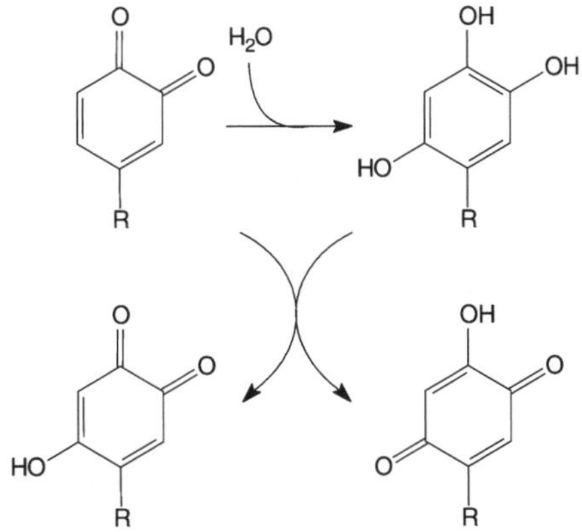

(6.16)

In animal tissues the only substrate is the amino acid tyrosine. The reaction in this case leads to the formation of the melanin pigments of skin and hair. The brown pigments that develop on the cut surface of fruit and vegetables may well be regarded as unsightly to consumers, but they do not detract from the flavour or nutritional value. To the plant the products of phenolase action, particularly the quinones, are very important as antifungal agents. The *o*-quinones interact with the tyrosine hydroxyl groups on the surface of fungal enzymes in much the same way as the polymerisation reactions (6.16) shown above. This can cause the inhibition of these enzymes and reduces the ability of fungi to penetrate tissues that have suffered minor physical damage. The relative resistance to fungal diseases of different varieties of onions and apples has been correlated with their phenolase activities and quinone contents.

The most important substrate for phenolase in apples, pears, and potatoes is chlorogenic acid (6.17) and in onions protocatechuic acid (6.18), but very little is known about the structures of the polymerised quinones that are derived from them.

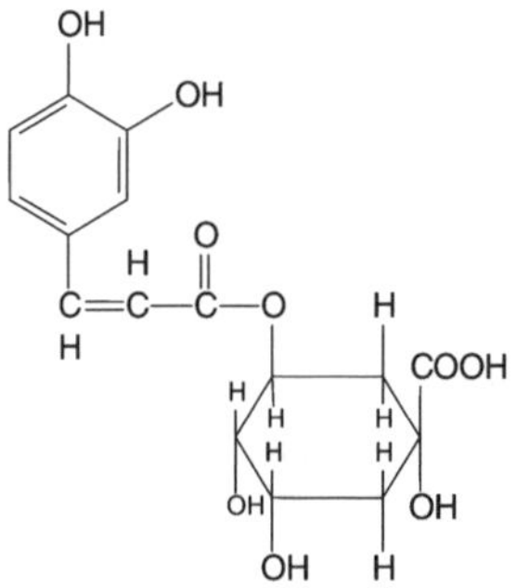

(6.17) 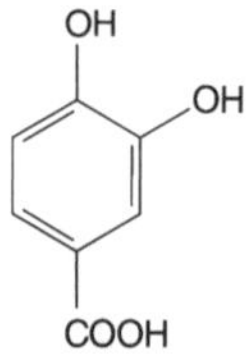   (6.18)

During fruit and vegetable processing (in the home as well as the factory) we try to prevent phenolase action. Making sure that the fruit and vegetables are blanched as soon as possible after (or even before) any tissue-damaging operations will reduce phenolase action to a minimum. Reducing contact with air by immersion in water is also common practice. Some of the most effective specific inhibitors of phenolase activity are chelating agents that act by depriving the enzyme of the copper ion that is a component of active site. However, substances used in analytical laboratories such as diethylthiocarbamate, CN^-, H_2S, or 8-hydroxyquinoline are hardly suitable for food use.

$$\begin{array}{cc}
\begin{array}{l}
\text{H}_2\text{C}\!\!-\!\!\text{COOH} \\
| \\
\text{HO}\!\!-\!\!\text{C}\!\!-\!\!\text{COOH} \\
| \\
\text{H}_2\text{C}\!\!-\!\!\text{COOH}
\end{array}
&
\begin{array}{l}
\text{COOH} \\
| \\
\text{H}\!\!-\!\!\text{C}\!\!-\!\!\text{OH} \\
| \\
\text{CH}_2 \\
| \\
\text{COOH}
\end{array}
\\
(6.19) & (6.20)
\end{array}$$

Though less specific and less potent, the organic acids citric acid (6.19) and malic acid (6.20) do have advantages. Not only are they common components of fruit juices but also their acidity helps to keep the pH well below the optimum for phenolase action (approx. pH 7). Immediately after being peeled (and especially if strong sodium hydroxide solution (lye) has been used for this), fruit such as peaches are often immersed in baths containing dilute solutions of these acids, frequently supplemented with ascorbic acid or sulfite.

Ascorbic acid is not only valued as a vitamin but it is also a good chelating agent and antioxidant. It interacts directly with the quinones to prevent browning and, if sufficient is used, will 'mop up' the oxygen in closed containers such as cans:

$$o\text{-diphenol} + O_2 \rightarrow o\text{-quinone} + H_2O$$

$$o\text{-quinone} + \text{ascorbate} \rightarrow o\text{-diphenol} + \text{dehydroascorbate}$$

Although it seems likely that the manufacturers of dehydrated potato powders include ascorbic acid in their products to ensure their whiteness rather than their vitamin C activity, the motive does not belittle the nutritional value of the addition for people, notably the elderly, who for one reason or another rely on such convenience foods in place of fresh fruit and vegetables.

Sulfur dioxide also has a specific, if not fully explained, inhibitory action against phenolase. A sulfite concentration sufficient to maintain a free SO_2 concentration of 10 p.p.m. will completely inhibit phenolase. Being a reducing agent, sulfite has the additional benefit of preserving the ascorbic acid level, but it is not without its disadvantages as it bleaches anthocyanins and also hastens can corrosion.

Tea

The enzymic oxidation of polyphenolic substances is a highly desirable feature of one particular foodstuff – tea. The tea we drink is a hot-water infusion of the leafy shoots of the tea plant, *Camellia sinensis*, which has been processed in various ways depending on the type of tea to be produced. The type of tea that most of us drink is known as 'black tea', to distinguish it from the 'green tea' often referred to as 'China tea'.* The leaves of the tea plant contain enormous quantities of polyphenolic materials – over 30% of the dry weight. Caffeine (6.21), the stimulant found also in coffee and 'cola'-based drinks, is some 3–4% of the dry weight of the leaf. The freshly plucked leaves (normally just the bud plus the first and second leaves, together known as the 'flush') are allowed to lose about one fifth of their water content, the production stage known as withering, after they are macerated. This physical damage to the leaves allows the phenolase in the cytoplasm of the leaf cells to come into contact with the polyphenolic substances contained in the cell vacuoles. The macerated leaves are then left at ambient temperatures for a few hours to 'ferment', during which time a high proportion of the polyphenolic substances are oxidised and much polymerisation occurs. The leaves are then 'fired', *i.e.* dried out at temperatures up to 75 °C, to inactivate the phenolase, bringing the fermantation reactions to a halt and giving us the material we know as tea. Green tea is manufactured by firing before the leaves are macerated so that the phenolase is inactivated before a a significant amount of fermentation can occur. An intermediate type 'Oolong' is fermented for only 30 minutes or so before firing.

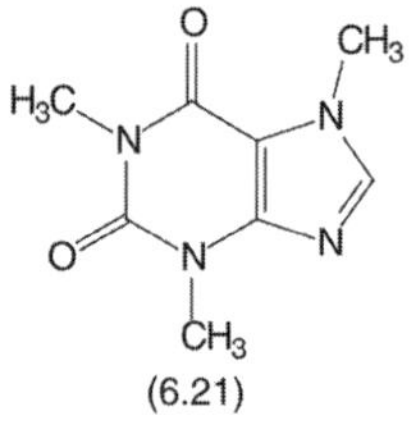

(6.21)

Almost all the polyphenolic substances in tea are flavonoids, either catechin or its derivatives as shown in Figure 6.12. Of these *epi*-gallocatechin gallate is by far the most abundant, some 9–13% of the dry weight.

The initial effect of phenolase action is in every case to convert the diphenols into the corresponding *o*-quinones (6.22 and 6.23).

---

* The tea trade does not appear to have noticed that Ceylon changed its name to Sri Lanka in 1972.

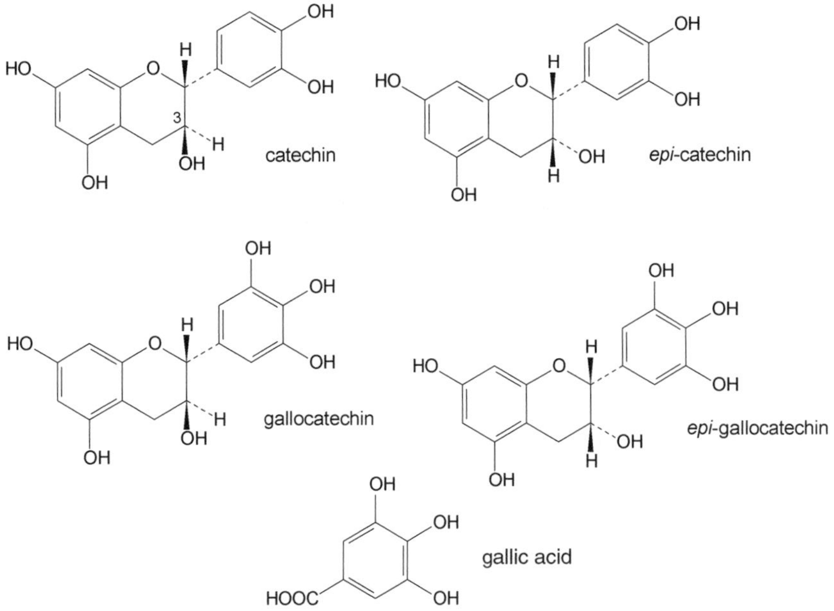

Figure 6.12 *Gallic acid and the principal flavonoids of the unfermented tea flush. The prefix 'epi-' refers to the optical configuration at the 3 position of of the flavonoid ring system. 'Gallo-' indicates three hydroxyl groups attached to adjacent carbon atoms in the aromatic ring. Both* epi-gallocatechin *and* epi-catechin *also occur as gallates, with gallic acid esterified at the 3-position.*

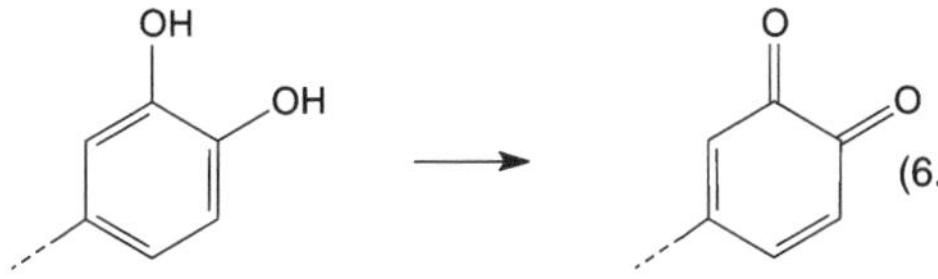

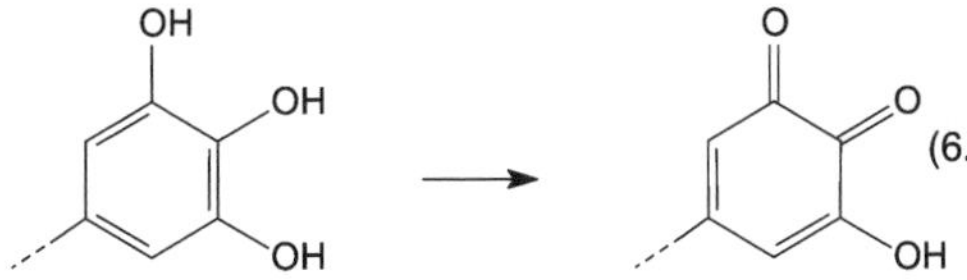

The quinones are highly reactive and readily form a wide variety of dimers. Most of these interact further to produce the polymeric thearubigins, but one class of dimers, the theaflavins (6.24), comprise about 4% of the soluble solids of black tea. Being bright red in colour, they make a major contribution to the 'brightness' of tea colour.

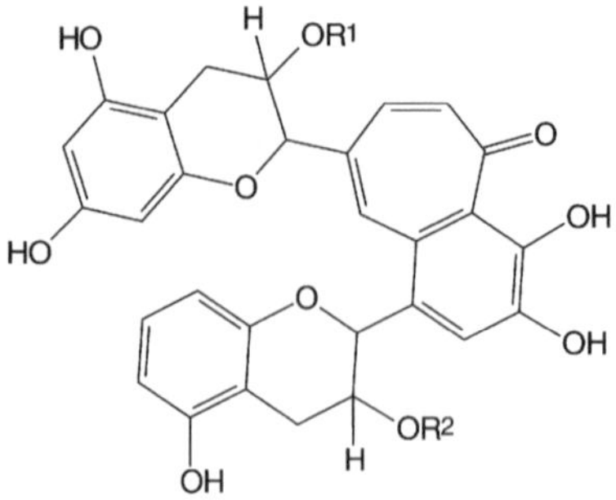

(6.24) R^1 and R^2 may be either
H or gallate residues

The brown thearubigins are the principal pigments of black-tea infusions. They contribute not only the bulk of the colour but also the astringency, acidity, and body. Thearubigins that are extracted with hot water have a wide range of molecular weights. Although their average molecular weight corresponds to oligomers of four or five flavonoid units, molecules of up to some 100 flavonoid units are found in tea infusions. Degradation studies have shown that all the principal flavonoids shown in Figure 6.12 are represented in these oligomers. Although there is no certainty as to the nature of the links involved in the polymerisation, a proanthocyanidin structure is most likely, as shown in Figure 6.13. The proanthocyanidins themselves are dimers which occur in small amounts in many fruit juices, where they contribute to the astringency.

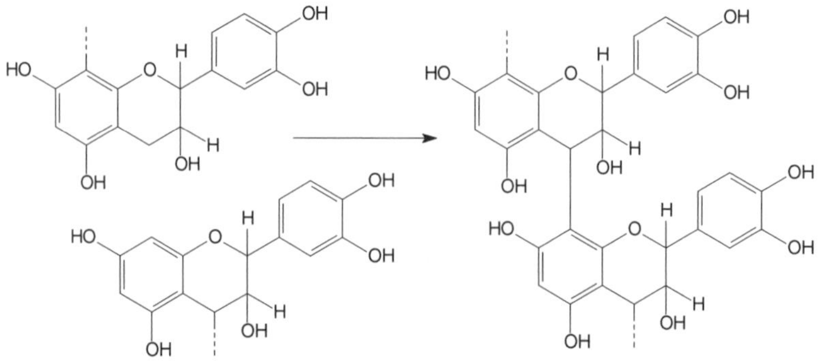

Figure 6.13   *The polymerisation of tea flavonoids forming a proanthocyanidin structure.*

Box 6.1   Flavonoids, Tannins and Health

The anthocyanins of fruit and the catechins of tea are important members of the flavonoid group of plant components. They are all based on the flavan nucleus (6.9) but differ in the state of oxidation and substituents of the central oxygen containing ring. Tannins are polymerised flavonoids. Flavonoids are difficult categorise in terms of their function in food (and therefore their placement in this book) since in different situations they can contribute colour, flavour, mouthfeel*. Significant degrees of anti-oxidant activity, anti-tumour activity and the converse, carcinogenicity, have also been attributed to them and it is these effects on human health that have brought them to wider attention.

As shown in Figure 6.14 flavonoids are grouped into classes which differ in the structure of the central ring. The A ring almost invariably carries hydroxyl groups at positions 5 and 7 and these frequently carry sugar residues. Similarly the B ring carries 1, 2 or 3 hydroxyl groups. When double

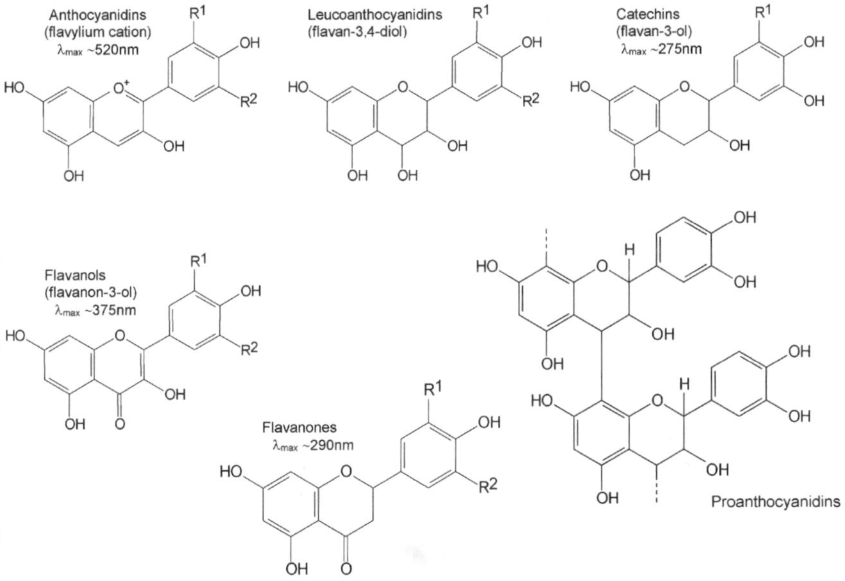

Figure 6.14   *The basic structures of the major groups of flavonoids. Where available typical values for the absorption maxima are shown. Except in the anthocyanins, where more details are shown in Figure 6.8, R¹ and R² are either −H or −OH. The proanthocyanidin here has two flavin-3-ol units but proanthocyanidins commonly include flavin-3,4-diol and anthocyanin units.*

*The word 'mouthfeel' is a slightly more scientific term used for the 'body' of drinks, but impossible to quantify.

continued on p. 200

Box 6.1 *continued*

bonds in the central ring provide a conjugated link between the aromatic systems of the A and B rings flavonoids absorb visible light, as indicated by the approximate absorption maxima (λ_{max}) included in Figure 6.14.

The definition of 'tannins' is imprecise and usually refers to polyphenolic substances, *i.e.* substances containing aromatic rings carrying multiple hydroxyl groups, that can form links with proteins. These links usually result in the protein being rendered insoluble, as in the tanning of leather. Most often the polyphenolics concerned are flavonoids, especially where two or more flavonoids have condensed together *via* their 4 and 8 positions. These oligomeric flavonoids are also known as 'condensed tannins' or more precisely proanthocyanidins. Most frequently proanthocyanidins are found to contain flavan-3,4-diols (leucoanthocyanins) but flavan-3-ols (catechins) and anthocyanins commonly occur. The oligomers may contain up to around ten flavonoid units, *i.e.* molecular weights in the range 500–3000.

In recent years the possible beneficial effects that flavonoids, including tannins, may have on the health of consumers has received considerable attention, with the focus on red wine and tea. A complication is that both adverse and beneficial effects have been reported and that epidemiological studies tend not give such encouraging results as laboratory or clinical experiments.

On the negative side the habit of chewing betel nuts, which have a very high condensed tannin content, has been implicated in the high incidence of mouth and oesophageal cancer in the Far East. Higher risk of oesophageal cancer in the Far East has also been linked to tea drinking but in this case it is now generally accepted that the culprit is not a component of the tea itself but repeated thermal cell damage caused by habitually swallowing the tea very hot ($>55\,^\circ$C). Different studies of stomach cancer have shown contradictory results, with both increased and decreased risks associated with tea consumption. Various flavonoids have been shown to possess anti-tumour activity in laboratory experiments but in the absence of unequivocal epidemiological data no useful conclusions can be drawn.

The other more clearly positive effects of dietary flavonoids are linked to arterial disease. The initial interest in their possible benefits arose from the observation that the French and Italians eat diets containing a lot of fat but suffer far less heart disease than expected. A reduced mortality rate from coronary heart disease (CHD) has been clearly linked to wine intake and wine is a good source of flavonoids ($1-3\,\mathrm{g\,l^{-1}}$ in red, $0.2\,\mathrm{g\,l^{-1}}$ in white). However, total mortality rate does not show the same association, presumably due the compensating effect of other factors such as the adverse effects of

continued on p. 201

> **Box 6.1** *continued*
>
> high alcohol intakes. The CHD rate is also low in the Far East where consumption of green tea is high.
>
> Two potential mechanisms have been implicated in these beneficial effects. Firstly flavonoids, including quercitin (6.25) found in wine, tea, onions and apples, and the catechins of tea, are recognised as effective antioxidants and many laboratory experiments and a few clinical studies indicate that they are effective inhibitors of the oxidation of low density lipoproteins (LDL), an important stage in the development of atherosclerosis. The transfer of catechins to the bloodstream has been shown to unimpaired by the presence of milk in tea. Wine flavonoids also inhibit platelet aggregation, another important stage in CHD. Quercitin is particularly effective, the catechins are not. Resveratrol (*trans*-3,5,4'-trihydroxystilbene, 6.26), a non-flavonoid present in wine ($\sim$4 mg l^{-1} in red, $\sim$0.7 mg l^{-1} in white), has a similar activity and has attracted lot of attention recently. 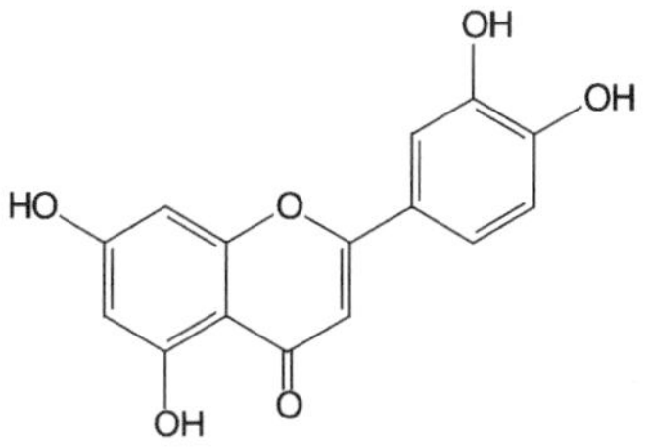
>
>
> (6.25) 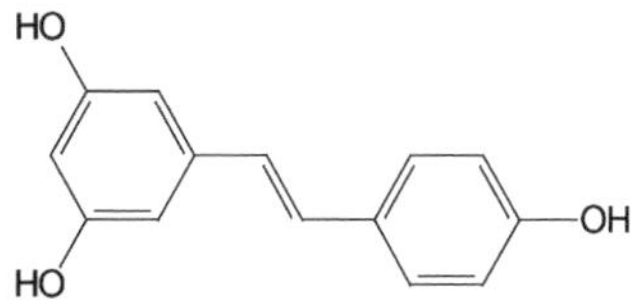 (6.26)

TURMERIC AND COCHINEAL

Most of the consideration of colour so far in this chapter has been devoted to the colour of a food imparted by the substances that it naturally contains. The applications of some of these natural colours as colorants for other foods has been mentioned and the important examples were listed in Table 6.1. However, two important food colorants are not members of the major pigment types and this section is concerned them.

Turmeric, the dried powdered roots of the turmeric plant *Curcuma longa,* has been an important spice for centuries, not only as a food colour but also for dyeing textiles. Nowadays it is best known for its role in oriental cookery, as an essential component of curry powder. It also contributes a characteristic earthy, slightly bitter taste. It is a traditional ingredient in the type of English pickle known as piccalilli. Turmeric contains the pigment curcumin (6.27).

Solvent extraction of the powdered spice gives a concentrated colouring material that can be purified and stripped of its flavour components. The result is a useful oil-soluble, bright-yellow colouring with a slight greenish cast. As synthetic dyestuffs have become increasingly unpopular as food additives curcumin is proving valuable for colouring in dessert products, ice-cream and pickles and other manufactured foods.

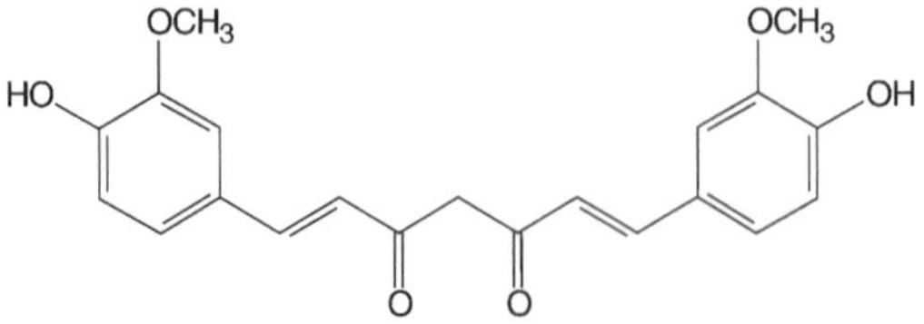

(6.27)

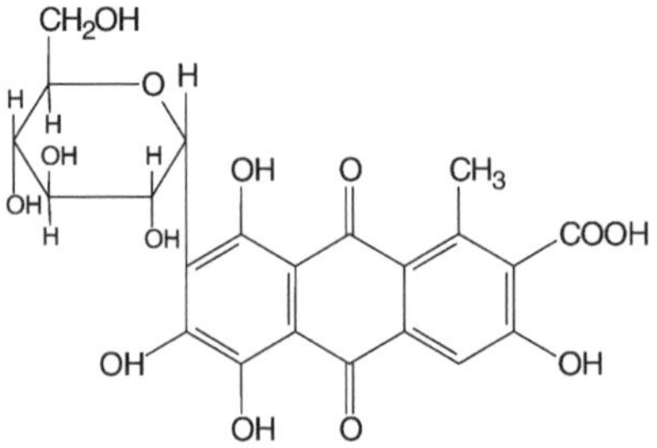

(6.28)

Plants do not have a monopoly in the supply of natural colouring materials. Cochineal is the name given to a group of red pigments from various insects, similar to aphids, notably *Coccus cacti*, which belong to the superfamily *Coccidoidea*. Cochineal, in the form of the dried and powdered female insects (about 100 000 per kg), has been an object of commerce since ancient times. The insects are harvested from cactus plants of the prickly pear or *Opuntia* family in various parts of the world including Central and south America and the Canaries. Besides its use as a food colour it has been used in a dyestuff for textiles and leather and also as a heart stimulant. Although varieties of cochineal have been 'cultivated' in many parts of the world, the bulk of world production now comes from Peru, where it is obtained from insects that are parasites on cacti.

The principal component of cochineal is carminic acid (6.28), an anthroquinone glycoside with an unusual link between the glucose and the aglycone. Cochineal itself gives a purple colour but textile dyers discovered that with tin salts it gave a strong scarlet dye. The deep-red colour pigment used as a food colour, variously referred to as 'carmine', 'carmines of cochineal' or simply 'cochineal' are obtained by first extracting the carminic acid from the crude cochineal powder with hot water. The extracts are then treated with aluminium salts to produce brilliantly coloured complexes of uncertain structure known as 'lakes'. These can be precipitated out with ethanol to give a water-soluble powder. In spite of cochineal's high cost, the increasing unpopularity of chemically synthesised dyestuffs is bound to lead to a revival in its use.

ARTIFICIAL FOOD COLORANTS

The use of artificial, unnatural colours in food has a long but far from glorious history. Wine was always vulnerable to the unscrupulous, and in the 18th and 19th centuries burnt sugar and a range of rather unpleasant 'vegetable' extracts were commonly used to give young red wines the appearance of mature claret. However, for sheer lethality the confectionery trade had no equal. In 1857 a survey of adulterants to be found in food revealed that sweets were commonly coloured by, for example, lead chromate, mercuric sulfide, lead oxide (red lead), and copper arsenite! Legislation and the availability of the newly developed aniline dyestuffs rapidly eliminated the use of metallic compounds as well as some of the more dubious vegetable extracts.

The new chemically synthesised dyestuffs had many advantages over 'natural' colours. They were much brighter, more stable, cheaper and offered a wide range of shades. It did not take long before the toxic properties of these dyes also became apparent, though mostly through their effects on those engaged in making them rather than on consumers.* Since that time there has been a steady increase in the range of dyestuffs available, but at the same time we have become increasingly aware of their toxicity. Thus in 1957 there were 32 synthetic dyestuffs permitted for use in food in Britain. By 1973 19 dyestuffs had been removed from the list and 3 added. In 1979, 11 of these were given identifying 'E' numbers and together with several others that were not permitted in the rest of the European Community. This situation has now been simplified and Table 6.3 lists all the permitted synthetic colours permitted in the EC together with their E numbers.

The structural formulae and full chemical names of a representative selection of permitted synthetic colours are shown in Figure 6.15. One complication in the providing chemical details of dyestuffs is that sometimes the name covers not a single substance but rather a mixture. An

---

* It appears that the laxative effects of phenolphthalein were first discovered, in Hungary in 1902, when wine to which it had been added to enhance the colour had quite unanticipated effects on the drinkers.

Table 6.3 *The synthetic food colours permitted in Europe. The US numbering system is only applied to synthetic dyestuffs. 'F', 'D' and 'C' indicate permitted use in food, drugs and cosmetics respectively with the prefix 'Ext' for 'External use only'. NB. In spite of the author's efforts to ensure its accuracy the data in this table should be treated as illustrative, not authoritative. As discussed in the text restrictions are imposed on the types of foods in which some of these colours may be used.*

Name	Class	EC No.	US No.	CI No.
Yellow:				
Tartrazine	Azo	E102	F,D & C Yellow No 5	19140
Quinoline Yellow	Quinoline	E104	D & C Yellow No 10	47005
Sunset Yellow FCF	Azo	E110	F,D & C Yellow No 6	15985
Red:				
Carmoisine	Azo	E122	Ext D & C Red No 10	14720
Amaranth	Azo	E123		16255
Ponceau 4R	Azo	E124	F,D & C Red No 2	16185
Erythrosine	Xanthene	E127	F,D & C Red No 3	45430
Red 2G	Azo	E128	Ext D & C Red No 11	18050
Allura Red AC	Azo	E129	F,D & C Red No 40	16035
Blue:				
Patent Blue V	Triarylmethane	E131		42051
Indigo Carmine	Indigoid	E132	F,D & C Blue No 2	73105
Brilliant Blue FCF	Triarylmethane	E133	F,D & C Blue No 1	42090
Green:				
(Food) Green S	Triarylmethane	E142		44090
Brown & Black:				
Black PN	Azo	E151		28440
Brown FK	Azo	E154		
Chocolate Brown HT	Azo	E155		20285

extreme example is Brown FK which has six components. Their elaborate ring structures ensures that they have rather intimidating chemical names. To make matters worse both the common and names chemical names vary between authorities and between Europe and the USA. For most dyestuffs, not just those used as food colours, the internationally agreed Colour Index (CI) numbers can provide an unequivocal identification. The reference numbers used in the USA (the equivalent of the European E numbers) are also shown in Table 6.3. The common names of many dyestuffs often include mysterious suffixes. The 'FK' of Brown FK is short for '*for kippers*', as Brown FK has been the dye of choice for enhancing the brown colour of this type of smoked herring.

All of these colours are freely soluble in water, a property generally conferred by their sulfonic acid groups. The chromophoric group of the

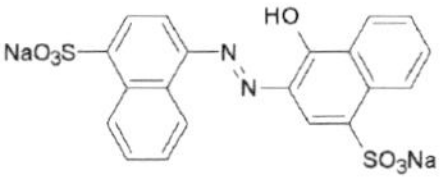

Carmoisine

disodium 4-hydroxy-3-(4-sulfonato-1-naphthylazo) naphthalene-1-sulfonate

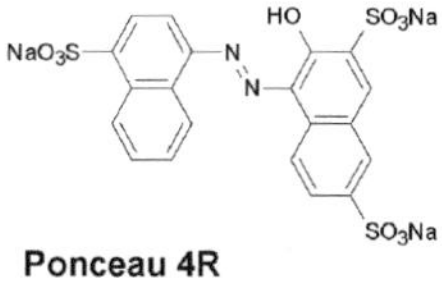

Ponceau 4R

trisodium 2-hydroxy-1-(4-sulfonato-1-naphthylazo) naphthalene-6,8-disulfonate

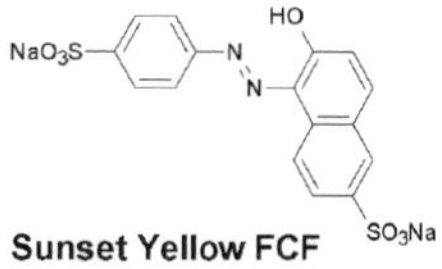

Sunset Yellow FCF

disodium 2-hydroxy-1-(4-sulfonatophenylazo) naphthalene-6-sulfonate

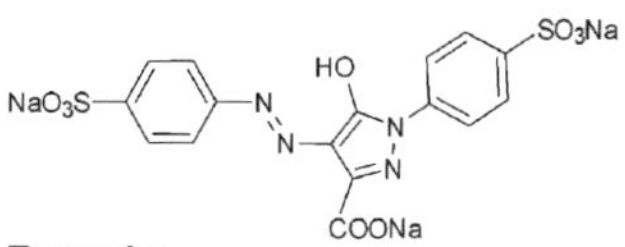

Tartrazine

trisodium 5-hydroxy-1-(4-sulfonatophenyl)-4-(4-sulfonatophenylazo)-H-pyrazole-3-carboxylate

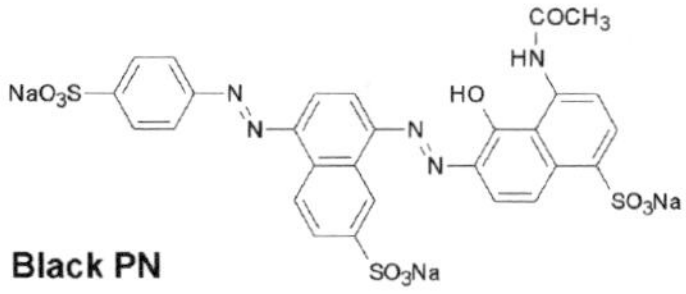

Black PN

tetrasodium 4-acetamido-5-hydroxy-6-[7-sulfonato-4-(4-sulfonatophenylazo)-l-naphthylazo] naphthalene 1,7-disulfonate

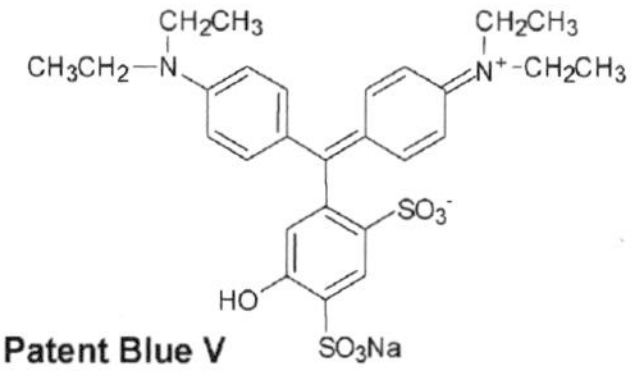

Patent Blue V

sodium [4-(α-(4-diethylaminophenyl)-5-hydroxy-2,4-disulfophenyl-methylidene) 2,5-cyclohexadien-1-ylidene] diethylammonium hydroxide

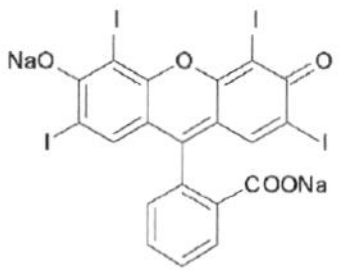

Erythrosine

disodium 2-(2,4,5,7-tetraiodo-3-oxido-6-oxoxanthen-9-yl) benzoate monohydrate

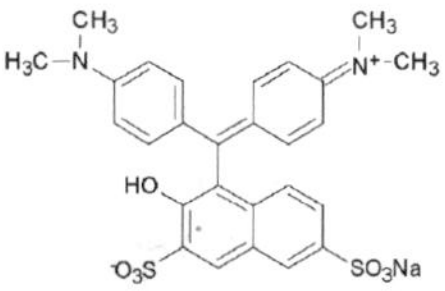

Green S

sodium N-{4-[4-(dimethylamino)phenyl][2-hydroxy-3,6-disulfo-1-naphthalenyl]-methylene]2,5-cyclohexadien-1-ylidene}-N-methylmethanaminium

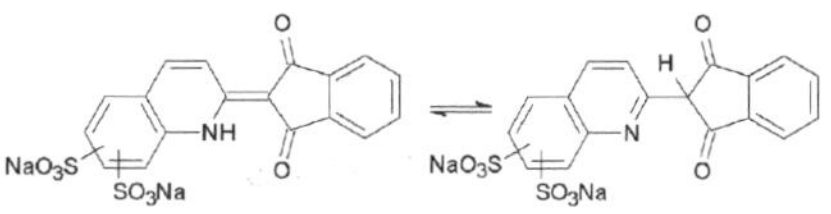

Quinoline Yellow

disodium salt of disulfonates of 2-(2- quinolyl)indan-1,3-dione

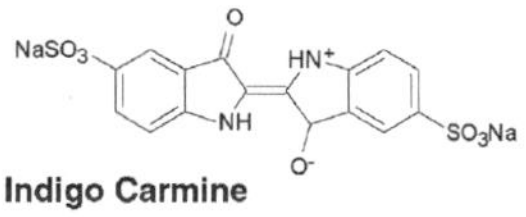

Indigo Carmine

disodium 3',3-dioxo-2,2'-bi-indolylyidene-5,5'-disulfonate

Figure 6.15 *The chemical structures and full chemical names of a selection of synthetic food colours. The names used here are those currently applied in official EC documentation. Indigo Carmine is also known as Indigotine.*

azo dyes is the azo group itself conjugated with aromatic systems on either side to give colours in the yellow, orange, red, and brown range. Three linked aromatic systems provide the chromophoric group of the triaryl-methane compounds, which are characteristically bright green or blue. The chromophoric group of xanthene compounds is similar to that of the triarylmethane compounds and imparts a brilliant red shade to erythrosine, the only permitted example. Resonance hybrids are important in the colours of both quinoline and indigoid compounds. Indigo Carmine in particular has many possible zwitterion arrangements besides the one shown in Figure 6.15. The absorption spectra of many of these dyes are shown in Figure 6.16.

The application of the synthetic food colours in food processing demands attention to many other factors besides the shade and intensity required. For example, in sugar confectionery and baked goods high cooking temperatures may pose problems. In soft drinks one may have to contend with light (Food Green S, erythrosine, and Indigo Carmine are particularly vulnerable), sulfur dioxide, low pH, and ascorbic acid. Probably the greatest use of food colours is in nominally fruit containing products including dessert products, confectionery, ice cream *etc.* Canned peas are coloured with a mixture of Food Green S and tartrazine. Brown colours in chocolate and caramel products are obtained either with Chocolate Brown HT or with mixtures such as Sunset Yellow and amaranth or Food Green S and tartrazine. Canned meats can be coloured with erythrosine, but products such as sausages or meat pastes that will be exposed to light are usually coloured with Red 2G. Some products,

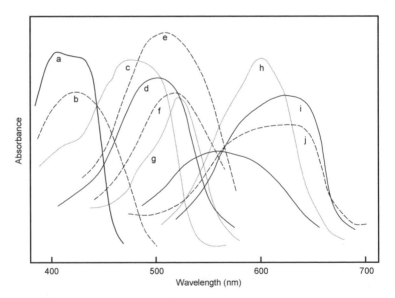

Figure 6.16   *The absorption spectra of the synthetic food colours (at pH 4.2 in citrate buffer):* a *Quinoline Yellow;* b *Tartrazine;* c *Sunset Yellow FCF;* d *Ponceau 4R;* e *Carmiosine;* f *Amaranth;* g *Erythrosine;* h *Indigo Carmine;* i *Patent Blue V;* j *Green S;* k *Black PN.*

particularly types of confectionery, contain very little free water in which the colour can be dissolved. In these cases insoluble lakes are used in which the colour has been absorbed onto a hydrated alumina substrate. The powdered lake is milled down to a particle size around 1 μm to ensure the absence of speckling. The intensity of colour obtained using a given amount of a lake colour increases as the particle size is reduced.

Current regulations in Britain forbid the addition of any colour to raw or unprocessed foods such as meat, fish, and poultry, fruit and vegetables, tea and coffee (including 'instant' products), wine, milk, and honey. The use of added colours in baby and infant food is also prohibited.

In general, synthetic food colours such as the azo dyes are added at a rate of 20–100 mg kg^{-1}. For foods with very little available water, rather higher levels of caramels (1000–5000 mg kg^{-1}) are required to achieve a useful degree of colouring. It has been calculated that in the UK there is a total annual consumption of some 9500 tons of food colours, of which all but 500 tons are caramels, which happily corresponds to daily consumption data for the various colours in all cases of less than 10%, and usually less than 1%, of the 'Acceptable Daily Intake' figures that are now being established by toxicologists.

There is one small group of food colours that do not fit easily into any classification. Listed in Table 6.4. these are inorganic substances that find occasional applications in food products.

Over the last few years a lot of concern has been expressed in the media over the safety of food additives generally and the synthetic food colours in particular. The stringency of the toxicity testing procedures that have been applied in Britain and abroad forces the conclusion that the food colours now permitted are at least as safe as any of the more 'natural' components of our diet. (The author is only too well aware of the diversity of interpretations that can be placed on this adjective when food is being discussed.) However, one question that conventional testing procedures

Table 6.4 *Inorganic food colours.*

Name	EC No.	
Carbon black	E153	Obtained from the limited combustion of vegetable material
Titanium oxide	E171	
Iron oxides & hydroxides	E172	
Aluminium	E173	Only permitted for the surface decoration of sugar confectionary
Silver	E174	
Gold	E175	

Box 6.2   Restrictions on the Use of Colours

While the so-called 'natural' colours have obvious attractions for consumers with concerns as to the safety of the synthetic food colours it should be noted that in the EC there are a considerable number of foods in which the only

Table 6.5 *Foods restricted in the EC to natural colours. Note that neither blue nor green colours may be added to the blue cheeses such as Stilton, Roquefort and Gorgonzola.*

Foods	Permitted colours
Beer, cider, vinegar, and spirits	Caramels (E150a-d)
Butter	Carotenes (E160a)
Margarine *etc.*	Carotenes (E160a), annatto (E160b), curcumin (E100)
Cheese	As for margarine plus paprika (E160c) and the chlorophylls (E140,141) for Sage Derby, cochineal (E120) for red marbled cheeses
Pickled vegetables	Chlorophylls, caramels, carotenes, betanin (E162), anthocyanins (E163)
Extruded, puffed &/or fruit flavoured breakfast cereals	Caramels, carotenoids (E160a-c), betanin (E162), anthocyanins (E163), cochineal, betanin, anthocyanins
Many types of sausages and pâtés	Carotenoids (E160a,c), curcumin, cochineal, caramels, betanin

---

(which are most often looking for evidence of carcinogenicity) do not address is that of intolerance and allergy. Of the food colours tartrazine is widely regarded as the most suspect, and there is some evidence that 0.01–0.1% of the population may show intolerance towards it. This intolerance usually manifests itself as one or more of the symptoms associated with allergic reactions (eczema, asthma, *etc.*) (*see* page 346) and is most often encountered in those who also show a reaction to certain other groups of chemicals, notably the salicylates (including aspirin) and benzoates. The possibility of food colours being involved in disorders such as 'hyperactivity' in children remains much more questionable and has been dismissed by most medical authorities. The attention given to this subject by the media is sadly out of all proportion to the number of properly conducted clinical trials.

The ever-increasing costs of safety testing of new food additives is making the appearance of new food colours extremely unlikely. It must be remembered that it is not only the food colour itself that must be shown not to cause death or disease but also any impurities that arise during its permitted colours are 'natural'. In this context it is usual to include the caramels with the natural colours.

There are corresponding restrictions on the range of foods in which some synthetic colours may be used, as listed in Table 6.6. Tables 6.5 and 6.6 are based on the current European Community regulations but do not claim to be authoritative, only illustrative.

Table 6.6 *Synthetic colours restricted in the EC to particular foods.*

Colour	Foods
Amaranth (E123)	Aperitif wines (vermouth), fish roe
Erythrosine (E127)	Cocktail and candied cherries
Red 2G (E128)	So-called breakfast sausages (*i.e.* the type eaten in the British Isles that include a significant cereal content) and burgers with a significant vegetable or cereal content
Brown FK (E154)	Kippers

manufacture, and breakdown products that might arise during food-processing operations, cooking, or digestion have to be cleared.

THE MOLECULAR BASIS OF COLOUR

Many readers will be happy to accept that certain substances are 'coloured' without needing to understand what it is about the molecular structure of such substances that imparts this particular property. These readers can ignore the next few pages with a clear conscience. Others, however, will want to know a little more, possibly so as to understand 'why' the colours of some food materials are what they are and 'how' they come to change.

Materials or objects appear coloured because they absorb some of the incident light that shines on or through them. In this context 'some' means light of some wavelengths is absorbed more than others. The light that subsequently reaches our eyes, having been reflected or transmitted, is therefore richer in light from some parts of the visible spectrum than from others. To understand the molecular basis of colour we need to understand firstly how molecules absorb light energy, and secondly what determines the wavelength, or colour, of the light they absorb.

Figure 6.17 shows the relationship between the wavelength of light and its colour. Although for most purposes physicists and chemists regard light as having the properties of waves it can also be thought of as a stream of energy containing particles. These particles are called 'quanta'. The amount of energy possessed by a quantum is related to the wavelength of the light. The shorter the wavelength, *i.e.* the nearer the violet end of the visible spectrum, the more energetic the quantum.

The energy, as light, that a coloured molecule absorbs cannot simply be soaked up and vaguely converted into heat. Each single quantum must be utilised individually to provide exactly the right amount of energy to facilitate a particular change (or transition) in the molecule. The quantum will only be absorbed if its energy content matches, is neither too large nor too small, the energy demand of a transition that the molecule can undergo. The transitions that match the range of energies associated with visible light quanta involve the electrons that form the bonds between the atoms of the molecules of the absorbing substance. Of the many theoretically possible types of transition only those associated with the π-electrons of conjugated double bonds are common amongst the coloured compounds in food.

Colour	Wavelength (nm)	Complementary colour
Violet	400	
		Yellow
Blue	450	
		Orange
Green	500	
		Red
Yellow	550	
		Violet
Orange	600	
		Blue
Red	650	
		Green
	700	

Figure 6.17 *The visible spectrum. Light of the wavelength shown appears as the colour shown in the left-hand column when it enters the eye. Substances that absorb light of that wavelength appear to be the colour shown in the right hand column. For example, the yellow carotenoid pigments strongly absorb light in the purple/blue region (420–480 nm; see Figure 6.6). We describe the response of the human eye to the mixture of all the other visible wavelengths that the carotenoids do not absorb (480–800 nm) as yellow or orange.*

A single bond between two carbon atoms consists of one pair of shared electrons which form one π-bond. In the normal, or ground, state of a σ- or π-bond the two electrons are regarded as spinning in opposite directions; shown as ($\uparrow\downarrow$). However, if an appropriate quantum of energy is absorbed by the molecule one of the electrons may undergo a transition to an excited, higher energy, state with both electrons spinning in the same direction ($\uparrow\uparrow$). Such transitions are known as σ to σ^* and π to π^*. The σ to σ^* transitions require quanta of much higher energy levels than those of visible or ultra-violet light, and are therefore irrelevant to the question of colour. However, π to π^* transitions require less energetic quanta and are relevant here. Once an electron has undergone transitions like these it normally returns immediately to its original state, the energy it absorbed being released as heat. In some molecules the return of the electron to its original state can lead to the release of a quantum with sufficient energy to appear as light, although obviously at a longer wavelength than that originally absorbed. This is the phenomenon of fluorescence.

The wavelength at which a substance absorbs most radiation, the position of the highest peak on the absorption spectrum, is referred to as the substance's λ_{max} value.

When the bonds linking a chain of carbon atoms are alternately single and double:

$$-C=C-C=C-C=C-$$

the double bonds are described as *conjugated*. All the coloured substances described in this chapter have an abundance of conjugated double bonds. The π-electrons of conjugated systems interact together and in doing so particular pairs of π-electrons lose their association with a particular carbon–carbon bond. In this state they are referred to as *delocalised* and require much less energy to undergo a π to π^* transition. In fact the more double bonds are conjugated together in a molecule the less energy is required for the π to π^* transition and therefore the radiant energy absorbed by the process is of increasing wavelength. This is clearly illustrated by the sequence of carotenoid structures shown in Figure 6.18. Only when the carotenoid has its full complement of conjugated double bonds, as in lycopene, is visible light absorbed. Of course the exact λ_{max} values we observe, and the precise details of the absorption spectrum, are affected by other structural elements in the molecule such as the involvement of the double bonds in rings and also by interactions with other substances such as solvents.

The need for at least seven conjugated double bonds for a substance to absorb visible light is also demonstrated nicely by the anthocyanins (*see*

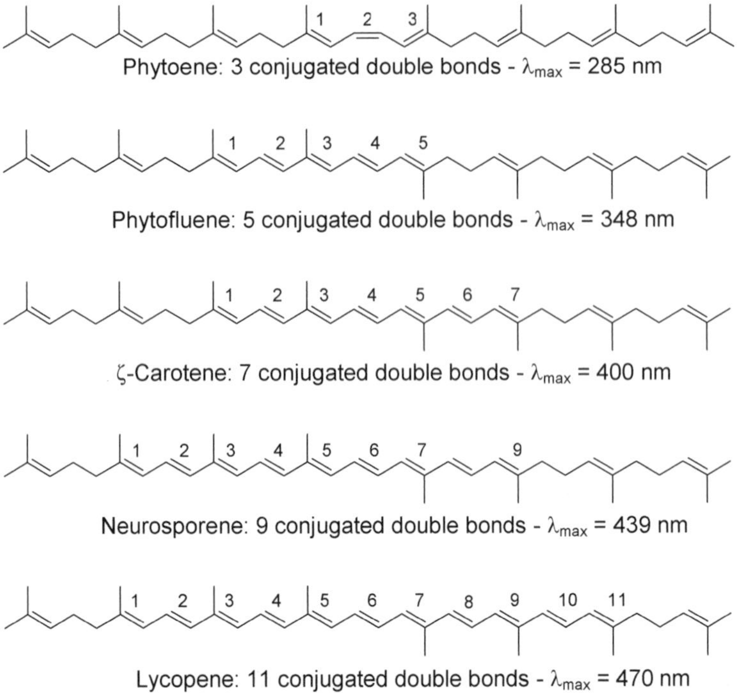

Phytoene: 3 conjugated double bonds - λ_{max} = 285 nm

Phytofluene: 5 conjugated double bonds - λ_{max} = 348 nm

ζ-Carotene: 7 conjugated double bonds - λ_{max} = 400 nm

Neurosporene: 9 conjugated double bonds - λ_{max} = 439 nm

Lycopene: 11 conjugated double bonds - λ_{max} = 470 nm

Figure 6.18   *The relationship between the structure of the carotenoid precursors and their*
λ_{max} values. This sequence, including the conversion of the central double
bond from trans phytoene into cis in phytofluene, happens to correspond to
the biosynthetic pathway for the formation of carotenoids in plants. β-
Carotene, and the other carotenoids with rings, are derived from neurospor-
ene.

Figure 6.8). The flavylium cation (red) has a total of eight but when a rise
in pH converts it into the carbinol pseudo-base the conjugation in the
middle of the system is lost. With at most only four double bonds then
conjugated together the result is colourless. The bleaching action of sulfur
dioxide has a similar basis. In contrast the ionisation or otherwise of
betanidine's carboxyl groups with change in pH causes no change in its
seven conjugated double bonds including the C-20 carbonyl group (*see*
Figure 6.11). The term 'chromophore' is often used to denote the region
of a molecule where delocalised electrons are responsible for light
absorption. For example the porphyrin ring system is the chlorophyll's
chromophore.

Some food pigments involve metals: the chlorophylls (*see* page 176),
myoglobin and its derivatives (*see* page 155), and vitamin B_{12} (*see* page
275). The structures of all of these are similar in that they consist of

elaborate systems of fused rings that provide large numbers of conjugated double bonds and hence, delocalised electrons. However, the electrons from the metal atom also get involved, with considerable effects on the colour. For example the loss of the magnesium from chlorophyll gives us brown pheophytin. The colour of myoglobin in meat depends on whether the iron is in the ferric or ferrous state and what other substances are also attached to the iron by coordinate bonds.

COLOUR MEASUREMENT

The measurement of colour is a major issue in many fields of food science and, strictly speaking, a subject for physicists rather than chemists. However, in the real world it almost invariably the analytical chemists that end up having to cope with this aspect of food science. What aspect of colour that is to be measured depends on the circumstances. Very often we need to be able to express the 'shade' of a food product, ingredient, or raw material in quantitative terms. To describe a tomato as 'red' may be adequate for some purposes but expressing the concise difference in colour between a tomato and a strawberry to someone who has never seen either fruit would place impossible demands on even our greatest poets. Besides the shade we may need to specify the intensity of the colour and its 'lightness' or 'darkness', for example in matters of fruit juice quality.

Chemists confronted by this problem would normally reach for the spectrophotometer to obtain an absorption spectrum, like those shown in this chapter (*e.g.* Figures 6.2, and 6.6). This approach starts to fail because food materials are usually solids, or more or less opaque liquids. This particular difficulty can be overcome by adapting the instrumentation to record the light reflected by the object rather than that which passes through it; so as to give a reflectance spectrum. However, this does not solve the real problem with absorption or reflectance spectra. This is that they present too much information.

The human eye uses the cone cells of the retina to perceive the colour of an object. There are three different types of cone cells each responding best to light of different wavelengths, *viz.* 445, 535, and 570 nm. There is considerable overlap between the 'response' spectra of the different cone types. This means that light of a given wavelength, or mixture of wavelengths, will stimulate at least two and possibly all three types of cone. Our perception of the colour of the light is based on the ratio between the three responses, combined with the response of the rod cells. The rods are some 10 000 times more sensitive than the cones and have their peak sensitivity at 505 nm.

It should therefore be possible to describe any colour, as perceived by the human eye, simply in terms of such ratios. In fact the wavelengths used in colour definition using scientific instruments are more evenly distributed across the visible spectrum and correspond to the primary colours of red, green, and blue.

It is well known that an appropriate mixture of red, green, and blue lights will appear white and it might be assumed that any colour could be similarly achieved. However, this turns out to be only nearly true. In 1931 the CIE (Commission International de l'Eclairage) introduced the concept of *theoretical* primary colours, known respectively as X, Y, and Z. Although no real lamps could produce them they are able, *theoretically*, to match any shade. Their spectra are shown in Figure 6.19. The 'red', 'green', and 'blue' curves are labelled X, Y, and Z respectively. An important characteristic of the Y curve is that it matches the response of the eye to the total amount of light energy; in other words it represents the response of the rod cells of the retina.

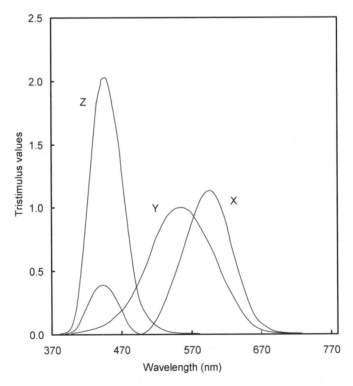

Figure 6.19   *The spectra of the theoretical primaries of the CIE system. The subsidiary peak on the X curve results from the mathematical jiggery pokery that was used to ensure that all shades could be theoretically matched, as explained by Billmeyer and Saltzmann (see Further Reading).*

The amounts of the three theoretical primaries required to produce a given colour can be calculated from its spectrum and the curves in Figure 6.19 and are known as the *tristimulus* values. The ratios between them are given by:

$$x = X/X + Y + Z, y = Y/X + Y + Z, \text{and } z = Z/X + Y + Z,$$

and, since *x*, *y*, and *z* must total one, any colour can actually be described by just two of them, such as *x* and *y*.

Figure 6.20 shows the famous horseshoe shaped CIE chromaticity diagram where *x* is plotted against *y*. On this plot quantities *x* and *y* for a colour are called its chromaticity co-ordinates. The theoretical red primary is located at *x* = 1, *y* = 0, blue at *x* = 0, *y* = 0; and green at *x* = 0,

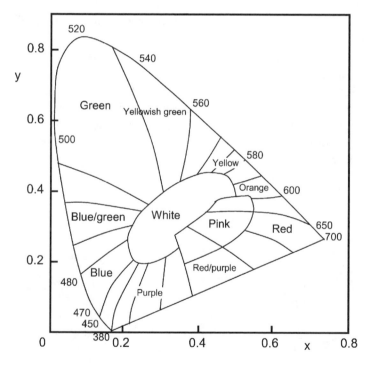

Figure 6.20 *The CIE Chromaticity diagram. The colour names shown on the diagram describe the colour perceived by a standard observer looking at a coloured surface illuminated by a standard illuminant (i.e. light source) with the spectral characteristics of ordinary daylight. A rather unpicturesque approach is usually adopted towards the naming of intermediate colours, largely dependant on the suffix 'ish'. All shades of grey are merely more or less bright white.*

$y = 1$. The outer curved line shows the positions of monochromatic colours at different positions in the spectrum, from 380 to 770 nm. The straight line closing the horseshoe represents the various purplish shades resulting from mixtures of red and blue. Towards the centre of the horseshoe colours approach white.

The CIE system is linked closely to theoretical concepts of colour perception but has a number of practical drawbacks. One of these is the status of 'black'. To handle this properly the diagram would have to be three-dimensional, with values for the tristimulus value Y, expressing the total amount of light, plotted perpendicularly to the plane of the paper. Another problem is that our perceptions of degrees of difference between colours are not uniform over the diagram. Both these problems are dealt with better by the Hunter and Munsell systems.

The Hunter L, a, b system is based on the concept of a colour space with the colour defined by three co-ordinates. The vertical co-ordinate (L) runs from $L = 0$ (black) through grey to $L = 100$ (white). The horizontal co-ordinate (a) runs from $-a$ (green) through grey to $+a$ (red). The other horizontal co-ordinate (b) runs from $-b$ (blue) to $+b$ (yellow). Purple and orange can be visualized as being located at $+a - b$ and $+a + b$ respectively, as shown in Figure 6.21.

The Munsell system, also shown in Figure 6.21, has a similar structure but describes colour in terms of the three attributes of hue, value, and chroma. Five basic hues are distributed around the circumference of a circle. The vertical axis of the circle defines the value or lightness from black below the circle, through grey at the centre, to white above the

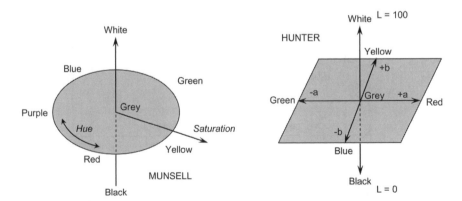

Figure 6.21 *The Munsell and Hunter colour spaces. The mathematical relationships between these and the CIE system are well established so that the parameters of a colour determined in one system can be mathematically related to those it would have in the others.*

circle. The distance from the centre to the circumference is given a figure for the chroma or saturation. It is most important to remember that it is extremely unlikely that the colours of two real objects would differ in only one of these parameters; in most cases all three would differ. Users of modern computers with colour monitors and printers will have encountered some of the practical aspects of these colour systems.

Armed with the theory one can then approach the actual task of recording the colour of a foodstuff. One simple approach is by visual matching with standard colour samples. Although apparently simple this approach still requires specialist instrumentation, in that a standardised light source is required. The sky, incandescent (tungsten) bulbs, and fluorescent tubes may all appear to give 'white light' but actually they differ considerably in their distribution of energy across the spectrum. Two samples may appear identical observed under a fluorescent lamp but quite different in daylight. This phenomenon is known as *metamerism*.

The instrumental alternatives are colorimeters and spectrophotometers. A colorimeter uses filters to generate three light beams whose spectra match as closely as possible the CIE curves. Photocells measuring the quantity of light reflected (or transmitted) by the sample will then give the tristimulus values directly. The inevitable inclusion of the microcomputer in modern instruments facilitates instant conversion into Hunter or Munsell values. More sophisticated spectrophotometers record the absorbance or transmittance of light at intervals, or even continuously, throughout the visible spectrum and then compute the tristimulus values.

FURTHER READING

'Anthocyanins as Food Colors', ed. P. Markakis, Academic Press, New York, 1982.

'Natural Food Colorants', 2nd edn., eds. G. A. F. Hendry and J. D. Houghton, Blackie, Glasgow, 1996.

F. W. Billmeyer and M. Saltzmann, 'Principles of Color Technology', 2nd edn., Wiley, New York, 1981.

J. Gross, 'Pigments in Vegetables. Chlorophylls and Carotenoids', Van Nostrand Reinhold, New York, 1991.

J. Gross, 'Pigments in Fruits', Academic Press, London, 1987.

'Natural Food Colorants', eds. G. A. F. Hendry and J. D. Houghton, Blackie, London, 1996.

J. B. Hutchings, 'Food Colour and Appearance', Blackie, London, 1994.

Chapter 7

Flavours

However much nutritionists and 'health food' enthusiasts may wish otherwise, it is the flavour and appearance of food rather than its vitamins or fibre that win the compliments at the dinner table. This attitude on the part of the diners is not as short-sighted as it may at first appear. The sense organs concerned in the detection of taste and odour (here collectively referred to as flavour) have evolved to perform an essential function, to establish what is and what is not suitable as food. As a general rule palatability can be equated with nutritional value, and most of the exceptions occur in relation to excess. For example, to the genuinely hungry sweetness is a clear indicator of the presence of energy-yielding sugars whereas to the mother of a well fed European child sweetness implies dental caries and obesity. Conversely, their bitterness and astringency will prevent us from eating many poisonous plants but these flavours are valued elements of the flavour of beer and tea.

We usually regard taste as the property of liquids or of solids and gases in solution that is detected in the mouth, not only by receptor cells in the taste buds of the tongue but also elsewhere in the oral cavity. Odour (or aroma, or smell), is similarly regarded as the property of volatile substances detected by the receptor cells of the olfactory systems of the nose. Very few flavours, and fewer complete foodstuffs, allow a clear distinction to be made between an odour and a taste.

The main interest in flavour for the food chemist is the identification of the particular substances responsible for particular flavour elements. One obvious motive for such work is that it enhances the possibilities for the simulation of natural flavours for use in processed food products. Related to this is the possibility of identifying which are the features of chemical structure that invoke a particular response from our sense organs. This line of enquiry was initiated some years ago by Theophrastus (*c.* 372–287 BC) who decided:

τὸν δὲ πικρὸν ἐκ μικρῶν καὶ λείων καὶ περιφερῶν τὴν
περιφέρειαν εἰληχότα καὶ καμπὰς ἔχουσαν. διὸ καὶ
γλίσχρον καὶ κολλώδη. ἁλμυρον δὲ τὸν ἐκ μεγάλων καὶ
οὐ περιφερῶν, ἀλλ' ἐπ' ἐνίων μὲν σκαληνῶν ....*

The research chemist faces a number of special difficulties in investigations of flavour. The first problem is that there is no physical or chemical probe that can be used for the specific detection of the substances of interest. There is no flavour equivalent of the spectrophotometer, which quantifies the light absorbing properties of substances that we perceive as colour. Here the success of an analytical procedure may depend on the skill of a trained taste panel or the nose of an experimenter. Another complication is that a food flavour rarely depends on a single substance. It is commonly found that when the mixture of substances involved in a particular flavour has apparently been elucidated and then reconstituted in the 'correct' proportions one still does not have the flavour one started with. Usually this is because some substances, present in exceedingly small proportions, without strong flavour on their own and therefore overlooked by the analyst, have great influence on the total flavour of the food. These difficulties will not be underestimated if it is realised just how sensitive the nose can be. For example, vanillin (7.1), the essential element of vanilla flavour, can be detected by most individuals at a level of 0.1 p.p.m.

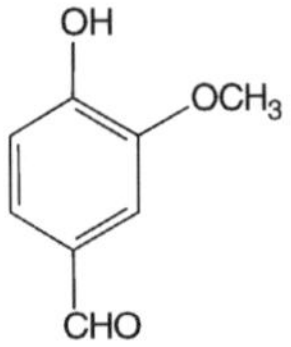

(7.1) 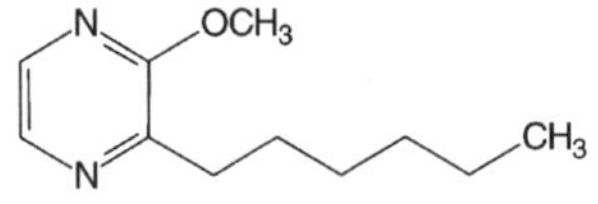   (7.2)

Many other substances are known to which we are far more sensitive. For example the odour threshold (*i.e.* the lowest concentration at which most humans can detect a substance) for 2-methoxy-3-hexylpyrazine (7.2) is reported to be 1×10^{-6} p.p.m., *i.e.* 1 part in 1 000 000 000 000! It is interesting to note that as the lengths of the pyrazine side chains shorten we become progressively less sensitive to this group of substances. The

---

* For those without classical Greek: 'Bitter taste is therefore caused by small, smooth, rounded atoms, whose circumference is actually sinuous; therefore it is both sticky and viscous. Salt taste is caused by large, rounded atoms, but in some cases jagged ones ...'

thresholds for the 2-methoxy-3-methyl-, 2-methoxy-3-ethyl-, and 2-methoxypyrazines are 4×10^{-4}, 4×10^{-3}, and 7×10^{-1} p.p.m. respectively.

Gas chromatography (GC) has proved to be the most valuable analytical technique available to the flavour chemist. Some of the most sophisticated modern instruments are capable of detecting as little as 10^{-14} g of a component in a mixture, and 10^{-10} g is within the capabilities of most commercially available gas chromatographs. The peaks on a gas chromatogram can only be tentatively identified by comparison of elution characteristics with those of standard known substances. Firm identification requires that substances emerging from the chromatography column are separately trapped (usually by condensation at very low temperature) so that they can be subsequently identified by techniques such as infrared spectroscopy, nuclear magnetic resonance, and mass spectrometry. Instruments are now available in which the link between GC and mass spectrometer is handled automatically and the substance identified automatically by comparison of its mass spectrum with a library of known spectra. The vast amounts of information that sophisticated GC methods may make available do not automatically make the food chemist much wiser. For example, a study of Scotch whisky in which 313 different volatile compounds were identified, including 32 alcohols and 22 esters, will probably be slow* to yield conclusions of real value to either manufacturers or drinkers.

Since it is the flavour, more than anything else, that enables us to distinguish between one foodstuff and another it would be surprising if the total number of different flavours was much smaller than the number of different foodstuffs and much harder to classify in any way that acknowledged that this is a textbook of chemistry rather than cookery. However, we can make a reasonable distinction between tastes and odours and this forms the basis of the organisation of this chapter. The scale of these subdivisions, as we shall see, tends to reflect the intensity of research and the volume of knowledge rather than the relative importance of one flavour over another, an impossible judgement to make.

TASTE

The flavour sensations detected in the mouth, and particularly by the tongue, are usually described as tastes. A number of different sensations are identified. Firstly there is the classic quartet of saltiness, sweetness, bitterness, and sourness mentioned in all school biology textbooks and often associated with different parts of the tongue. Then we have three

---

* Not least because the best Scotch whiskies spend many years maturing between distillation and bottling.

other tastes, astringency, pungency (*i.e.* hotness), and meatiness (otherwise known as *umami*). The substances involved in all these sensations have in common a number of characteristics that distinguish them from substances commonly associated with odours. Taste substances are usually polar, water-soluble, and non-volatile. Besides their necessary volatility, odour substances are generally far less polar and elicit a much broader range of flavour sensations.

Sweetness

Of the tastes detected by the taste buds of the tongue sweetness has received by far the most attention from research workers. We usually think of sweetness as being the special characteristic of sugars; their taste is certainly a major reason for their incorporation into so many foods. What is frequently overlooked is that most sugars are much less sweet than sucrose and many are not sweet at all. Furthermore, our pursuit of low calorie diet foods has increased our awareness of the fact that sweetness is in no way the prerogative of sugars: saccharin is a household word. The measurement of the sweetness of a substance is peculiarly problematical. There are no laboratory instruments to perform the task, no absolute or even arbitrary units of sweetness except 'one lump or two?' in a teacup. Instead we have to rely on the human tongue and the hope that if we average the findings of large numbers of tongues we can obtain useful data. This is one area of biological research where experiments on laboratory animals will never achieve very much. The data we obtain will still not be in absolute units but will be expressed relative to some arbitrary standard, usually sucrose. Table 7.1 gives some relative sweetness data for a number of sugars and other substances. It is a revealing class exercise for students to devise for themselves procedures for comparing the sweetness of, say, sucrose and glucose and then compare the results of individual tongues, and the 'class average' with the data given here. Any practical work in this field should be carried out in the workrooms of the home economists, not chemistry laboratories!

A vast amount of work has been devoted to identifying the particular molecular feature of sweet-tasting substances that is responsible for the sensation of sweetness. One approach is simply to survey as many sweet-tasting substances as possible with a view to identifying a common structural element. A second approach has been to prepare large numbers of derivatives of sweet-tasting substances in which potentially important groups are blocked or modified. This second approach led Shallenberger's group to propose, in 1967, a general structure for what they termed the 'saporous unit'. They suggested the AH,B system shown in Figure 7.1,

Table 7.1 *The relative sweetness of sugars and sugar alcohols. As discussed in the text these values must be regarded as typical rather than absolute.*

Sugar	Relative sweetness		Sugar alcohol	Relative sweetness	
	On a weight basis	On a molarity basis		On a weight basis	On a molarity basis
Sucrose	1.00	1.00	Xylitol	1.00	0.44
D-Fructose	1.52	0.80	Maltitol	0.90	0.90
D-Glucose	0.76	0.40	Sorbitol	0.60	0.32
D-Galactose	0.50	0.26	Isomalt	0.55	0.55
Lactose	0.33	0.33	Mannitol	0.50	0.27
Maltose	0.33	0.33	Lactitol	0.37	0.37
Trehalose	0.25	0.25			
D-Ribose	0.15	0.06			

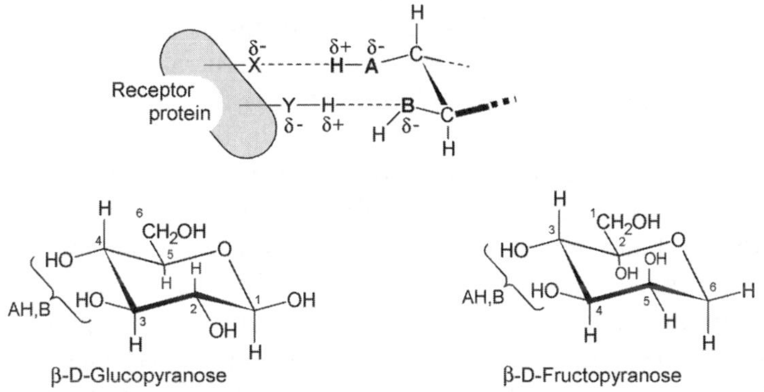

Figure 7.1 *Shallenberger's 'saporous unit'. By analogy with the 'chromophores' of coloured substances (see page 212) the term 'glycophore' is now commonly used in place of 'saporous unit'. The recognised AH,B elements of the glucose and fructose are also shown. Although it is not immediately obvious this portrayal of β-D-fructopyranose is entirely compatible with that on page 2.10 (2.33); it is shown here as seen from an unusual viewpoint.*

where A and B represent electronegative atoms (usually oxygen) and AH implies a hydrogen-bonding capability.

It has been shown that for sweetness the distance between the electronegative atom B and the hydrogen atom H on A must be close to 3 Å (*i.e.* 0.3 nm). Not only can it be assumed that the corresponding groups on the surface of a receptor protein (X and YH) of the taste bud epithelial cells are a similar distance apart but also that if AH and B were any closer they would form an intermolecular hydrogen bond of their own and not with the receptor protein. A likely candidate for the role of receptor protein (*i.e.* one having the ability to bind substances in proportion to their sweetness) has been isolated from the epithelium of the taste buds of a cow. It is assumed that when a sweet tasting molecule binds to the receptor protein it undergoes a small change in shape. (Students of biochemistry will recognise this as similar to the changes in regulatory enzymes when a ligands bind at allosteric sites.) Different theories (way beyond the scope of this book) have been proposed to describe subsequent events but the most popular view is that the initial change in the receptor protein leads to a cascade of changes in other neighbouring proteins in the cell membrane, much as one person fidgeting about in a packed train compartment can lead to passengers some distance away having to adjust their position. Ultimately these effects lead to the cell membrane briefly opening up to the passage of ions such as Ca^{2+} or Na^+ and a nerve impulse being dispatched to the brain.

Studies of numerous sugar derivatives have established that in aldo-hexoses such as glucose the 3,4-α-glycol structure is the principal AH,B system. The conformation of the sugar ring is of crucial importance to the question of sweetness. For the α-glycol AH,B system to have the correct dimensions the two hydroxyl groups must be in the *skewed* (or *gauche*) conformation. If the α-glycol is in the eclipsed conformation, the two hydroxyl groups are sufficiently close for the formation of an intramolecular hydrogen bond, which excludes the possibility of interaction with a receptor protein. In the *anti* conformation the distance between the hydroxyl groups is too great for interaction with the receptor protein. Inspection of the conformation of *C*1 β-D-glucopyranose (*see* page 28) shows that all the hydroxyl groups attached directly to the ring (to C-1, C-2, C-3, and C-4) are potential 'saporous units'. In contrast 1*C* β-D-glucopyranose has all of its α-glycol structures in the anti conformation, making the distance between the hydroxyl groups too large for binding to the receptor. If D-glucose could be persuaded to take up the 1*C* conformation (it can't) we can guess that in this form it would not be sweet.

The presence of an AH,B system of the correct dimensions is clearly not the only factor. For example, the *axial* C-4 hydroxyl of D-galactose is able to form an intramolecular hydrogen bond with the ring oxygen, thus explaining why D-galactose is less sweet than D-glucose. The furanose form of β-D-fructose is believed to lack sweetness, which can be explained by the ease with which a hydrogen bond can form across the ring between the hydroxyl groups on C-2 and C-6. The observation that the disaccharide trehalose is not twice as sweet as D-glucose tells us that, although a sugar may have more than one saporous unit, only one at a time actually binds with the receptor. It should not then be surprising that high-molecular-weight polymers of glucose, such as starch, are quite without taste. The relative sweetness of a range of sugars is shown in Table 7.1

There are clearly vast numbers of substances that contain an AH,B system of the correct dimensions that are actually tasteless or even bitter. These include the L-amino acids where it is the ionised carboxyl and amino groups that constitute a well defined AH,B system. The fact that the corresponding D-amino acids are sweet made a third binding site, and the concept of a three-dimensional receptor site inevitable. At the third site, known as 'γ', the binding is depends upon the hydrophobic character of both the γ site of the receptor protein and the corresponding position on the saporous unit. Figure 7.2 shows the most probable dimensions of the receptor and shows how this principle applies to leucine, of which only the D isomer can bind.

Of course sugars are not the only substances that taste sweet and we are now familiar with saccharine, aspartame *etc.* as 'non-sugar sweeteners' that have the advantage of sweet taste without the calorific value. A small number of sweet tasting proteins, notably thaumatin and monellin, have also been identified. The relative sweetness values of some of the

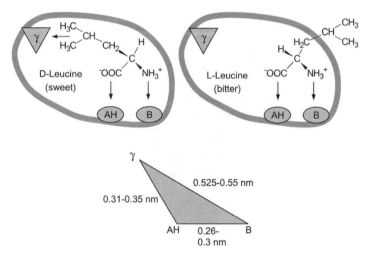

Figure 7.2 *The third, γ, binding site proposed by Kier in 1972, as applied to D- and L-leucine. Opinions vary as to the exact dimensions of the triangle formed by AH,B and γ. It is assumed that other elements of the receptor site get in the way of the L-leucine side chain. Glycine, which does have a weak taste sweet has just a single hydrogen atom in place of leucine's (CH₃)₂CHCH₂– and therefore can get into the receptor site even though it does not bind to the γ site.*

important non-sugar sweeteners are shown in Figure 7.3. Relative sweetness values are inevitably rather variable. This is partly because it has to be measured subjectively using human subjects but also because the sweetness of a substance in solution is not linearly related to its concentration. Thus if one prepares solutions of different sweeteners that have equal apparent sweetness diluting both solutions by the same factor will not necessarily result in a new pair of equally sweet solutions.

The AH,B,γ concept has been extended to many of the non-sugar sweeteners that are listed in Figure 7.3. The differences in relative sweetness are ascribed to both closeness of fit in geometrical terms and also the strength of hydrogen bonds and hydrophobic interactions at the γ site. Identifying the AH, B and γ sites within a molecule is far from straightforward. For example many authorities still record the AH and B sites of D-fructopyranose to be the C-2 and C-1 hydroxyl groups in spite of the evidence provided by Birch in favour of C-3 and C-4, as shown in Figure 7.1. The γ site appears to be of much less significance in the binding of sugars compared with the intensely sweet non-sugar sweeteners whose chemical structures are included in Figure 7.4.

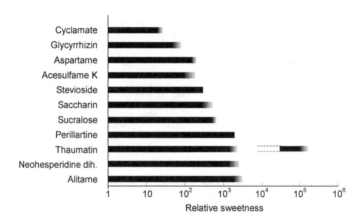

Figure 7.3   *The relative sweetness of a range of non-sugar sweeteners. Relative sweetness is expressed relative to sucrose (=1), shown on a logarithmic scale. Data is obtained by diluting a solution of the test substance until its sweetness is judged to be equal to that of a dilute sucrose solution and then relating the concentration to that of the sucrose standard. The blurring of most of the bars in this chart reflects the variations between the results of different laboratories. The apparent precision of the data for stevioside and perillartine is merely reflection of the shortage of data. For the protein, thaumatin the concentration ratios were also calculated on a molarity basis, giving the extremely high relative sweetness values plotted to the right.*

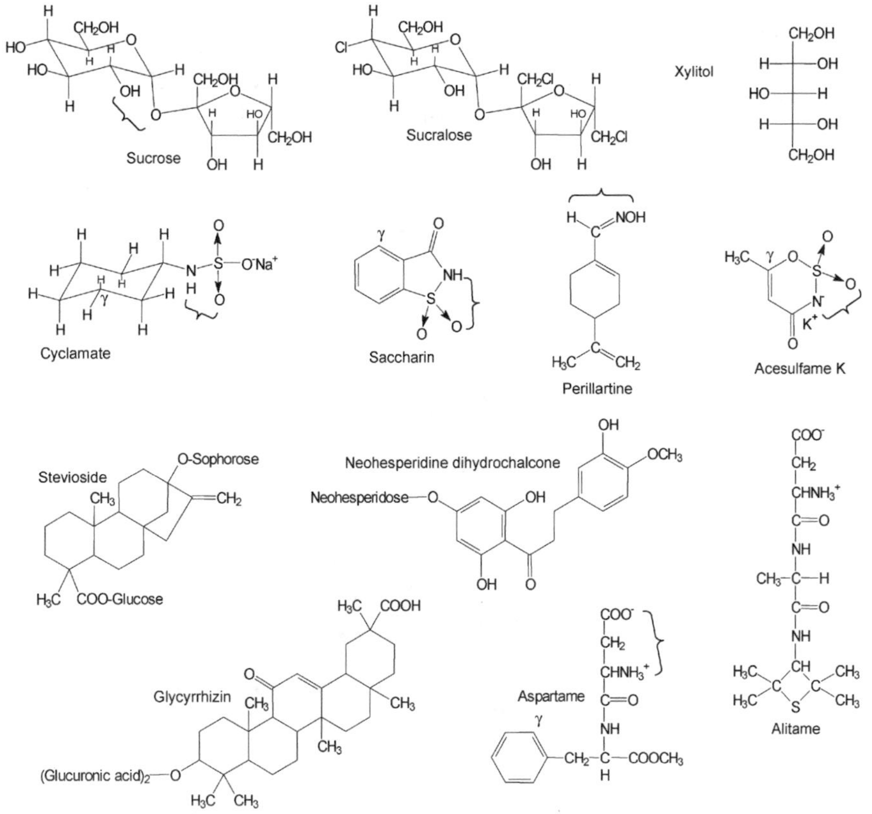

Figure 7.4 *The chemical structures of sweeteners showing, where they have been assigned with reasonable confidence, the AH, B and γ sites. The AH and B sites are shown by braces: }. Neohesperidose is α-L-rhamnose-1,2-β-D-glucose. Sophorose is β-D-glucose-1,2-β-D-glucose.*

Sugar derivatives form an important class of sweeteners. Sugar alcohols such as xylitol (2.20), sorbitol (2.21), mannitol (7.3), lactitol (7.4), and maltitol (7.5) are obtained from their parent mono- or dissaccharide by hydrogenation. In contrast isomalt, an increasingly popular member of this group, is made by a novel process involving immobilized cells of the soil bacterium *Protaminobacter rubrum*. These cells carry an enzyme which is able convert sucrose to its α-1,6 isomer, isomaltulose (7.6), which is then chemically hydrogenated to a mixture of 6-α-D-glucopyranosyl-(1,6)-D-sorbitol (7.7) and the corresponding mannitol derivative that is sold as isomalt:

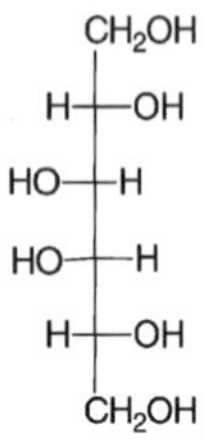

(7.3) 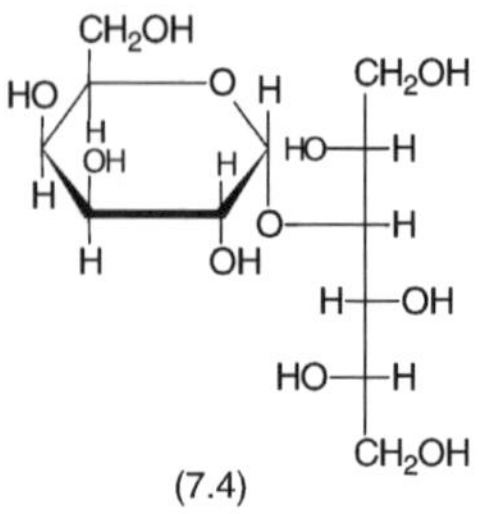   (7.4) 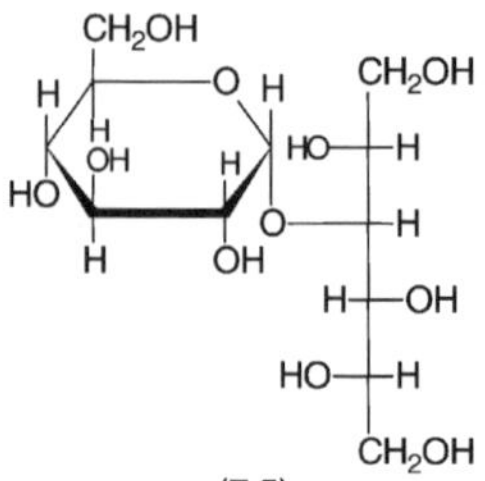   (7.5)

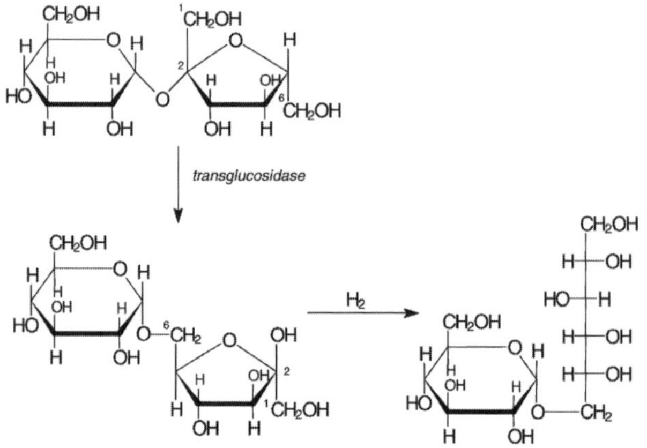

(7.6)   (7.7)

The hydrogenated sugars are gaining importance as bulk, low-calorie sweeteners. In general they are a little less sweet than sucrose. The term bulk refers to the fact that they can fill the other roles, *i.e.* besides sweetness, of sucrose in flour and sugar confectionery but being largely unabsorbed by the small intestine do not contribute to one's calorific intake. However, they also carry the water binding properties of sugars on into the large intestine and have the potential, if enough is consumed, to cause osmotic diarrhoea. They are frequently used in the formulation of commercial confectionery products *etc.* marketed as suitable for diabetics although many diabetes specialists disapprove of this approach to diabetes management. Another approach to low-calorie bulk sweeteners is typified by sucralose (*see* Figure 7.4), a chlorinated sucrose derivative. This is permitted in the USA and many other countries and is currently (2002) being considered by the EC.

Non-sugar sweeteners have a much longer history than one might think. The attractive sweetness of a saccharin was noticed within weeks of its first synthesis in 1878, and it was quickly exploited as a sweetening agent. By the turn of the century some 200 tons were being produced annually by a German company and in 1901 the Monsanto Chemical Company was set up to manufacture saccharin in the USA. Shortages of sucrose during both World Wars led to its widespread use both domestically and for commercial products such as soft drinks. The advantages of saccharin include its stability to processing procedures and the essentially clean bill of health it has received from toxicologists. It is also very cheap, in terms of sweetening power some 2% of the cost of sucrose. To many consumers it offers the advantage of a sweetener that contributes no 'calories' to either the human diet or that of the bacteria that cause dental caries. Its main disadvantage is an unpleasant metallic after-taste that often leads to its use along with sucrose to reduce, rather than eliminate, the calorific value of a product.

In more recent years, several other non-sugar sweeteners have been introduced, some entirely the product of chemical synthesis but others wholly natural in origin (Table 7.2). Two of these, sodium cyclamate and acesulphame K (*see* Figure 7.4), have some structural features similar to those of saccharin. Both of these have a superior taste profile to saccharin, lacking the metallic after-taste, and have all the other dietary advantages. However, cyclamate fell foul of a single carcinogenicity test in 1969. It was immediately banned in the USA and many other countries followed suit. However, subsequent biological testing did not support this conclusion and it has been reinstated in most countries around the world and is currently being reconsidered for use in the USA.

Like saccharin, the sweet taste of aspartame, L-aspartyl-L-phenylalanine methyl ester (*see* Figure 7.4), was discovered by chance, following a lapse of laboratory hygiene on the part of a chemist working on a quite different problem. Aspartame has a clean, sucrose-like flavour that has made it extremely popular with soft-drink manufacturers. Its main drawback is its instability outside the pH range 3–6, especially at temperatures above

Table 7.2 *Non-sugar sweeteners permitted in Europe. Most of these are subject to restrictions on the levels and types of food in which they may be used. In particular none are permitted in infant foods. The status of these and other sweeteners varies widely around the world.*

Sweetener	E number	Sweetener	E number
Acesulphame K	E950	Sorbitol	E420
Aspartame	E951	Mannitol	E421
Cyclamate	E952	Isomalt	E953
Saccharin	E954	Maltitol	E965
Thaumatin	E957	Lactitol	E966
Neohesperidine DHC	E959	Xylitol	E967

70 °C. Its decomposition follows two different paths: (i) simple hydrolysis to its components aspartic acid, phenyalanine, and methanol and (ii) methanol loss followed by cyclisation to a diketopiperazine (7.8).

(7.8)

   Fortunately the diketopiperazine proved to present no toxicological problems, but aspartame itself did cause some concern. The initial attraction of aspartame was that its natural hydrolysis in the gastrointestinal tract would not result in products whose 'naturalness' at reasonable levels could be doubted. However, the possibility arose that sufferers from the inherited disease phenylketonuria, whose capacity to cope with an excess of phenylalanine is dangerously impaired, might be at risk. In fact the dietary regime of sufferers from the disease is so exacting that avoiding food and drinks that contain aspartame is only a minor tribulation. The risk to symptomless carriers of the defective gene has been shown to be insignificant. Alitame (*see* Figure 7.4) emerged from the intense research on potentially sweet peptides that followed the discovery of aspartame. It is permitted in some countries around the world and under consideration for approval in the USA and EC.

   A number of naturally occurring intense sweeteners have attracted attention. Although neohesperidine dihydrochalcone has not been found in nature, close chemical relatives have been found to be the source of the strong sweet taste in a number of traditional food plants used around the world. It is obtained by chemical conversion of naringin (7.9) (the source of the bitterness of grapefruit, *Citrus paradisii*) into neohesperidine followed by hydrogenation to the chalcone. Neohesperidine can also be obtained directly from the peel of the bitter, or Seville, orange (*Citrus aurantium*). It is not normally used alone in food products (it is quite expensive) but it has been found to enhance the potency of other sweeteners, of all types. One unusual property is that it masks the bitterness of other substances, making it an attractive ingredient for some types of pharmaceutical products.

   Glycyrrhizin (*see* Figure 7.4) is the natural sweet component of liquorice root (*Glycyrrhiza glabra*). Extracts containing glycyrrhizin are widely used in the Far East for flavouring and sweetening foods and drinks, medicines and tobacco. Nowadays liquorice based confectionery products appear to be less popular in Britain than they were. Unfortunately glycyrrhizin is well known to have a range of biological activities. While its anti-allergenic, anti-inflammatory and expectorant properties could be regarded as beneficial its effects on kidney function and blood pressure could not. For this reason it is widely permitted as a flavouring in confectionery, soft drinks *etc.* where the total consumption is unlikely to be high but not as sweetening agent. The well known laxative action of liquorice may or may not be caused by glycyrrhizin but this does ensure that consumption is, to a degree, self-limiting.

Stevioside (*see* Figure 7.4) is the principal terpenoid glycoside extracted from the leaves of *Stevia rebaudiana*, a member of the sunflower family cultivated in parts of the Far East and South America. Extracts containing stevioside are approved for food use in some countries but doubts over its safety (like glycyrrhizin it shows a wide range of biological activities) mean that it is not permitted in the USA or the EC. Sweetness is not the prerogative of relatively small molecules. Thaumatin is the best known example of a sweet tasting protein. It is intensely sweet, especially if compared to sucrose on a molecule for molecule basis (*see* Figure 7.3). It is obtained from the fruit of a West African tree, katemfe (*Thaumatococcus danielli*) and marketed under the name Talin®. It is unusually resistant to thermal denaturation, making it suitable for cooked products, but totally hydrolysed to amino acids by the human digestive system. It has been approved for use in a number of countries including the USA and the EC but is likely to find most application as a flavour enhancer rather than as a sweetener in its own right.

Bitterness

Romantics have often linked bitterness and sweetness in terms of human emotions but they are also closely associated, almost opposite sides of the same coin, where taste is concerned. Several classes of the substances that taste bitter have, as we shall see, close structural relationships with sweet tasting substances. However, such an association is not found with the inorganic salts. Only a few salts actually taste salty, most are very bitter. Studies with the alkali metal halides have suggested that the structural criterion that distinguishes saltiness from bitterness is simply size. Where the sum of the ionic diameters is below that of potassium bromide (0.658 nm), which tastes both salty and bitter, then the salty taste predominates. Sodium chloride, at 0.556 nm, is the obvious example; potassium iodode, however, at 0.706 nm, tastes bitter. Magnesium salts such as the chloride, $MgCl_2$ (0.850 nm) also have a bitter taste.

Phenolic substances in the form of flavonoids are important sources of bitterness in fruit juices, particularly citrus juices. The best known is naringin (7.9), a glycoside of the flavanone naringenin with the disaccharide neohesperidose, which occurs in grapefruit and Seville oranges. Its bitterness is, such that it can be detected at a dilution of 1 in 50000. Another bitter element that can occur in citrus juices is limonin (7.10), which is sometimes formed spontaneously from tasteless precursors during commercial juice extraction.

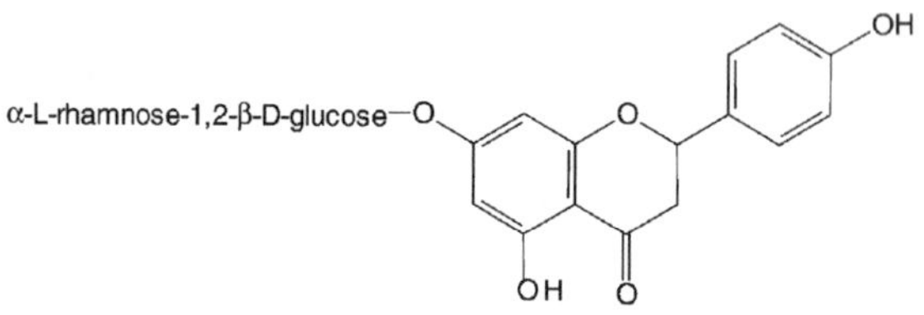

(7.9) 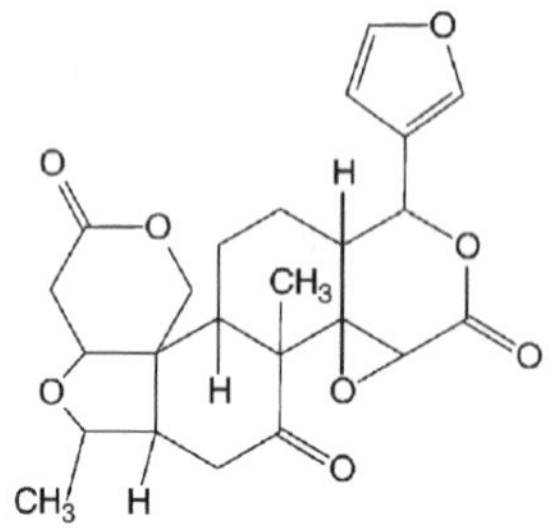   (7.10)

Bitterness is a sought-after characteristic in many types of beers, particularly those brewed in Britain. Bitterness is achieved by adding hops to the wort, *i.e.* the sugary extract from the malt, before it is boiled and then cooled in the stage preceding the actual fermentation. Hops are the dried flowers of the hop plant, *Humulus lupulus*, and they are rich both in volatile compounds that give beer its characteristic odour and in resins that include the bitter substances. The most important bittering agents are the so-called α-acids. As shown in Figure 7.5, these isomerise during the 'boil' to give forms which are much more soluble in water and much more bitter. Different varieties of hops have different proportions of the three α-acids and the structurally related, but less important, β-acids.

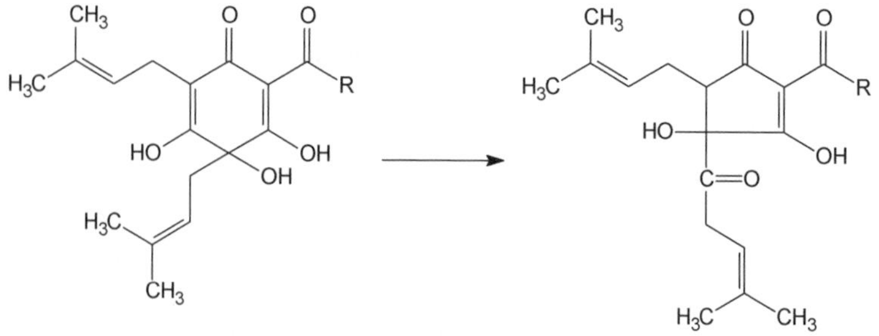

Figure 7.5   *The isomerisation of the α-acids of hops. R is $(CH_3)_2CHCH_2-$ in humulone, $(CH_3)_2CH-$ in cohumulone, and $C_2H_5(CH_3)CH-$ in adhumulone.*

The ability to perceive bitterness almost certainly evolved in order to protect man, and presumably other animals, from the dangers posed by the alkaloids present in many plants. Alkaloids are defined as basic organic compounds having nitrogen in a heterocyclic ring. Although we have little idea as to the function of the alkaloids in plants, many have highly undesirable pharmacological effects on animals, besides their extremely bitter taste; nicotine, atropine, and emetine are all alkaloids. In view of its medicinal use, quinine (7.11) is one of the best known, and it is much used as a bittering agent in soft drinks such as 'bitter lemon' and 'tonic water'.

(7.11)

Until recently the bitter taste of many amino acids and oligopeptides was of only academic interest. However, the need for more efficient use of the proteins in whey, waste blood, *etc.* has focused attention on the properties of the peptides that are obtained by partial hydrolysis (by enzymes or acid) of these proteins. These protein hydrolysates are being developed as highly nutritious food additives with many useful properties relating to food texture. Examination of the tastes of the amino acids that occur in protein hydrolysates shows that bitterness is exclusively the property of the hydrophobic L-amino acids: valine, leucine, isoleucine, phenylalanine, tyrosine, and tryptophan. Whether a peptide is bitter or not depends on the average hydrophobicity of its amino acid residues. A great deal of work on the conformation of peptides, in relation to their taste, has demonstrated that the structural requirements for bitterness mirror those required for sweetness in sugars and other compounds. There is a similar requirement for a correctly spaced pair of hydrophilic groups (AH and B in sugars), one basic and one acidic, together with a third, hydrophobic group in the correct spatial relationship to the hydrophilic groups. The ability to predict the likely taste of a peptide can be linked to our knowledge of amino acid sequences to predict the appearance of bitter, and therefore unwelcome, peptides in hydrolysates of particular proteins. For example, caseins and soya proteins are rich in hydrophobic amino acids and therefore tend to yield bitter peptides. One possible solution is to ensure that the hydrolysis is limited to producing fragments with molecular weights above about 6000, which are too large to interact with the taste receptors.

Close relatives, in terms of chemical structure, of many non-sugar sweeteners are being found to taste bitter. For example the isomer of aspartame in which the L-phenylalanine residue is replaced by D-phenylalanine tastes bitter. The relatives of saccharin, sucralose and perillartine shown here (7.12, 7.13 and 7.14 respectively) all taste bitter. Since there is, at best, only a very limited market for novel bittering agents the most likely outcome of these observations is greater insights into the structure of the sweet taste receptor proteins.

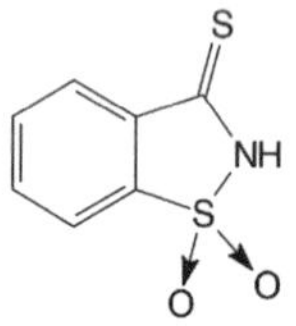

(7.12) 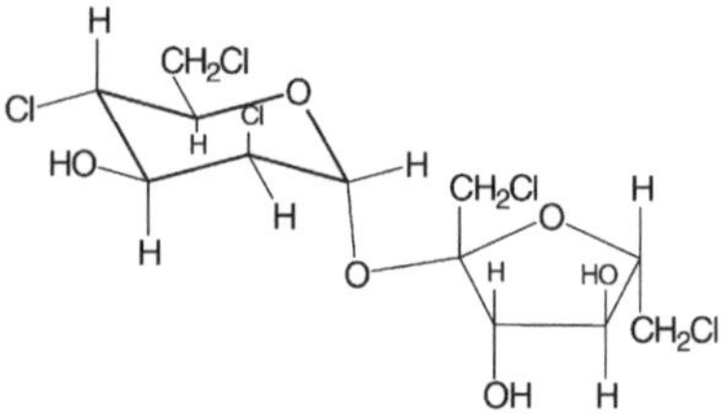   (7.13) 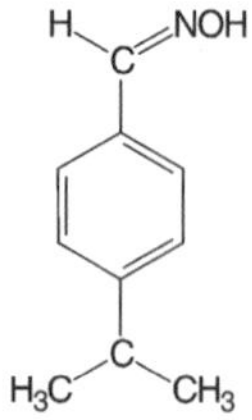   (7.14)

Saltiness

Saltiness is most readily detected on the sides and tip of the tongue and is elicited by many inorganic salts besides sodium chloride. However, the taste of salts other than sodium chloride has very little relevance to food. The contribution of sodium chloride to food flavour is often under-estimated, as anyone who has experienced the awfulness of bread made without salt will testify. We frequently add salt to both meat and vegetables to enhance their flavour, but the sodium ions naturally present in many foods have a major role in their flavour. For example, when sodium ions were eliminated from a mixture of amino acids, nucleotides, sugars, organic acids, and other compounds known to mimic the flavour of crab meat successfully (as well as corresponding closely to the composition of water-soluble components of crab meat), the mixture was totally lacking in crab-like character. Besides this tendency to stimulate meaty flavours, salt tends to decrease the sweetness of sugars to give the richer, more rounded flavour that is required in many confectionery products. The relationship between molecular size and the saltiness or bitterness of salt was dealt with in the bitterness section.

Sourness

Sourness is always assumed to be a property of solutions of low pH, but it appears that the hydrogen ion H^{+*} is much less important for taste than the undissociated forms of the organic acids that occur in acidic foodstuffs. In most fruit and fruit juices, citric acid (7.15) and malic acid (7.16) account for almost all the acidity. Tartaric acid (7.17) is characteristic of grapes, blackberries are especially rich in isocitric acid (7.18), and rhubarb is of course well known for its possession of oxalic acid (7.9), which occurs together with substantial amounts of malic and citric acids. Typical organic-acid contents of citrus and grape juices are shown in Table 7.3.

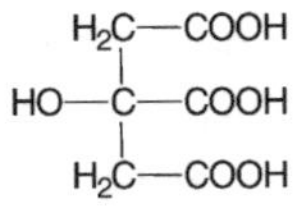

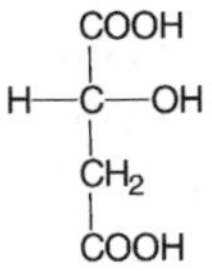

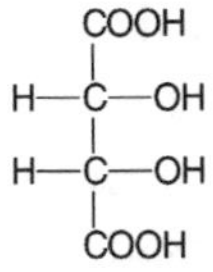

Ethanoic (acetic) acid, at levels around 5%, gives vinegar and products derived from it their sourness. In many pickled products such as pickled cabbage (*sauerkraut* in Germany) and cucumbers it is lactic acid, derived from the sugar in the vegetable by bacterial fermentation, that is

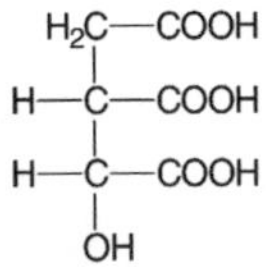

Table 7.3 *The organic acid content of some fruit juices. The values here are typical, but wide variations occur between different varieties and especially different degrees of ripeness. Grapes grown in relatively cold climates such as northern Europe, and the wines produced from them, have much higher acid contents than those from warmer, sunnier climates.*

	Acid content/mmol dm^{-3}		
	Malic	*Citric*	*Tartaric*
Orange	13	51	–
Grapefruit	42	100	–
Lemon	17	220	–
Grape	7	16	80

* In aqueous solutions hydrogen ions are hydrated to form hydronium, ions, H_3O^+ so that strictly speaking it is these whose role in taste is under consideration here.

responsible for the low pH and sourness. One should not overlook the contribution that acetic acid makes to the odour of foods, for example when fried fish and chips are dowsed in vinegar. Lactic acid derived from lactose by fermentation occurs in cheese at around 2%, contributing some of the sharpness. Lactic acid, in the form of its sodium salt, is also used to give a vinegar taste to 'salt and vinegar' flavoured potato crisps.

Astringency

Astringency is a sensation that is clearly related to bitterness but is registered within the oral cavity generally as well as on the tongue. We tend to register astringency in terms of 'mouth-feel' and though it may therefore be incorrect to regard this sensation as a *taste* we should not let semantics spoil our appetite. The reaction of polyphenolic substances with enzyme proteins was mentioned in the previous chapter (*see* page 6.10) and it is believed that the sensation of astringency results from similar reactions involving proteins in the saliva or possibly the surfaces of the tongue, palate *etc.*

Astringency is usually regarded as a desirable characteristic of fruit and cider, but it is in red wine and tea that it is most important. In both these beverages it is associated with the high content of the polyphenolic substances, which are also involved in colour, and were described in Chapter 6. In wine and tea the polyphenols responsible in part for flavour are collectively described as tannins. In black tea most of the polyphenolic material is considered to contribute to the overall astringency,* but it is the gallyl groups of the epi-gallocatechin gallate (*see* Figure 6.12) that contribute most of the astringency. *epi*-Gallocatechin itself is merely bitter, with no discernable astringency, whereas the astringency of epicatechin gallate is apparent at concentrations above 50 mg per 100 ml. There is evidence for an interaction between these groups and the caffeine (which alone is merely bitter) in producing astringency. When we drink black tea we normally seek to moderate the astringency. Two techniques are available, the choice being driven by custom and geography. In the USA, and in other countries where coffee rather than tea is the primary source of caffeine, lemon juice is added. This lowers the pH, the ionisation of the gallate carboxyl groups is repressed and astringency reduced. The alternative approach, deeply ingrained in the culture of the British Isles, is to add milk. The milk proteins successfully compete with the proteins of the mouth for the polyphenolics. This technique is limited by the availability of pasteurised milk. Milk that has been subjected to more

---

* The astringency of tea is referred to 'briskness' in the trade.

aggressive heat treatments such UHT or sterilization contains flavour compounds, presumably generated by the Maillard reaction, that clash with the tea's delicate flavour with unpleasant results. Coffee, whose flavour is more dependent on the Maillard reaction, is less sensitive to the heat treatment of any milk that is added to it.

In wine the situation is more complex as there are many more elements in the flavour, but all the evidence suggests that it is flavanoids similar to those found in tea that are responsible for the astringency. Besides their obvious anthocyanin content, which does contribute a little to the taste, red wines generally have up to 800 mg dm^{-3} of catechins (as in Figures 6.12 and 6.14) compared with no more than 50 mg dm^{-3} in white wine.

Pungency

Pungency, often described 'hotness', is another sensation experienced by the entire oral cavity and elsewhere, notably the anal mucosa. It is common nowadays to link pungency with astringency and the cooling effect of some compounds (*see* page 6.12) and the cooling effect of some compounds under the general heading of 'chemesthesis', by analogy with 'somesthesis', the sensory system in the skin which responds to mechanical stimulus *i.e.* touch. The response to chemesthetic stimulants may be better described as chemical irritation and has much in common with the perception of pain.

Pungency is the essential characteristic of a number of important spices, but it is also found in many members of the family Cruciferae. The genus *Capsicum* includes the red and green chillies, the large *C. annuum* being much less pungent than the ferocious little fruit of *C. frutescens*. The active components in *Capsicum* species are known as capsaicinoids, capsaicin (7.20) and dihydrocapsaicin (which has a fully saturated side chain) being the most abundant and the most pungent.

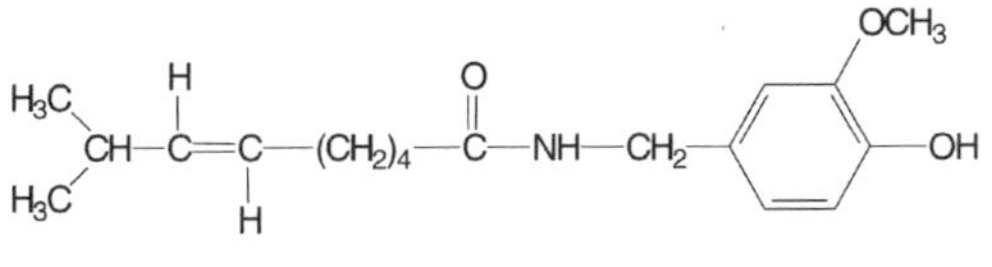

(7.20)

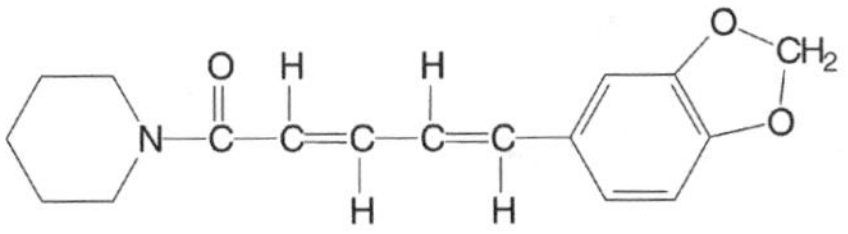

(7.21)

Pepper, both black and white, comes from *Piper nigrum*. About 5% of the peppercorn is the active pungent component, piperine (7.21), together with small traces of less pungent isomers that have one or both of the double bonds *cis*. The active components of ginger, *Zingiber officinale*, another pungent spice, show a striking structural similarity to capsaicin and piperine. These are the gingerols (7.22) and shogaols (7.23). Gingerols and shogaols with *n* equal to 4 predominate, but higher homologues do occur in small amounts. Different forms of ginger, *e.g.* green root ginger and dry powdered ginger, have different proportions of gingerols and shogaols as a result of the readiness with which gingerols dehydrate to the corresponding shogaols. The vanillyl side chain is also found in eugenol (7.24), the essential pungent component of cloves.

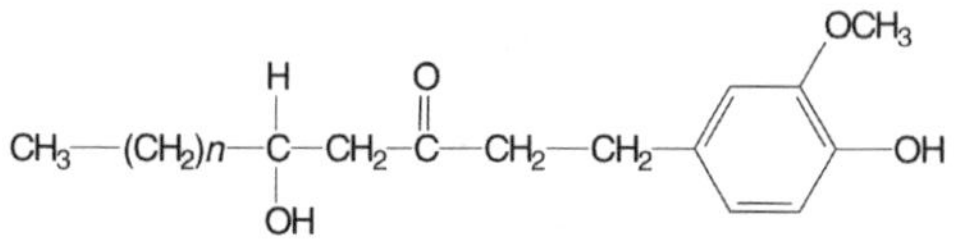

(7.22)

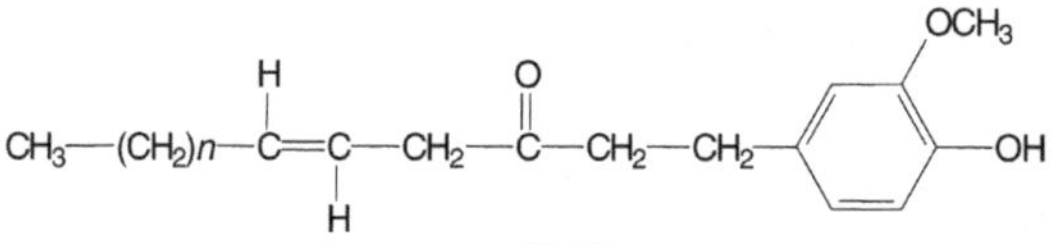

(7.23)

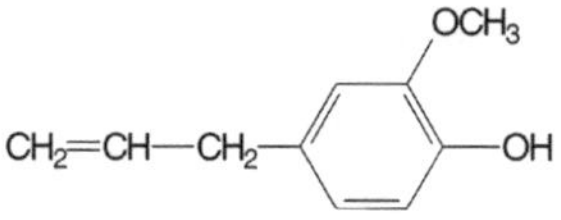

(7.24)

The final group of pungent food components to be considered are the glucosinolates, found in plants belonging to the family Cruciferae, which includes the brassicas (cabbage, Brussels sprouts, broccoli) and many plants renowned for their pungency when raw, such as horse-radish, black and white mustards, and radish. They are normally found at levels (of total glucosinolates) up to around 2 mmol (or $\sim$2 g kg^{-1}) fresh weight with the highest amounts in the most pungent species or varieties. The glucosinolates themselves are not themselves pungent but all the plants that contain them also contain the enzyme *myrosinase*. This occurs in the cytoplasm of the plant's cells but it only encounters its substrates, the glucosinolates, when the plant tissue is disrupted when we, or grazing animals, chew it, it suffers insect or other damage in the field, or it is chopped up in the kitchen. As shown in Figure 7.6 the enzyme catalyses the hydrolysis of the unusual carbon sulfur bond liberating thiohydroxamate-*O*-sulfates. These compounds are unstable and depending on the conditions spontaneously form isothiocyanates or nitriles. The volatile isothiocyanates are the most likely products and it is these that have the pungent taste. Collectively isothiocyanates are often referred to as mustard oils. The glucosinolates of

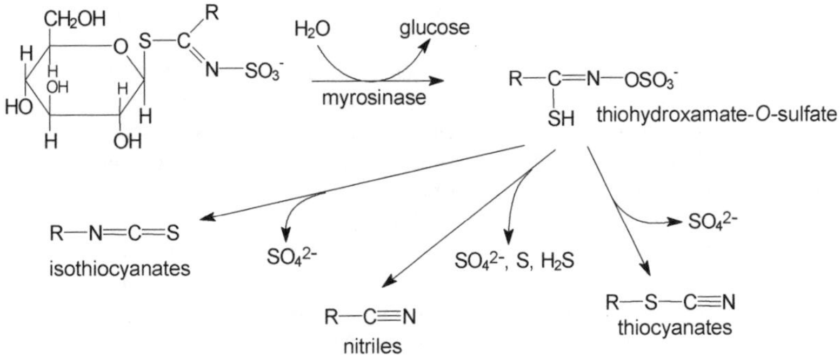

Figure 7.6 *The breakdown of glucosinolates to form isothiocyanates and other products. Nitrile production is favoured by low pH such as encountered, for example, in the production of sauerkraut.*

a range of plants are shown in Figure 7.7. In the cases of sinigrin and glucotropaeolin the thiocyanates are formed by enzymic isomerisation of the isothiocyanates.

Nitrile formation is a significant problem in the utilisation of rape-seed meal, *i.e.* the residue from rape-seed oil extraction. Another problem with rape-seed meal is the tendency of isothiocyanates that have a hydroxyl group in the *β*-position, like that derived from progoitrin, to cyclise spontaneously to oxazolidinethiones (7.34). These compounds, also known as goitrins, interfere with iodine uptake by the thyroid gland resulting in goitre, *i.e.* massive enlargement of the thyroid gland and deficiency of the hormone thyroxine. For these reasons plant breeders have made great efforts to develop strains of rape with very low levels of glucosinolates. These efforts are not without their downside. Breakdown products of glucosinolates are recognised as important in the defence of plants against insect and fungal pathogens. The suggestion that milk from cattle fed largely on rape-seed meal and other brassicas might be harmful should not be taken seriously. The only time goitrins have proved to be a problem for human consumers was in the winter of 1916–17 in Germany. Food shortages caused by the 1st World War led to turnips replacing potatoes and cereals as staple foods. Combined with other effects of malnutrition the incidence of goitre rose sharply.

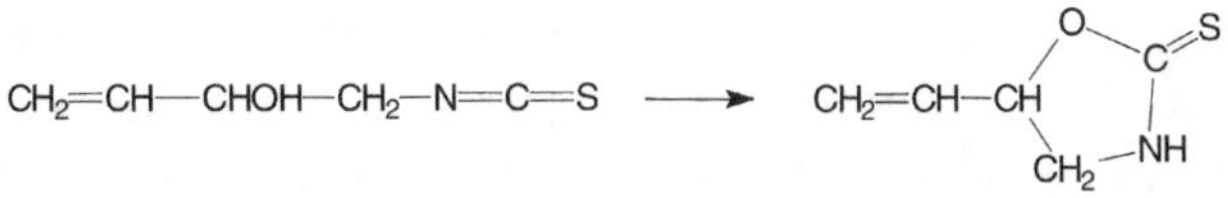

(7.34)

Principal or characteristic glucosinolates

Cabbage	*Brassica oleracea*	sinigrin (7.25), glucobrassicin (7.26), progoitrin (7.27)
Brussels sprouts	*B. oleracea* var. *gemmifer*	glucobrassicin, neoglucobrassicin (7.28)
Broccoli	*B. oleracea* var. *italica*	glucobrassicin, glucoraphanin, (7.29)
Cauliflower	*B. botrytis* var. *cauliflora*	sinigrin, glucobrassicin, neoglucobrassicin
Black mustard	*B. nigrum*	sinigrin
White mustard	*B. alba*	sinalbin (7.30)
Rape	*B. napus*	progoitrin, glucobrassicin, neoglucobrassicin
Horse-radish	*Amoracia lapathiofolia*	sinigrin
Radish	*Raphanus sativus*	glucoraphasatin (7.31)
Watercress	*Nasturtium officinalis*	gluconasturtiin (7.32)
Cress	*Lepidum sativum*	glucotropaeolin (7.33)

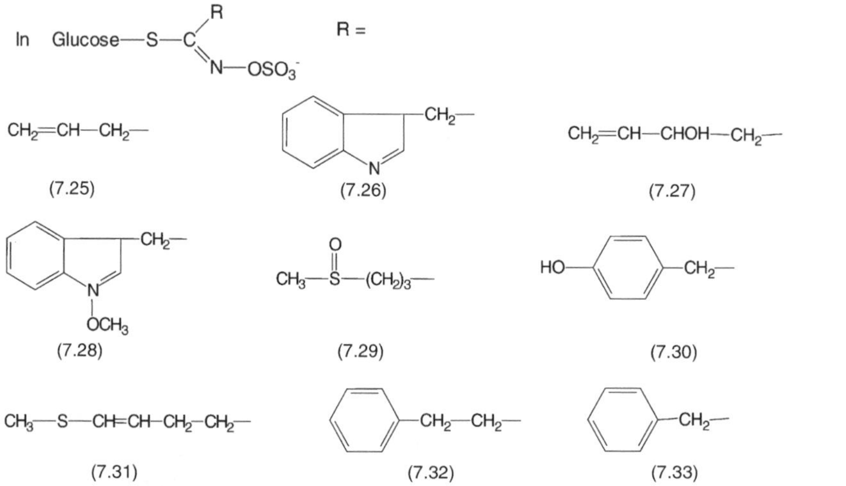

Figure 7.7 *The glucosinolates of members of the Cruciferae family.*

The lower incidence of cancer in populations that consume large amounts fruit and vegetables is now well recognised and at least some of the beneficial effects, especially in relation to cancers of the large colon, can be related to the consumption of glucosinolates in cruciferous vegetables. In experiments with laboratory animals and tissue cultures of human cells allylisothiocyanate (7.35) (from sinigrin) and sulforophane (7.36) (from glucoraphanin) at realistic concentrations have been shown to kill cancerous cells. Sulforophane appears to cause cells to increase production of glutathione transferases, a group of enzymes involved in the neutralization of carcinogens.

$$CH_2{=}CH{-}CH_2{-}N{=}C{=}S$$
(7.35)

$$CH_3{-}\overset{\overset{\displaystyle O}{\|}}{S}{-}(CH_2)_3{-}CH_2{-}N{=}C{=}S$$
(7.36)

When vegetables in this group are cooked by pouring boiling water over them, followed by a few minutes at 100 °C the myrosinase is destroyed before tissue disruption brings the enzyme and substrate into contact; the isothiocyanates are not formed. Instead a wide range of other sulfur compounds arise which give over-boiled cabbage and other similar vegetables their characteristic unattractive odour. The alternative procedure of immersing vegetables in cold water and bringing them slowly to 100 °C tends to promote myrosinase activity. The difference in flavour between frozen and fresh Brussels sprouts after cooking has been shown to reflect the different rates of myrosinase destruction during commercial blanching and domestic cooking.

Cooling is a very minor taste response that hardly justifies a section of its own. It gets a mention here simply because, like pungency, it is regarded as another manifestation of chemesthesis, and is also the natural opposite of heat. Cooling sensations are associated with some confectionery and and also products such toothpaste, mouthwash *etc.* Just as the burning taste of pungent substances is believed to be registered by the same, or similar, nerve endings as those concerned with high temperature the cooling taste is assumed to invoke a response in the nerves that register low temperature. The best known cooling substance is (−)-menthol (7.37). It is the dominant component of peppermint oil, extracted from *Mentha piperiia* and *M. arvensis,* that is used in confectionary (menthone, (7.38) is the other major component). Menthone and all the optical isomers of (−)-menthol share the characteristic minty odour but only (−)-menthol induces a cooling sensation in the mouth. There are other substances, notably xylitol, that have a cooling effect but this is usually attributed to their large negative heat of solution. As the solid xylitol included in chewing gum dissolves in the saliva its temperature falls, briefly.

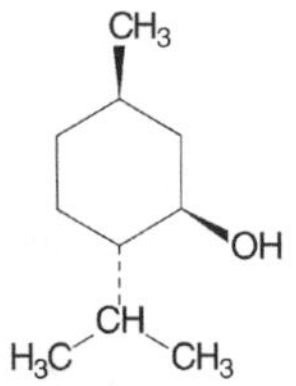

(7.37) 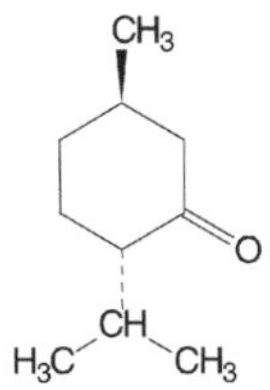 (7.38)

Meatiness

Of all the taste classes the most recent to attract attention is meatiness. Although the flavour of meat is obviously an amalgam of our responses to both volatile and non-volatile compounds, meaty taste has come to be associated with two particular compounds, inosine monophosphate, IMP, (7.39) and monosodium glutamate, MSG, (7.40). Both these substances were first identified as the active components in popular food ingredients used to enhance the flavour of traditional Japanese dishes. The Japanese word *umami* is used to identify the particular taste of these two compounds and others related to them. MSG was found, in 1908, in a variety of seaweed and has been produced commercially ever since, in recent years by a bacterial fermentation. IMP was similarly isolated from a dried fish preparation. Fish muscle served as the source for commercial preparations of all the 5′-nucleotides, until recently, when enzymic hydrolysates of RNA became available.

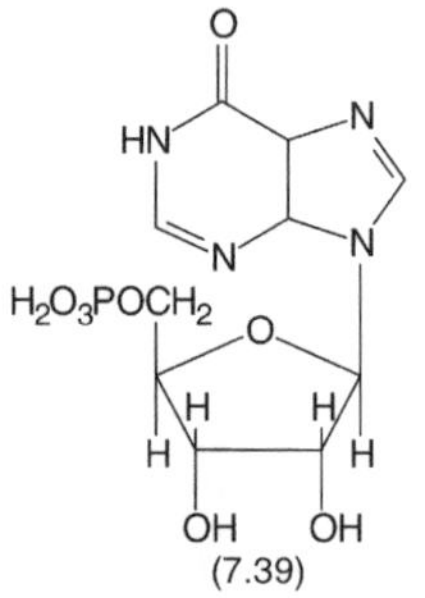

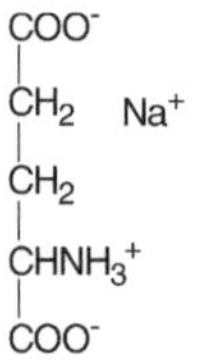

Alone, neither of these substances has a particularly strong taste. They need to be at a concentration above about $300 \, mg \, dm^{-3}$ before they make much impact. However, when present together they show a remarkable synergism. A mixture of equal proportions of IMP and MSG tastes some 20 times stronger than the same total amount of one alone. Although the mechanism of this synergism is not understood, its importance to food manufacturers is obvious. The addition of only a small amount of one of these substances (usually MSG) to a food product that naturally contains modest amounts of the other will have a dramatic effect on the flavour. Thus MSG is used as a flavour enhancer rather than flavouring in dried soups and similar products, at levels to give about $1 \, g \, dm^{-3}$ in the final dish. Although they are much more expensive, preparations of IMP, often blended with guanosine monophosphate (7.41), are now being used by some food manufacturers.

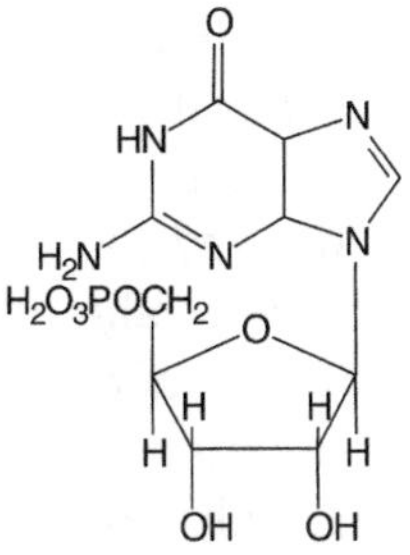

(7.41)

Glutamic acid, as MSG, will naturally be present at low levels in muscle tissue, but in meat it will be derived mostly from the protein breakdown that occurs during the ageing process. IMP is a breakdown product of the adenosine monophosphate, AMP, that accumulates as adenosine triphosphate, ATP, is utilised in the muscle post-mortem:

$$ATP \xrightarrow{\text{-phosphate}} ADP \xrightarrow{\text{-phosphate}} AMP \xrightarrow{\text{-NH}_3} IMP$$

Some typical data for the levels of *umami* substances are shown in Table 7.4. It is not difficult to correlate these data with one's personal impression of the flavour of many of these foods. The use of tomatoes to 'bring out' the flavour of meat and parmesan as garnish for minestrone soup can be

Table 7.4 *Typical levels of umami substances in a range of foodstuffs. The gaps in the table indicate lack of data rather than zero concentration.*

	IMP	GMP	MSG
	mg per 100 g		
Kelp			2240
Parmesan cheese			1200
Sardines	193		280
Tuna	188		
Tomato juice			260
Shiitake mushroom		30	67
Prawns	92		43
Pork	122	3	23
Cod	44		9
Salmon			20
Human milk			22
Cow's milk			2
As a food additive			20–80

readily justified by this data. The difference in MSG levels between human and cow's milk is also interesting and one could speculate that it reflects the different dietary priorities of the herbivorous cow and the omnivorous human. Any protein-rich food will, by the time it is eaten, have degraded sufficiently to have accumulated detectable levels of *umami* substances so that our taste for them provides a good test for the likely presence of protein. Results obtained by the petfood industry suggest that cats, being obligate carnivores and using high protein foods to supply all their nutritional needs, are much more sensitive to *umami* substances than humans*.

While IMP and MSG provide the basic 'meatiness' of meat and meat products, the subtle difference in flavour between different meats depends on variations in their proportions and also involves many other flavour compounds that occur in trace amounts. For example, the IMP contents of beef and pork are roughly similar (100–150 mg per 100 g), but beef has significantly more free amino acids than pork, including twice as much MSG. Lamb is especially rich in MSG but has markedly low levels of dipeptides such as carnosine (7.42), which also contribute to the meat flavour. An increase with maturity in the level of the free amino acids has been suggested as the difference between the flavour of veal and beef.

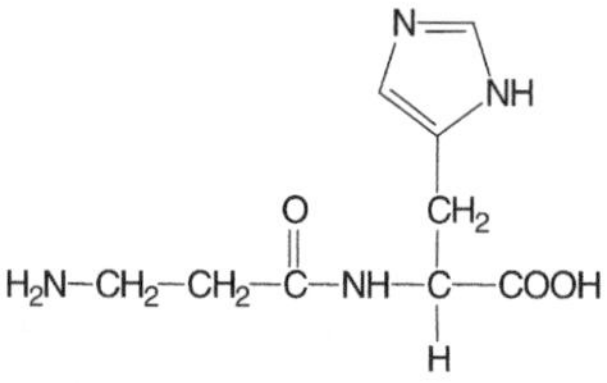

$$H_2N-CH_2-CH_2-\overset{\overset{\displaystyle O}{\|}}{C}-NH-\underset{\underset{\displaystyle H}{|}}{\overset{\overset{\displaystyle CH_2}{|}}{C}}-COOH$$

(7.42)

Although some dishes do depend on raw meat (*e.g.* steak tartare), cooking is obviously the origin of the innumerable volatile compounds that are so important in our appreciation of meat. These are discussed in the second half of this chapter.

ODOUR

While the sense of taste enables us to make some broad assessment of the nature of a potential foodstuff it is the odour that fills in the detail. Counting the number of different substances that we can detect is impossible, as is any attempt to list all the different odours. It is also

---

* This argument also provides an explanation for the apparent indifference of cats to sweetness, human's indicator taste for of the presence of carbohydrates, our major energy supply.

impossible to draw a line between food odours and the host of other odours, more often referred to as smells, that we encounter every day that have nothing to do with food. Yet another complication is that natural odours, as we perceive them, are almost never generated by single pure substances. The complexity of Scotch whisky was mentioned at the beginning of this chapter; a typical fruit may well have as many as 200 different volatile components, all making a contribution to its overall odour, but even in total these may comprise only a few parts per million of the total fruit. Attempts have been made to classify odours and relate the different classes to elements of chemical structure. Seven primary odours were described by Amoore in 1972: camphoraceous, ethereal, musky, floral, minty, pungent and putrid. However, if these correspond to seven types of receptor in the nose then many substances must be able to interact with more than one receptor. Since most of the odours we experience in the real world outside the laboratory are caused by a mixture of a great number of different substances with several types of receptors responding simultaneously the brain is left to make sense of an amalgam of nerve impulses (*cf.* colour vision?). By analogy with what is now well understood in relation to sweet and bitter tastes our olfactory organs must be equipped with cell membranes carrying receptor proteins that can discriminate between molecules of similar structure, include optical isomers. For example, of these two isochroman derivatives one (7.43) is reported to have an intense musk odour but the other (7.44) is odourless.

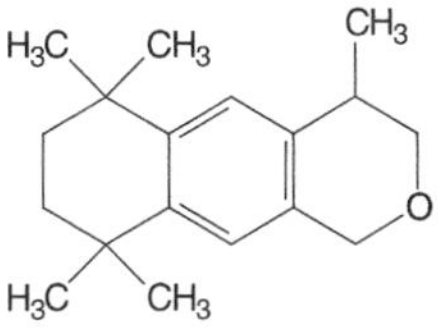

(7.43) 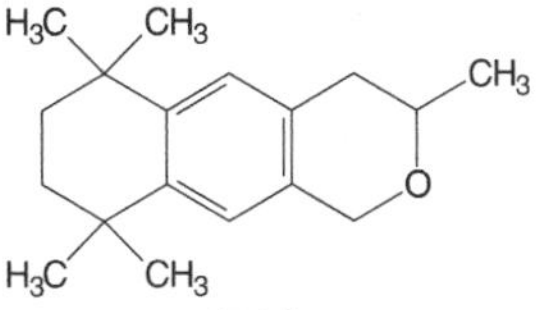   (7.44)

The organisation of the following sections is therefore based on a few broad groups of foodstuffs and makes no claim to be comprehensive.

Meat

The first half of this chapter concluded with an account of the *taste* of meat and it seems reasonable to emphasise the dependence of flavour on both taste and odour by immediately returning to the same foods. There are three major groups of compounds that contribute to the flavour of meat and one of these, the *umami* substances and dipeptides, was covered in the previous section. The other two groups are:

1 Heterocyclic compounds derived from amino acids, nucleotides, sugars and thiamine *via* the Maillard reaction (*see* Chapter 2) during cooking.

2 Unsaturated fatty acid breakdown products (*see* Chapter 4) formed during cooking.

The Maillard reaction can yield a great diversity of carbonyl compounds from the residual sugars (from glycogen and nucleotides) in meat, including 2-oxopropanal (2.36), 2,3-butanedione (2.43), hydroxypropan-2-one (2.44), hydroxymethylfurfural (Figure 2.9). The Strecker degradation (Figure 2.12) of these with most amino acids yields the corresponding aldehydes but, as shown in Figure 7.8, cysteine gives rise to both hydrogen sulfide, and ammonia, which can contribute to the formation of a variety of heterocyclic compounds. Thiazoles and oxazoles are considered important in roasted meat flavour. Reactions of hydrogen sulfide with aldehydes can lead to ring compounds with multiple sulfur atoms that are thought to be particularly important in the flavour of boiled (as oppose to roasted) meat. Several other sulfur compounds that contribute to cooked

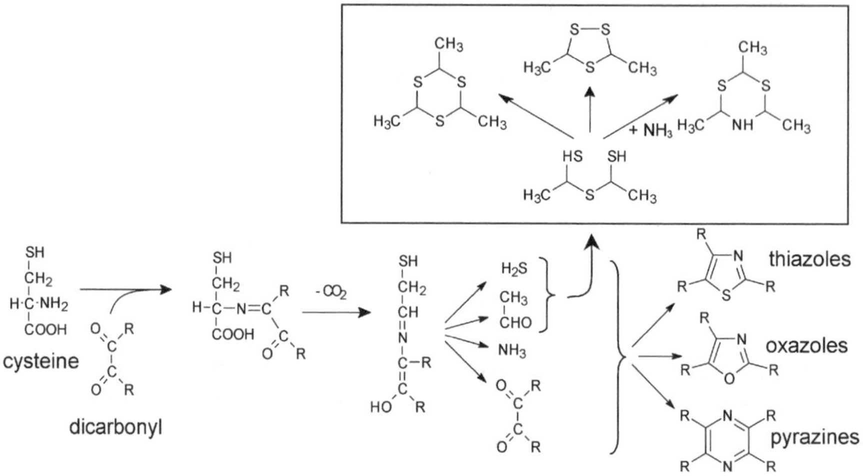

Figure 7.8 *The formation of heterocyclic meat flavour compounds following the Strecker degradation of cysteine. For simplicity the complex sequence of condensation reactions postulated to lead to the ring compounds has not been shown. R represents any of the terminal fragments of sugar molecules that can arise in the Maillard reaction. It is likely that aldehydes and dicarbonyls from the degradation of other amino acids will join in the sequence and provide even greater diversity of end products. The boxed reactions show the formation, from acetaldehyde and hydrogen sulfide, of the cyclic sulfides that have been identified in boiled meat flavour.*

meat flavour have been traced back to the thermal decomposition of the vitamin thiamin.

Thermal decomposition of ribonucleotides (*e.g.* IMP, 7.39) has been shown to lead to compounds, such as methylfuranolone (7.45), that not only have a pronounced meaty odour themselves but can also be precursors of sulfur-containing compounds (the sulfur is derived from the breakdown of cysteine and methionine), such as methylthiophenone (7.46), that also have a strong meaty odour.

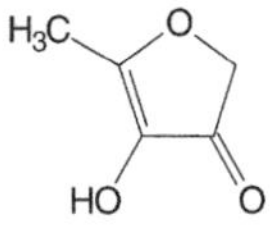

(7.45) 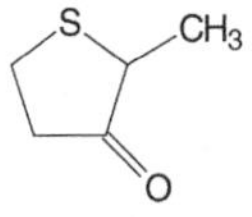   (7.46)

Fatty-acid breakdown, along the lines of the autoxidation reactions described in Chapter 4, is also an important source of the characteristic odours of meat products. The low content of polyunsaturated fatty acids in the depot fat and intra-muscular fat of beef, lamb and pork points to the polar membrane lipids of muscle cells as the most likely source of flavour significant autoxidation products. The arachidonic acid has been shown to break down on heating to give four aldehydes which together have a distinctive cooked-chicken odour:

$CH_3(CH_2)_4CH = CHCH_2CHO$ 3-*cis*-nonenal

$CH_3(CH_2)_4CH = CHCH_2CH_2CHO$ 4-*cis*-decenal

$CH_3(CH_2)_4CH = CHCH_2CH = CHCHO$ 2-*trans*, 5-*cis*-undecadienal

$CH_3(CH_2)_4CH = CHCH_2CH = CHCH = CHCHO$ 2-*trans*, 4-*cis*,
                                                   7-*cis*-tridecatrienal

Aldehyde fatty acid fragments such as these can react with hydrogen sulfide or ammonia to provide yet another source of odorous heterocyclic compounds, as illustrated here (7.47).

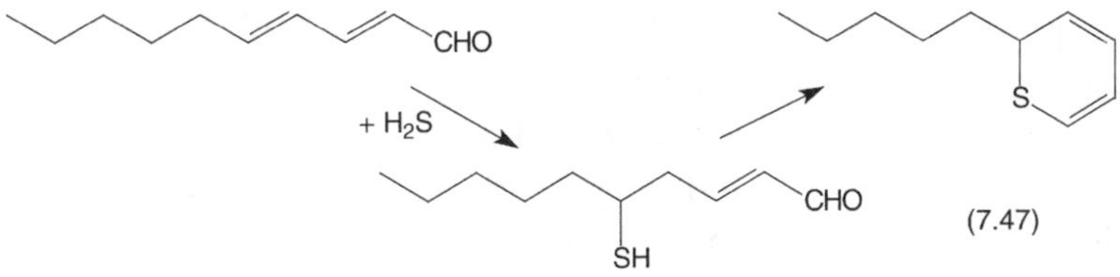

+ H_2S

CHO

SH

CHO

(7.47)

This is an appropriate point at which to mention boar taint. Male animals tend to be more lean than females and in theory this should make them preferred for meat production. Unfortunately this presents a difficulty for pig rearers. The uncastrated male pig, as it reaches about 200 lb (90 kg) live weight,* accumulates 5-androst-16-ene-3-one (7.48) in its body fat. This by-product of normal steroid metabolism is the causative substance of boar taint. Being volatile, this steroid is driven off when the meat is cooked, and its odour may give offence – not everyone can smell it! Over 90% of women but only about 55% of men can detect it at typical cooking concentrations, and of those that can smell it far more women than men find it unpleasant. One can only speculate as to the implications for human behaviour or physiology of this evidence of gender differences between olfactory organs. The odour is described as having much in common with both urine and sweat, and it is therefore hardly surprising that the meat industry would like to find a method of dissuading boars from synthesising it.

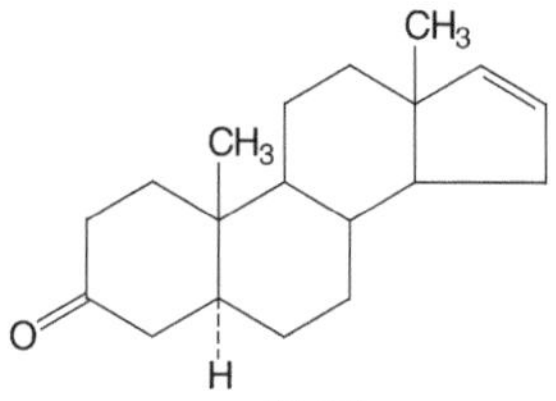

(7.48)

Fruit

The flavour of fruit is very definitely a blend of taste and odour. The taste, an aggregate of sweetness due to sugars, the sourness of organic acids and sometimes the astringency of polyphenolics, provides the basic identity that confirms that one is consuming 'fruit'. A simple solution of just sucrose and citric acid actually tastes 'fruity' to most people. However, to distinguish between different fruit, we rely heavily on the distinctive odours of the fruit's volatile components. If one has a cold, which will effectively eliminate one's sense of smell, it is extremely difficult to distinguish between raspberry and strawberry flavours.

A typical fruit may well have as many as 200 different volatile components, but even in total these may comprise only a few parts per million of the total fruit. As a generality it is observed that citrus fruit odours are dominated by terpenoids whereas in other fruit** terpenoids (*see* page 178) are much less in evidence. Long lists of compounds have

---

* 200 lb is the typical weight for a bacon pig; younger animals at around 140 lbs are used for pork.
** The term 'berry-type' aromas is sometimes used to cover all the remainder, a satisfactory term to cover raspberries and the like but highly stretched to cover apples and bananas.

been assembled for many fruit which show that acids, alcohols, the esters resulting from their combination, aldehydes, and ketones predominate numerically. Considering apples as a fairly typical example, we find:

at least 20 aliphatic acids ranging from formic to n-decanoic,
at least 27 aliphatic alcohols covering a corresponding range of chemical structures,
over 70 esters involving the predominant alcohols and acids,
26 aldehydes and ketones also with structures corresponding to the acids and alcohols, and
smaller numbers of miscellaneous ethers, acetals, terpenoids, and other hydrocarbons

a total of 131! In a consideration of aroma the first point to remember is that by no means all of these substances make a significant contribution: volatility does not equal aroma. Another point is that a very high proportion of these compounds are common to many different fruits. For example, of the 17 esters identified in banana volatiles only five are not found in apples. The particularly common occurrence of aldehydes, alcohols, and acids with linear or branched (*e.g.* isopropionyl) chains of up to seven carbon atoms, often with an occasional double bond, suggests common origins. In fact there are two sources which are both represented in many fruits. During ripening, cell breakdown is accompanied by the oxidation of the unsaturated fatty acids of membrane lipids catalysed by the enzyme lipoxygenase. This converts fatty acids with a *cis,cis* methylene-interrupted diene system (as in linoleic acid) into hydroperoxides. These break down spontaneously or under the influence of other enzymes to give aldehydes, as shown in Figure 7.9 in the context of vegetable flavour. The pattern of reaction products is very similar to that found in the non-enzymic autoxidation reactions of fatty acids. Amino acids are the other sources of aldehydes. Enzymic transamination and decarboxylation of free amino acids often accompany ripening. For example, leucine gives rise to 3-methylbutanal (7.49).

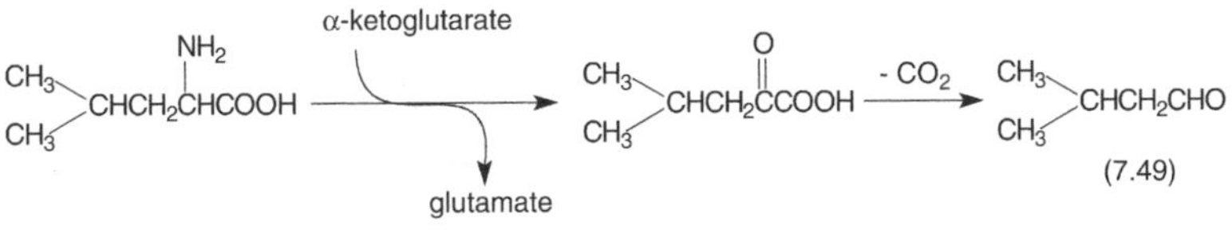

(7.49)

The aldehydes are not only important in fruit and vegetables. During alcoholic fermentations, yeast enzymes catalyse similar reactions with the amino acids from the malt or other raw materials. These aldehydes, and the alcohols derived from them, are important as flavour compounds in many alcoholic drinks, particularly distilled spirits (where they are described as fusel oils), in which they have a special responsibility for the headaches associated with hangovers. Other enzyme systems in plant tissues are responsible for the reduction of the aldehydes to alcohols or their oxidation to carboxylic acids and the combination of these to form esters.

In many cases the distinctive character of a particular fruit flavour is dependent on one or two fairly unique 'character impact' substances. For example, isopentyl acetate is the crucial element in banana flavour, although eugenol (7.24) and some of its derivatives contribute to the mellow, full-bodied aroma of ripe bananas. Benzaldehyde is the character impact substance in cherries and almonds. It is difficult to establish to what extent hydrocyanic acid (HCN), which is also present at the low levels of a few p.p.m. in cherries, contributes to their aroma. Hexanal ($CH_3(CH_2)_4CHO$) and 2-hexenal (7.50) provide what are described as 'green' or 'unripe' odours in a number of fruits and vegetables. In at least one variety of apple they have been shown to balance the 'ripe' note in the aroma given specifically by ethyl 2-methylbutyrate (7.51). The olfactory thresholds of these three substances are, respectively, 0.005, 0.017, and 0.0001 p.p.m.! The distinctive character of raspberry aroma is due mostly to 4-(*p*-hydroxyphenyl)-2-butanone (7.52), often referred to as 'raspberry ketone', but the fresh grassy aroma which is characteristic of the fresh fruit, and notably lacking in many raspberry-flavoured products, is principally contributed by *cis*-3-hexenol (7.53) supplemented by α- and β-ionone (*see* page 179). Furaneol, 2,5-dimethyl-4-hydroxy-2,3-dihydrofuran-3-one (7.54), is a major component of strawberry flavour.

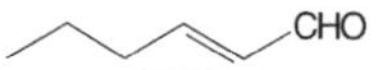

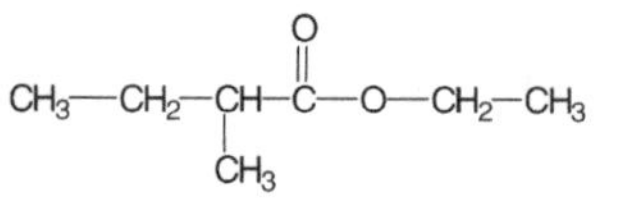

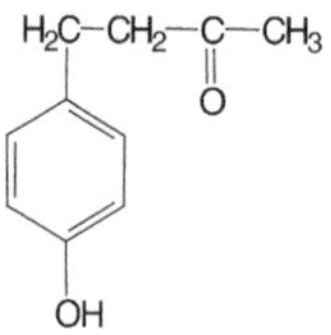

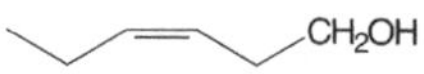

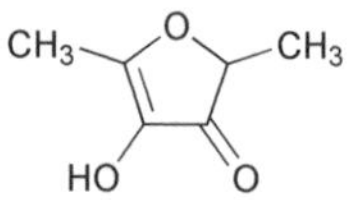

Wine presents special challenges to the flavour chemist. Different varieties of red wine, *e.g. Merlot, Grenache, etc.* generally share most of the same flavour compounds, merely in different proportions. Wine experts are well known for their apparent skill in characterising the aroma of a wine in terms of other fruit. For example the presence of 2,6,6'-trimethyl-2-vinyl-4-acetoxytetrahydropyran (7.55) in a red wine is said to impart a blackcurrant aroma. Similarly the peach and pineapple aromas of several red wines have been attributed to γ-decalactone (7.56) and 'wine lactone', (3a,4,5,7a-tetrahydro-3,6-dimethylbenzofuran-2(3*H*)-one, 7.57) respectively. *cis*-Rose oxide (4-methyl-2-(2-methyl-1-propenyl)tetrahydropyran, 7.58) has been identified as the source of the lychee aroma characteristic of *Gewürztraminer,* a white wine.

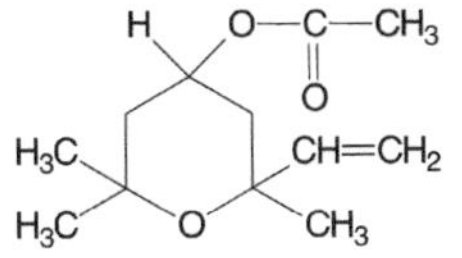

(7.55) 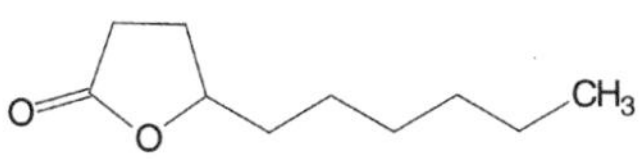 (7.56) 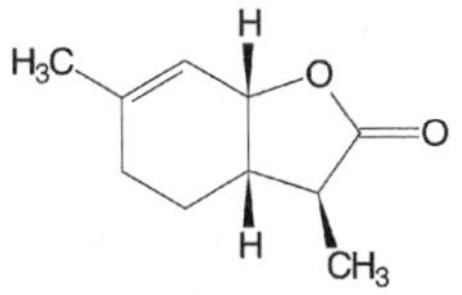 (7.57) 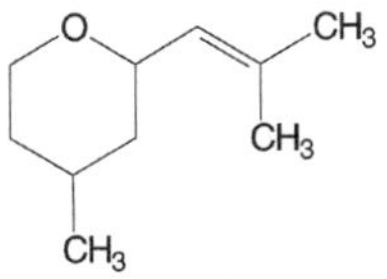 (7.58)

The character impact substances in many fruit, especially the citrus group, are terpenoids. Terpenoids are often isolated from plant materials by means of steam distillation. The oily fraction of the distillate is known as the essential oil and these are often used as components of flavourings. The essential oils of citrus fruits (*i.e.* of the entire fruit including the peel) consist mainly of terpenes, of which the monoterpene (+)-limonene (7.59) is usually at least 80%. The optical isomer of (+)-limonene, (−)-limonene (7.60), is reported to smell of lemons rather than oranges. However, (+)-limonene is not nearly as important as a flavour component as the oxygenated terpenoids which occur in the oil in much smaller amounts. For example, fresh grapefruit juice contains about 16 p.p.m. of (+)-limonene carried over from the peel during juice preparation, but the characteristic aromatic flavour of grapefruit is due to the presence of much smaller amounts of (+)-nootkatone (7.61). The characteristic bitterness of grapefruit juice was mentioned earlier in this chapter. The character impact substance in lemons is citral, which is more correctly named as a mixture of the isomers geranial (7.62) and neral (7.63). Most varieties of oranges do not appear to possess a clearly identified character impact substance although the mandarin orange, *Citrus reticulata*, is characterised by αsinensal (7.64). The various varieties of so-called sweet or Valencia orange, *C. sinensis*, contain β-sinensal (7.65).

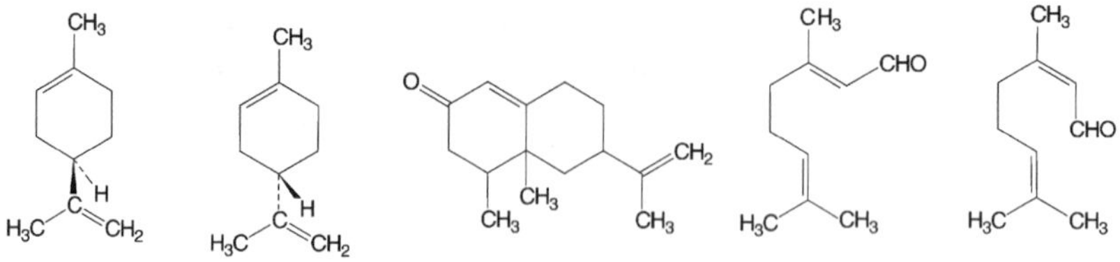

(7.59) (7.60) (7.61) (7.62) (7.63)

(7.64) (7.65)

Vegetables

Compared with fruit, vegetables tend to have much milder more subtle flavours where aroma is much less significant than taste. The aromas of fruit have evolved to entice insects, birds and other consumers into joining in seed distribution and related activities. This hardly applies to the leaves of a cabbage or the tubers of a potato. The list of vegetables whose flavour is sufficiently desirable to prompt food manufacturers to seek out artificial flavourings is very short, mushrooms and onions, and certainly does not include broccoli or beans. The character impact substance of the common mushroom, *Agaricus bisporus,* is 1-octen-3-ol, $CH_3(CH_2)_4CHOHCH=CH_2$. Many vegetables owe their flavour to the enzyme *lipoxygenase*. The action of this enzyme is shown in Figure 7.9. 2-*trans*-Nonenal, the end product shown in Figure 7.9, is an important contributor to the aroma of cucumbers although 2-*trans*-6-*cis*-nonadienal is recognised as the character-impact substance. Lipoxygenase activity is not always a good thing. High levels of the enzyme occur in peas and beans and if, during harvesting, they are physically damaged the enzyme is able to react with unsaturated fatty acids from the cell membranes and gives rise to off-flavours, notably 'delay off flavour' in peas. The 'delay' is an unintended few hours between the peas being mechanically removed from the pod by the vining machine and blanching. Blanching, usually consisting of a few seconds in a blast of steam, is designed to inactivate lipoxygenase and other enzymes.

Onion, *Allium cepa*, and garlic, *A. sativum*, are by far the most studied vegetable flavours. All members of this genus contain S-alkyl cysteine sulfoxides (*see* Figure 7.10). The 1-propenyl derivative is particularly associated with onions and the allyl derivative with garlic. Just as in the

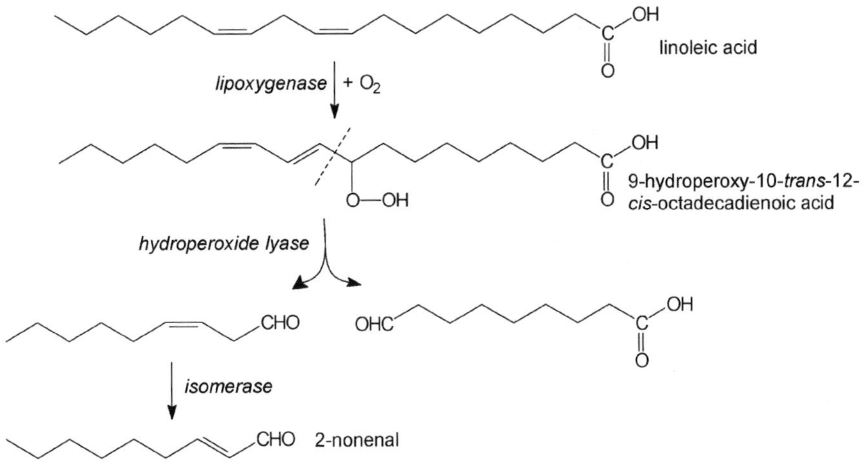

Figure 7.9 *The formation of* trans-2-nonenal *following the breakdown of linoleic acid initiated by lipoxygenase action. Lipoxygenases in some plant species react at the other end of the conjugated diene system, giving, for example, 9-hydroperoxy-10-*trans-12-cis-*octadecadienoic acid which leads to a quite different range of end products.*

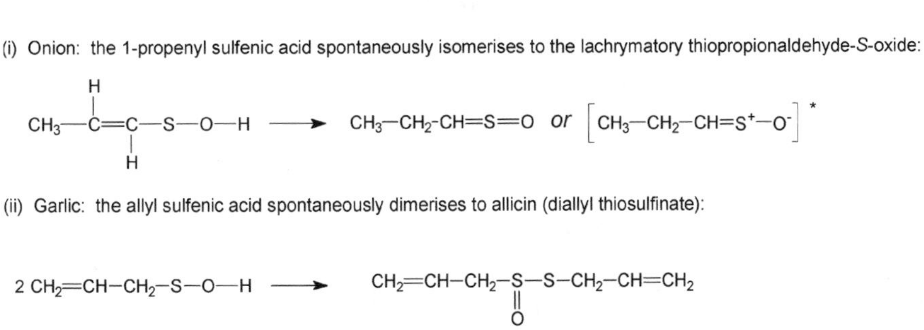

(i) Onion: the 1-propenyl sulfenic acid spontaneously isomerises to the lachrymatory thiopropionaldehyde-S-oxide:

(ii) Garlic: the allyl sulfenic acid spontaneously dimerises to allicin (diallyl thiosulfinate):

Figure 7.10 *The breakdown of S-alkyl cysteine sulfoxides in onion and garlic. The asterisked structures shown in square brackets are alternatives favoured by some authorities.*

formation of isothiocyanates in cruciferous plants, the compounds that occur in the raw, undamaged plant tissues are quite odourless. It is only when the garlic is crushed or the onion is sliced that the enzyme *alliinase* gains access to its substrates and the flavour becomes obvious. The fate of the unstable sulfenic acid depends on the location of the double bond in the side chain. In onions the double bond is sufficiently close to interact with the sulfenic acid group allowing isomerisation to thiopropionaldehyde *S*-oxide, the source of onion's legendary lachrymatory effects.

$$R\!-\!\underset{\underset{O}{\|}}{S}\!-\!CH_2\!-\!\underset{\underset{\underset{\underset{\underset{COOH}{|}}{CHNH_2}}{|}}{\underset{(CH_2)_2}{|}}}{\underset{\underset{CO}{|}}{\underset{NH}{|}}}\!CH\!-\!COOH$$

(7.66)

In garlic, by contrast, the reactivity of the sulfenic acid group of the allyl sulfenic acid is undiminished by such interactions leading to dimerisation to *allicin*, well known for its antibacterial properties. In the biosynthesis of *S*-alkyl cysteine sulfoxides the immediate precursors are the γ-glutamyl derivatives (7.66) and considerable amounts of these compounds are present in the fresh plant material. They are not susceptible to the action of alliinase but they are available to participate in the many exchange reactions, dimerisations and condensations that take place when onions and garlic are cooked. For example sulfonates (7.67) and disulfides (7.68) are formed, the latter going on to form mono- (7.69) and trisulfides (7.70) which all contribute to the flavour of the cooked vegetable.

$$2\,R\!-\!\underset{\underset{O}{\|}}{\overset{\overset{O}{\|}}{S}}\!-\!S\!-\!R \longrightarrow R\!-\!\underset{\underset{O}{\|}}{\overset{\overset{O}{\|}}{S}}\!-\!S\!-\!R + R\!-\!S\!-\!S\!-\!R$$

(7.67)   (7.68)

$\Big\downarrow$ x 2

$$R\!-\!S\!-\!R + R\!-\!S\!-\!S\!-\!S\!-\!R$$

(7.69)   (7.70)

Di- and trisulfides tend to be the most significant sulfur compounds present when cooking has taken place in water. However, in fried foods (including stir fried dishes) other compounds are more significant. These include (*E*)-ajoene* (7.71) and 2-vinyl-[4H]-1,2-dithiin (7.72).

---

* The (*E*) refers to the *trans* configuration of the central double bond. (*Z*)-ajoene with the double bond *cis* also occurs. The Spanish word for garlic is 'ajo'.

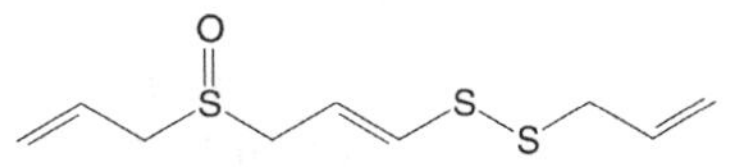

(7.71)

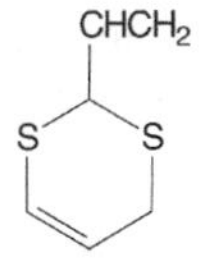

(7.72)

A brief mention should also be made of the medicinal properties of garlic. In spite of the long history of its medicinal use, results in clinical trials with human subjects have been rather variable. Most studies show reductions in blood cholesterol levels between 10 and 15% when 0.5–1.0 g of garlic powder (equivalent to 1.5–3.0 g fresh garlic) is consumed daily for 2 or 3 months. Besides its antibacterial properties allicin is a potent antithrombotic agent, preventing the formation of platelets. However, it remains questionable whether the health benefits of these effects can compensate for the social impact of this level of garlic consumption. The 'odour free' capsules of garlic oil popular as 'alternative' medicines are suspected of containing only very small amounts of these biologically active substances.

Asparagus is another vegetable in which sulfur compounds make an important contribution to flavour. However, the aroma of the vegetable itself is of less interest than the unpleasant smell it imparts to the urine of consumers. The most likely compounds in urine causing the 'rotten cabbage' smell are methyl mercaptan (methanethiol, CH_3SH), *S*-methyl-thioacrylate (7.73) and *S*-methyl-3-(methylthio)thiopropionate (7.74). These are assumed to be derived from asparagusic acid (7.75) which, unlike most of the other sulfur compounds in asparagus, is not found in any other vegetables. Only some 40% of asparagus eaters can detect this smell in their urine but whether this is due to a failure (?) in the majority's sense of smell or their sulfur metabolism is yet to be resolved.

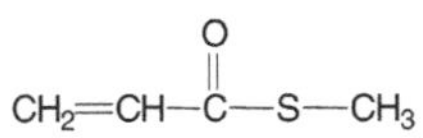

(7.73)

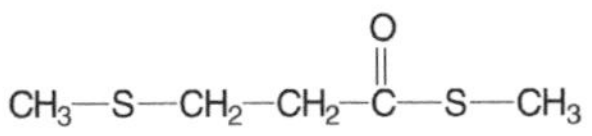

(7.74)

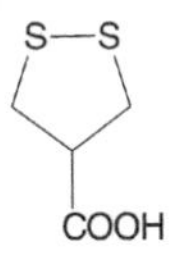

(7.75)

Herbs and Spices

These important sources of food flavour are usually lumped together, not least because the distinction between them is not clear. Both terms describe plant materials used specifically as flavourings rather than for any nutrients they may contain. The culinary use of 'herb' tends to follow on from its botanical use (a plant with no woody tissues, as in 'herbaceous border') and refer to leafy materials, such as mint, basil, rosemary and sage. By contrast 'spices' are usually derived from non-leafy plant tissues, such as seeds (*e.g.* nutmeg), flower buds (*e.g.* cloves), roots (*e.g.* ginger) or tree bark (*e.g.* cinnamon). The principal compounds responsible for the flavour of a selection of herbs and spices are shown in Figure 7.11.

Principal aroma compounds

Basil	*Ocimum basilicum*	methylchavicol (7.76)
Rosemary	*Rosmarinus officinalis*	verbenone (7.77), 1,8 cineole (7.78)
Sage	*Salvia officinalis*	α-thujone (7.79), camphor (7.80)
Origano	*Origanum spp.*	carvacrol (7.81), thymol (7.82)
Peppermint	*Mentha piperita*	(-)-menthol (7.83), menthone (7.84), menthofuran (7.85)
Spearmint	*M. spicata*	(-)-carvone (7.86)
Caraway	*Carvum carvi*	(+)-carvone (7.87)
Nutmeg	*Myristica fragrans*	(+)-sabinine (7.88)
Cardamom	*Elettaria cardomomum*	α-terpinyl acetate (7.89)

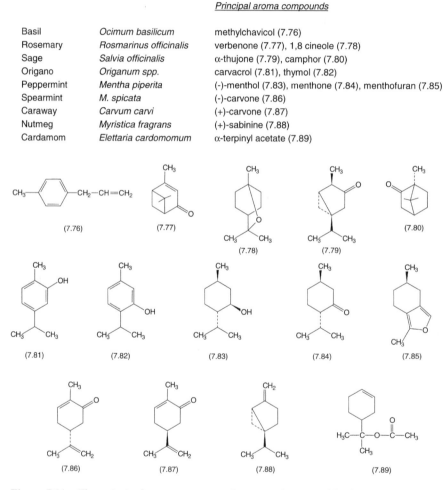

Figure 7.11 *The principal aroma compounds in a selection of herbs and spices. The compounds shown are not necessarily the most abundant but the ones believed to impart the characteristic flavour notes. The carvones (7.86 and 7.87) demonstrate the common observation that optical isomers frequently differ markedly in aroma. (For clarity the* −CH₃ *groups carried by the bridging carbon atoms have been omitted from structures 7.77 and 7.80.)*

Synthetic Flavourings

A principal objective for the flavour chemist is the provision of synthetic flavourings for use by both the domestic cook and the food manufacturer. An important trend in recent years has been the replacement of simple esters, with their sweet and unsubtle odours (*e.g.* amyl acetate in pear drops), with lighter, fresher, and very much more expensive flavourings based more closely on our knowledge of the natural flavour components. Very little information on the composition of their products is ever released by flavour manufacturers, but the patent literature does give some insights. Table 7.5 provides an example of a formulation for use in confectionery and other products. Mixtures such as these will normally be only the basis for a commercial flavouring.

Flavour manufacturers will 'customise' products for different food manufacturers (and retailing chains selling 'own label' products) by making small changes in the composition and adding small amounts of other substances. Even though most of the components are pure, synthetic, compounds there is still, as shown here, a place for natural essential oils that are complex mixtures in their own right. It is also important to remember that most of these compounds have been identified as components of the original fruit whose flavour is being mimicked. To arrive at the formulation of a flavouring the flavourist will be guided initially by the identities and apparent proportions of the natural volatiles, as revealed by gas chromatography linked to mass spectrometry. However, the final composition will owe as much to art as it does to science as the experienced, and highly trained, nose of the flavourist takes over.

In recent years a demand has built up in the food industry for flavourings outside the range of the traditional fruit flavours. For example,

Table 7.5 *A synthetic banana flavouring. The concentrations in the final product of even the most abundant of these components will still be of the order of parts per million.*

Major components	Minor components	Trace components
Amyl acetate	Amyl valerate	Acetaldehyde
Amyl butyrate	Benzyl propionate	Butyl acetate
Ethyl butyrate	Cyclohexyl propionate	2-Hexenal
Isoamyl acetate	Ethyl caproate	Isoamyl alcohol
Isoamyl butyrate	Geranyl propionate	Lemon oil
Linalool	Heliotropin	α-Ionone
	Vanillin	Methyl heptanone
		Orange oil

the very high price of cocoa has stimulated demand for good-quality chocolate flavourings. The natural flavour of chocolate has been extensively but inconclusively investigated. Two groups of compounds have been implicated, sulfides (*e.g.* dimethyl sulfide and other sulfur compounds) and pyrazines. A total of 57 different pyrazines have been identified in cocoa volatiles! A patented chocolate-flavouring formulation containing a representative pyrazine and sulfide is given in Table 7.6. In view of the understandable desire of most people to minimise the number of 'chemicals' in our food, perhaps we should consider the replacement of natural ingredients with synthetic ones!

In recent years snack foods have appeared in a remarkable range of savoury flavours. 'Smoky bacon' and 'barbecue beef' are typical. Synthesising flavours like these by replicating the range of volatiles found in the original foods presents special problems. Firstly the flavour compounds are not present in the original raw materials; they arise during cooking, mostly *via* the Maillard reaction or the thermal degradation/oxidation of unsaturated fatty acids. Secondly such Maillard reaction products usually have highly complex molecular structures that place them outside the scope of commercial synthetic organic chemistry. A third problem is that in isolation such compounds would be highly unlikely to gain the necessary clearances, with regard to safety, from regulatory authorities. The solution has been the development of 'process flavours'.

Process flavours are produced by heating precisely defined mixtures of proteins, amino acids, sugars, and fats together under closely controlled conditions. The raw materials must be completely acceptable, from a toxicological point of view, as food ingredients and heating conditions (time, temperature, and pH) also resemble those used in normal cooking. Comparison of the composition of a food product's raw materials with its range of flavour volatiles when cooked will often enable the chemist to

Table 7.6 *A synthetic chocolate flavouring (US Patent No. 3619210). The proportions of carriers, solvents etc. have been discounted.*

	Parts
Dimethyl sulfide	1
2,6-Dimethylpyrazine	3324
Ethylvanillin	143
Isovaleraldehyde	100

predict what mixture of amino acids *etc.* is most likely to produce a good process flavour simulation.

Off-flavours and Taints

Undesirable flavours can arise in foods as a result of contamination (taints) or by distortion of the normal processes that occur before it is consumed (off-flavours).

Many off-flavours in fruit and vegetables result from unwanted enzyme action in the post-harvest period. The enzyme lipoxygenase is frequently the culprit (as was mentioned on page 249), especially in legume seeds where it is especially abundant. The unwanted formation of the extremely bitter compound limonin in citrus juices was referred to earlier in this chapter (page 222) and is another example of an enzyme, in this case a *lactonase*, becoming active during post-harvest processing. Such an enormous diversity of substances can give rise to taints in food, from a similar diversity of sources, that there is only space for a few examples here. The most important sources are storage areas, packaging materials, and cleaning agents used on processing plant. In many cases tracing the source of the contamination is extremely difficult. For example a case has been described[*] in which a packaged food had an off-odour caused by trichloroanisole (7.90). This has a musty, mouldy odour that is detectable down to 3 parts in 10^{16}! Some polyethylene packaging was found to be the source of the trichloroanisole and it was eventually shown that a single polyethylene bag containing polyethylene granules, the raw material for manufacture of the packaging material, had been stored on a wooden pallet. The wood had been previously treated with chlorinated phenolic compounds as preservatives. Micro-organisms are known that are capable of converting chlorinated phenolics into the corresponding anisoles.

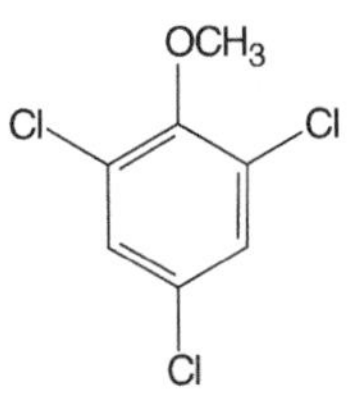

(7.90)    (7.91)

Although this was perhaps a rather extreme case phenolic disinfectants are a frequent cause of problems. Chlorophenol wood preservatives may

* J. Ewender *et al.*, in 'Food and Packaging Materials – Chemical Interactions', ed. P. Ackermann, M. Jägerstad and T. Ohlsson, The Royal Society of Chemistry, Cambridge, 1995, p.33.

find their way into fibreboard used for cartons. Disinfectants applied to buildings or to bedding for livestock or poultry can ultimately reach milk and meat at low, but detectable, levels. *p*-Cresol (7.91), a common component of disinfectants, is detectable down to 2 parts in 10^7. Trichloroanisole contamination is also the cause of cork taint in wine.

FURTHER READING

C. Fisher and T. R. Scott, 'Food Flavours. Biology and Chemistry', The Royal Society of Chemistry, Cambridge, 1997.

'Alternative Sweeteners', ed. L. O'Brien Nabors, 3rd edn., Dekker, New York, 2001.

'The Biochemistry of Fruits and Their Products', ed. A. C. Hulme, Academic Press, London, 1970.

'Flavor of Meat, Meat Products, and Seafoods', ed. F. Shahidi, 2nd edn., Blackie, London, 1998.

'Food Flavourings', ed. P. R. Ashurst, 3rd edn., Aspen, Gaithersburg, Md., 1999.

'Understanding Natural Flavours', eds. J. R. Piggott and A. Paterson, Blackie, London, 1994.

'Food Taints and Off-flavours', ed. M. J. Saxby, 2nd edn., Blackie, London, 1996.

T. W. Nagodawithana, 'Savory Flavors', Esteekay, Milwaukee, 1995.

R. S. Jackson, 'Wine Science', Academic Press, New York, 2000.

Chapter 8

Vitamins

The vitamins are an untidy collection of complex organic nutrients that occur in the biological materials we consume as food (and those we do not). In terms of chemical structure they have nothing in common, and their biological functions similarly offer no help in their definition or classification. What does draw them together is that:

(i) they are essential components of the biochemical or physiological systems of animal life (and frequently plant and microbial life),
(ii) as animals evolved they lost the ability to synthesise these substances for themselves in adequate amounts and
(iii) they tend to occur in only tiny amounts in biological materials, and
(iv) their absence from the tissues (whether by absence from the diet or by failure of absorption from the diet) causes a specific deficiency syndrome.

These specifications rule out:

(v) trace metals and other minerals since these are not 'organic',
(vi) essential fatty acids and essential amino acids which are required in larger amounts,
(vii) hormones, which are synthesised by the body as required and cannot be supplied by the diet,
(iix) substances that might well be beneficial in the treatment of some illness or condition, but whose absence does not invoke a disorder or disease in the otherwise healthy.

Animals are distinguished from most other forms of life by their dependence on other organisms as food. Directly or indirectly, plants are the fundamental source of basic nutrients, and it is hardly surprising that animals have come to rely on plants for the supply of other substances as well. The risks of vitamin deficiencies occurring in an otherwise adequate diet would have been slight when this dependence evolved, and a small price to pay for the loss of the burden of maintaining the synthetic machinery for such a diverse range of complex substances. Several of the diseases caused by vitamin deficiency have been recognised since classical times. For example accurate descriptions of *scurvy*, including its association with seamen living on preserved food, and caused by a deficiency of ascorbic acid, go back to *c*. 1150 BC (the Eber papyrus) and Hippocrates (*c*. 420 BC). Similarly *beriberi*, caused by a deficiency of thiamin, was described in Chinese herbals dating back to around 2600 BC. However, although the links to diet were noted the idea that the problem with these and other diseases was the *absence* of substances from diets was not widely accepted until well into the 20th Century. *Pellegra*, caused by a deficiency of niacin was associated with the consumption of maize but was assumed to be caused by a toxin present in spoilt maize.

Even when the association of the disease with a dietary component had been made progress in isolation and identification of the substance could only be made if a suitable animal model was available. The discovery of thiamin is a classic illustration of this. In 1886 Christian Eijkman was sent to Indonesia to discover the cause, presumed to be a bacterial infection, of beriberi. He failed to find a causative micro-organism but did notice a paralytic illness resembling beriberi in the chickens the laboratory kept. No connection with their nutrition or the human disease was made until the disease suddenly disappeared. This was quickly associated with a change in diet for the chickens: polished rice caused the disease, brown* rice prevented it. Eijkman guessed that the chickens' symptoms were very close to those of beriberi in humans and from that moment Eijkman had his experimental model. He quickly confirmed the link when he found the incidence of beriberi amongst Javanese prisoners fed polished rice diets was 2.8% but only 0.09% amongst those on brown rice diets. Water or alcoholic extracts from rice husks had the same effects, confirming that the idea that the anti-beriberi activity belonged to a chemical substance. Subsequently (in 1912) Casimir Funk came close to isolating the anti-beriberi factor from rice husks, using chickens to test his extracts. Suspecting that the factor was, in chemical terms, an amine, he used the term 'vitamine', from 'vital amine' for it. His later observation '*I must admit that when I chose the word "vitamine" I was well aware that these*

---

* Brown rice is the equivalent of wholemeal wheat, complete with the outer husk layers and the embryo, or germ. Polishing removes these, leaving just the starchy endosperm, the more familiar form used in most cookery.

substances might later prove to not all be of an amine nature. However it was necessary for me to use a name that would sound well and serve as a "catch word".' has a modern ring to it. When it became apparent that other factors were not turning out to be amines the final 'e' was dropped.

Studies in the years up to 1914 with laboratory rats fed highly purified diets of proteins, fats, starch and minerals demonstrated the necessity of 'growth factors' missing from such diets. McCullom and Davis suggested that there were two of these, one associated with milk fat and egg yolk that they called 'fat-soluble A' and the other with wheat, milk and egg yolk, 'water-soluble B'. Vitamin A became identified with the prevention of night-blindness. But in the 1920s another fat soluble factor, distinct from vitamin A, that was involved in the prevention of the bone disease *rickets* was isolated. By this time the term 'vitamin C' was being used to describe the, as yet unidentified, factor in fruit and vegetables that prevented scurvy which meant that with 'A', 'B' and 'C' now spoken for the 'anti-rachitic' factor acquired the letter 'D'.

Over the years 'water-soluble B' has been resolved into a host of different vitamins, often referred to as the 'vitamin B complex' although they are not physically bound to each other in the way that the word 'complex' implies. Through the 1920s and 1930s plant and animal foodstuffs yielded an ever increasing number of water soluble and fat soluble factors that could be linked to the growth of laboratory animals or deficiency diseases. For a time this led to a proliferation of apparent 'vitamins' that used up most of the alphabet. Once chemistry had advanced sufficiently to provide concise chemical definitions and structural formulae many of these disappeared. Some turned out to be identical to previously identified vitamins, mixtures of previously identified vitamins or to lack the activity initially attributed to them. Others were substances which, as far as reputable nutritionists, biochemists *etc.* are concerned, fail to meet the criteria set out above. Nowadays scientists try to use names wherever possible. Where a number of closely related substances, *e.g.* the tocopherols, have vitamin activity it is convenient to refer to, as in this example, 'vitamin E', to avoid the suggestion that all tocopherols are equally potent vitamins. One should not underestimate the difficulties that vitamin isolation and identification presented to the original researchers. For example the first isolation of pantothenic acid required a quarter of a ton of sheep liver as the starting material! Although most vitamins had had their chemical structures determined by the end of the 1930s that of cobalamin, better known as B_{12}, was only elucidated in 1955 by Dorothy Hodgkin, at Oxford University.

For the modern food chemist this brief detour into history may seem irrelevant but it is important for us to appreciate how vitamin terminology evolved so that we can recognise when it is being abused. Even today it is common, especially in toiletry products and on the less responsible fringes of the food supplements industry, to resurrect long obsolete terms and sections of vitamin alphabet.*

The first tasks for chemists were the isolation of vitamins in a pure form and the determination of their chemical structures. Once these were known, chemists were then called upon to provide sensitive and accurate methods for the determination of vitamin levels in food materials. Initially only biological assays had been available, based on the relief of deficiency diseases in laboratory animals, but their high cost and great slowness made them unsuitable for routine measurements.† Nowadays almost all vitamin analysis is carried out by chromatographic methods, notably high performance liquid chromatography (HPLC). A second objective for vitamin chemists has been the provision of commercially viable chemical syntheses of vitamins for clinical use and as dietary supplements. Strictly chemical routes are available for the commercial synthesis of most vitamins but in some cases, notably riboflavin and cobalamin, bacterial fermentation is used. Although the tocopherols can be synthesised chemically they are commonly extracted form vegetable oils.

THIAMIN (VITAMIN B₁, ANEURINE)

Thiamin occurs in foodstuffs either in its free form (8.1a) or as its pyrophosphate ester (8.1b) complexed with protein. No distinction is made between these forms by the usual analytical techniques or in tables of food composition. Although it is extremely widespread in small amounts, only a few foodstuffs can be regarded as good sources. As a general rule it is present in greatest amounts (0.1–1.0 mg per 100 g) in foodstuffs that are rich in carbohydrate and/or associated with a high level of carbohydrate metabolism in the original living material. Examples are legume seeds and the embryo component of cereal grains (the germ), which are involved in high rates of carbohydrate metabolism during the germination of the seed. Although meat, both muscle tissue and liver, contains very little carbohydrate in the living animal, carbohydrate is the chief source of energy for these tissues. The reason why pork should contain about ten times the amount of thiamin (about 1 mg per 100 g) that beef, lamb, poultry, and fish contain has yet to be established.

---

* The author's bathroom cabinet contains shaving gel that claims to be enriched with 'Vitamin F' a long dead term that was once applied to thiamin activity and also to essential fatty acids – definitely not vitamins.
† In the earlier years of the last century the moral question of the use of laboratory animals was sadly less prominent than it is today.

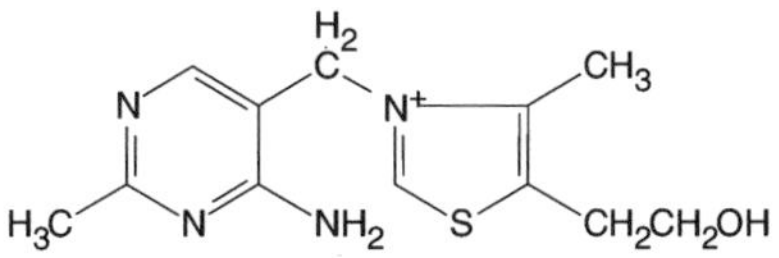

(8.1a)

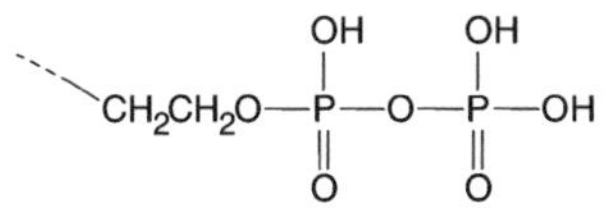

(8.1b)

The association of thiamin with carbohydrate is related to its role in metabolism. Thiamin pyrophosphate (TPP) is the prosthetic group* of a number of important enzymes catalysing the oxidative decarboxylation of α-keto acids, including pyruvic acid (8.2) and α-ketoglutaric acid (oxoglutaric acid, 8.3), which occur as intermediates in the Krebs cycle (the key metabolic pathway of respiration).

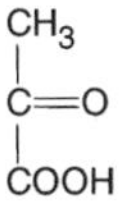

(8.2) 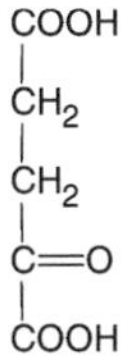   (8.3)

The role of TPP in oxidative decarboxylation is shown in Figure 8.1. Although details, such as these, of enzyme reaction mechanisms are not normally the province of food chemists, this is a valuable illustration of why vitamin structures are so often complex. It is also worth noting that the essential component of coenzyme A, pantothenic acid, is also a vitamin. There is some evidence that thiamin, probably as the pyrophosphate, has a quite separate role in nerve function, but details are not established.

Although thiamin, like the other B vitamins, is nowadays usually determined by HPLC it is actually one of the few that can be fairly easily determined by chemical methods; even so, the method requires the use of sophisticated techniques. First, the vitamin is extracted with hot dilute acid. Then the extract is treated with enzyme phosphatase to convert any TPP into thiamin and 'cleaned up' by column chromatography. This extract is then treated with an oxidising agent, usually hexacyanoferrate(II) (ferricyanide), to convert the thiamin into thiochrome (8.4), whose concentration is measured fluorimetrically.

---

* A prosthetic group is a non-protein component of an enzyme which is involved in the reaction catalysed and remains bound to the enzyme throughout the reaction sequence.

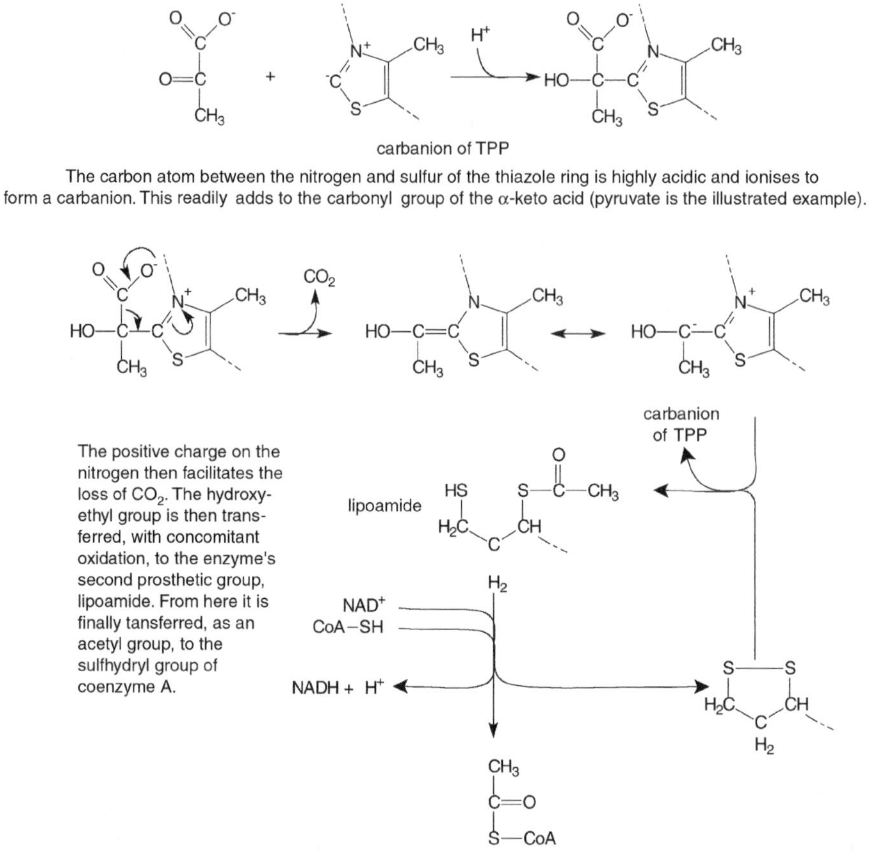
H+
HO—C—C
carbanion of TPP
The carbon atom between the nitrogen and sulfur of the thiazole ring is highly acidic and ionises to form a carbanion. This readily adds to the carbonyl group of the α-keto acid (pyruvate is the illustrated example).
CO2
HO—C=C
carbanion of TPP
The positive charge on the nitrogen then facilitates the loss of CO2. The hydroxy-ethyl group is then trans-ferred, with concomitant oxidation, to the enzyme's second prosthetic group, lipoamide. From here it is finally tansferred, as an acetyl group, to the sulfhydryl group of coenzyme A.
lipoamide
NAD+
CoA–SH
NADH + H+
S—CoA

Figure 8.1 *The mechanism of oxidative decarboxylations involving thiamin. Lipoamide is the other prosthetic group required in this type of decarboxylase. NAD^+ and NADH are, respectively, the oxidised and reduced forms coenzymes based on the vitamin niacin, described later in this chapter.*

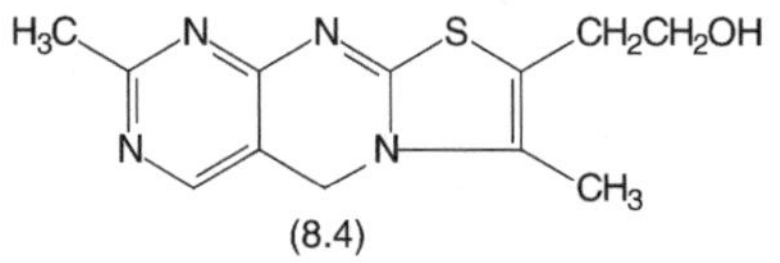
H3C
CH2CH2OH
CH3

(8.4)

Thiamin is one of the most labile of vitamins. Only under acid conditions, *i.e.* below pH 5, will it withstand heating. The reason for this is the readiness with which nucleophilic displacement reactions occur at the atom joining the two ring systems. Both OH^- and, more seriously sulfite, HSO_3^- split the molecule in two (8.5):

$$(8.5)$$

Thus sulfite added to fruits and vegetables to prevent browning will cause total destruction of the thiamin. Fortunately these foods are not important sources of dietary thiamin. Heat treatments such as canning, especially when the pH is above 6, cause losses of up to 20% of the thiamin, but only in the baking of bread, where up to 30% may be lost, are these losses really important. The greatest losses of thiamin, in domestic cooking as well as commercial food processing, occur simply as a result of its water solubility rather than any chemical subtleties. It is impossible to generalise on the extent of such losses, depending as they do on the amount of chopping up, soaking, and cooking times *etc.*, but they can be minimised by making as much use as possible of meat drippings and vegetable cooking water – perhaps gravy is nutritious after all!

Polishing rice, *i.e.* removing the outer layers of the grain and the germ, enhances its keeping properties by denying some the nutrients from invading moulds. Although this practice is a major cause of the thiamin deficiency that afflicts those parts of the world where rice is a staple the difficulties of storing rice, especially in warm, humid climates, make it essential. One way in which its effects can be minimised is by 'parboiling', or partial boiling. The rice is steeped in water, steamed, and then dried. Ostensibly this is done to make polishing easier but in fact it allows the thiamin to diffuse from the germ into the endosperm so that it is not then lost when the outer layers are removed. Niacin's poor solubility means that parboiling is of no benefit with regard to this vitamin.

The question of how much of any particular vitamin is required to maintain health in various circumstances is much better discussed in the more physiological context offered by textbooks of nutrition, but thiamin is a somewhat unusual case. Its involvement in carbohydrate metabolism has already been referred to, and it is recognised nowadays that the evaluation of the thiamin content of a single staple foodstuff should have regard for the energy that it also contributes. Authorities recommend a minimum thiamin intake of 96 μg MJ^{-1} (*i.e.* 0.4 mg kcal^{-1}) for all humans. Armed with this figure, it can be easily demonstrated that vegetarian diets dominated by refined cereals (from which the germ has been removed) will automatically result in thiamin deficiency unless they are supplemented with some whole-cereal products or legumes.

RIBOFLAVIN (VITAMIN B₂)

The structure of this vitamin is usually presented as that of riboflavin itself (8.6), the isoalloxazine nucleus with just a ribitol side chain attached, but in most biological materials it occurs predominantly in the form of two nucleotides, flavin mononucleotide, FMN (8.7), and flavin-adenine dinucleotide, FAD (8.8). These occur both as prosthetic groups in the group of respiratory enzymes known as flavoproteins and also free, when they are referred to as coenzymes.

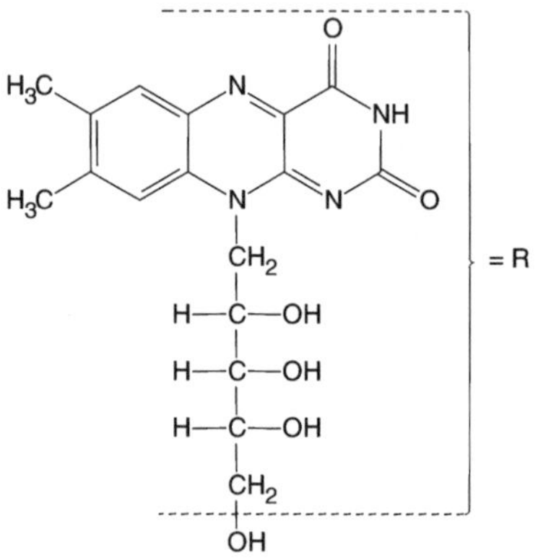

(8.6)

R—OPO₃H₂

(8.7)

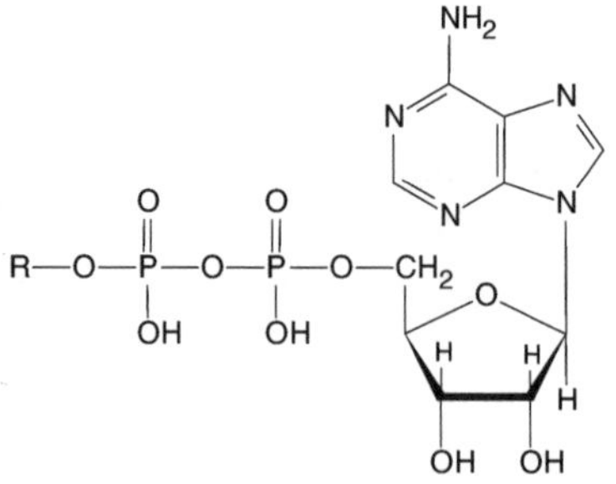

(8.8)

The distribution of riboflavin in foodstuffs is very similar to that of thiamin, in that it is found at least in small amounts in almost all biological tissues, and is particularly abundant in meat (0.2 mg per 100 g), especially liver (3.0 mg per 100 g). In contrast to the case of thiamin, cereals are not a particularly rich source, but milk (0.15 mg per 100 g) and cheese (0.5 mg per 100 g) are valuable sources. Dried brewer's yeast and yeast extracts contain large amounts of riboflavin and many other vitamins, but it is interesting that the thiamin does not leak out of the yeast during fermentation whereas the riboflavin does (to give a concentration in beer up to 0.04 mg per 100 g).

The flavoproteins of which riboflavin forms part of the prosthetic group are all involved in oxidation/reduction reactions. It is the isoalloxazine nucleus that receives electrons from substrates that are oxidised and donates electrons to substrates that are reduced (8.9):

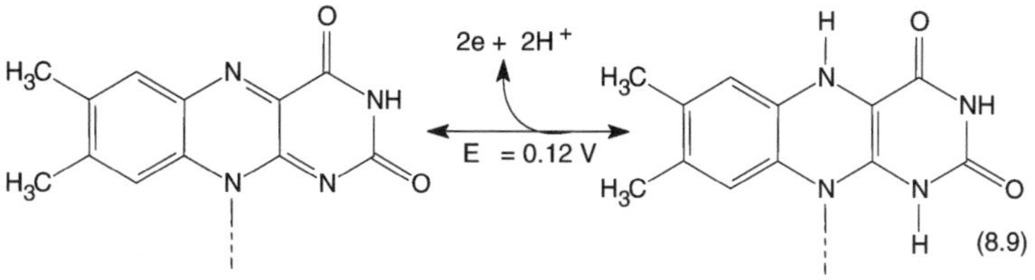

Most flavoprotein enzymes are involved in the complex respiratory processes that occur in the mitochondria of living cells, but some are involved in other aspects of metabolism. The enzyme glucose oxidase, which was mentioned in Chapter 2, is one example. This enzyme transfers two hydrogen atoms from C-1 of the glucose to a molecule of oxygen *via* the FAD prosthetic group, resulting in the formation of hydrogen peroxide.

Riboflavin is one of the most stable vitamins. The alkaline conditions in which it is unstable are rarely encountered in foodstuffs. The most important aspect of riboflavin's stability is its sensitivity to light. While this is not significant in opaque foods such as meat, in milk the effect can be dramatic. It has been reported that as much as 50% of the riboflavin of milk contained in the usual glass bottle may be destroyed by two hours' exposure to bright sunlight. The principal product of the irradiation is known as lumichrome (8.10); at neutral and alkaline pH values some lumiflavin (8.11) is also produced. This breakdown of riboflavin has wider significance than the simple loss of vitamin activity. Both breakdown products are much stronger oxidising agents than riboflavin itself and catalyse massive destruction of ascorbic acid (vitamin C). Even a small drop in the riboflavin content of the milk can lead to a near-total elimination of the ascorbic acid content. Fortunately milk is not an important source of vitamin C in most diets. The other important side effect of riboflavin breakdown is that in the course of the reaction, which involves interaction with oxygen, highly reactive states of the oxygen molecule are produced, in particular singlet oxygen. Although these have only a transient existence, they are able to initiate the autoxidation of unsaturated fatty acids in the milk fat, as discussed in Chapter 4. The result is an unpleasant 'off' flavour. When milk is to be sold in supermarkets, whose chilled cabinets are often brightly lit with fluorescent lamps, it is obviously essential that opaque containers rather than the traditional glass bottles are used.

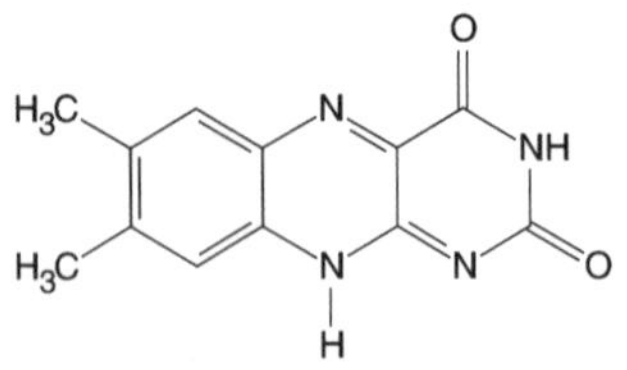

(8.10) 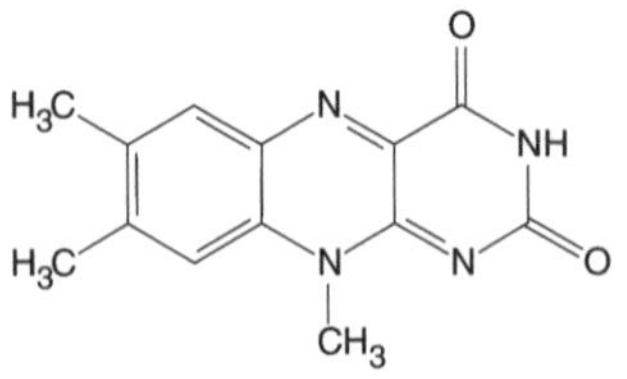   (8.11)

As vegetables are not generally important as sources of riboflavin, losses by leaching during blanching or boiling in water are not significant. Other cooking operations, including meat roasting and cereal baking, have negligible effects.

Riboflavin is not readily determined in the laboratory unless one has access to HPLC. The most commonly applied chemical technique requires an initial treatment with dilute HCl at high temperatures to liberate the riboflavin from the proteins to which it is normally bound. After clean-up procedures the extract is examined fluorimetrically. The absorbance by riboflavin at 440 nm is accompanied by an emission at 525 nm.

Until HPLC instrumentation became widely available laboratories that routinely determined the levels of many different vitamins in food frequently used microbiological assays. These assays depend on the existence of special strains of bacteria that have lost the capability to synthesise one particular vitamin for themselves. These strains are obtained by subjecting more normal (*i.e.* wild-type) members of the species to irradiation or mutagenic chemicals and isolating mutants that have the required characteristics. Specially prepared growth media providing all the nutrients the test strain required except the vitamin to be determined were used. The amount of bacterial growth in a culture using this medium depends on how much of the vitamin is added, either from a food sample extract or from a standard solution used for comparison. Although they were difficult to carry out their sensitivity, in experienced hands, was very impressive. The assay for riboflavin using a strain of *Lactobacillus casei* covers the range 0–200 ng per sample, but this is one of the least sensitive. The vitamin B_{12} assay using *Lactobacillus leichmannii* has an assay range of 0–0.2 ng per sample!

PYRIDOXINE (VITAMIN B$_6$, PYRIDOXOL)

The form of this vitamin that is active in the tissues, again as the prosthetic group of a number of enzymes, is pyridoxal phosphate (8.12). However, when the vitamin is identified in foodstuffs, it is in one of three forms, all having lost the phosphate group: pyridoxine (also more correctly pyridoxol) (8.13), pyridoxal, and pyridoxamine (8.14). In foodstuffs of plant origin the first two of these usually predominate whereas in animal materials it is the last two that predominate. At least two-thirds of the vitamin in living tissues is probably present as enzyme prosthetic groups, *i.e.* tightly bound to protein. In muscle tissue one particular enzyme, phosphorylase, has a major share of the total. Milk is one exception, only about 10% being protein bound.

(8.12)   (8.13)   (8.14)

Pyridoxine (this name is used to describe all the active forms of the vitamin besides pyridoxol) is considered to be widely distributed amongst foodstuffs, but the problems posed by its analysis are such that really reliable data are scarce. It has been found in at least small amounts (around 10 μg per 100 g) in almost all biological materials in which it has been sought. Meat and other animal tissues, egg yolk, and wheatgerm are particularly rich sources (around 500 μg per 100 g); milk and cheese have about 50 μg per 100 g).

Compared with other vitamins/prosthetic groups, pyridoxal phosphate is quite versatile in terms of the types of reactions in which it is involved. Almost all the enzyme-catalysed reactions in which it participates have amino acids as substrates; phosphorylase is one exception. The reaction sequence of a transamination, a typical pyridoxal phosphate-requiring reaction, is shown in Figure 8.2.

The question of the stability of pyridoxine during food processing is complicated by the tendency of the different forms to differ in stability and also to interconvert during some processing operations. Pyridoxol is very stable to heat within the pH range usually encountered in foodstuffs. The other two forms are a little less stable, but, as with the other water-soluble vitamins, leaching is the major cause of losses during cooking and processing. Although there is little loss of vitamin activity during milk processing, the more extreme methods, such as the production of

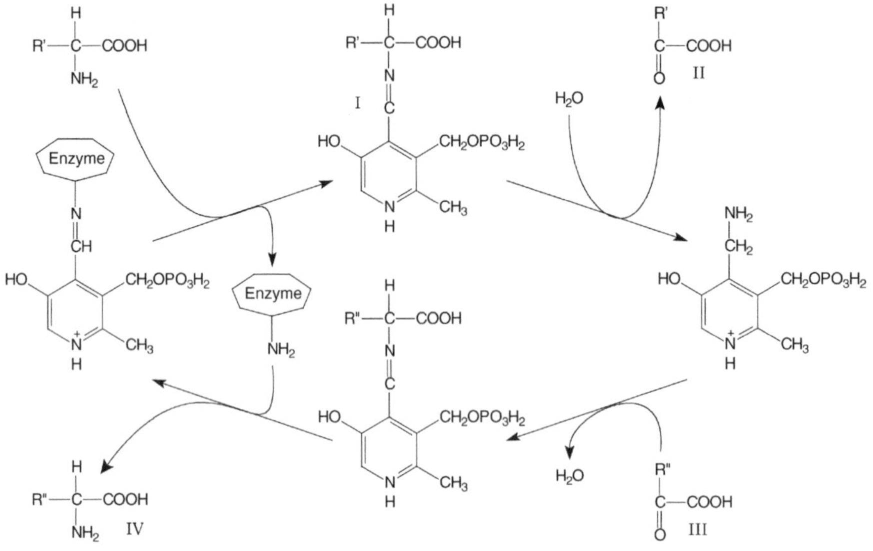

Figure 8.2   *The mechanism of a transamination reaction. The ε-amino group of a lysine in the active site of the enzymes carries the pyridoxal phosphate. This is displaced by the incoming amino acid to form a Schiff base (I). The hydrolysis of the weakened N–C bond then liberates the α-keto acid (II). The sequence is then reversed: as the second α-keto acid (III) enters, the second amino acid (IV) is formed. The result is an overall reaction such as:*

$$\textit{Glutamate} + \textit{oxaloacetate} \rightleftharpoons \alpha\textit{-ketoglutarate} + \textit{aspartate}$$

evaporated milk, cause extensive conversion of pyridoxal, the predominant form in raw milk, into pyridoxamine. An identical change occurs during the boiling of ham. When milk is dried, extensive losses of pyridoxal can occur because of interactions with sulfhydryl groups of proteins. Interaction with free amino groups of proteins leads to pyridoxamine formation – which does not cause loss of vitamin activity.

The analysis of pyridoxine levels in foods is not practical using chemical analytical techniques, and therefore HPLC or microbiological methods must be used. A special source of difficulty with microbiological assays for the vitamin is that the different forms of the vitamin differ in potency for micro-organisms, although apparently they do not differ where animal nutrition is concerned.

NIACIN (NICOTINIC ACID AND NICOTINAMIDE)

Niacin is the collective name given to nicotinic acid (8.15) and its amide, nicotinamide (8.16). Niacin is regarded as a member of the B-vitamin group (as are all the water-soluble vitamins except ascorbic acid), but it's occasional designation as vitamin B_3 is now obsolete. In living systems the pyridine ring occurs as a component of nicotinamide adenine dinucleotide, NAD* (8.17), and its phosphate derivative, NADP* (8.18). NAD does occur as a prosthetic group in a few enzymes such as glyceraldehyde-3-phosphate dehydrogenase, but most of the NAD and NADP in living cells functions as an electron carrier in respiratory systems.

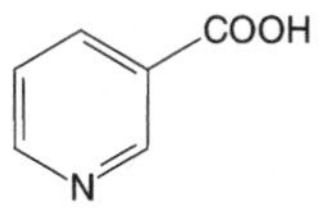

(8.15) 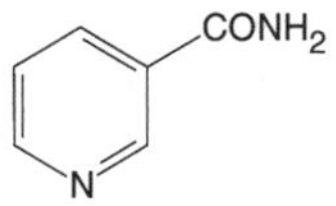   (8.16)

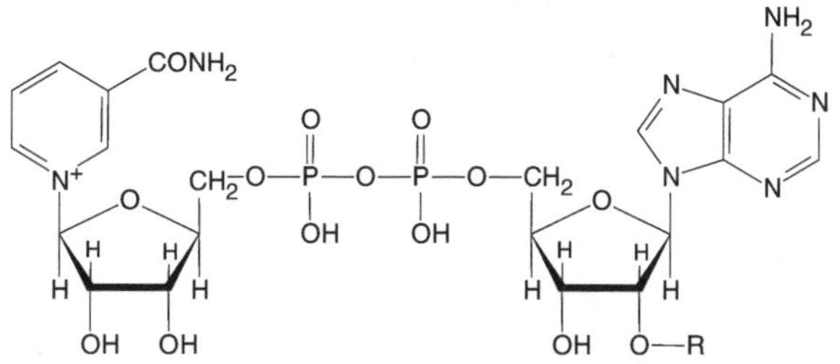

(8.17) R = H
(8.18) R = PO$_3$H$_2$

For example, in most of the Krebs cycle reactions (the key metabolic pathway of respiration) the electrons removed from the organic acid substrates are transferred initially to NAD (8.19). From NAD these are then used to reduce oxygen to water in the oxidative phosphorylation system of the mitochondrion. The route from NAD involves first flavoproteins and then porphyrin ring-containing proteins called cyto-chromes. The most striking feature of the reduction of NAD, is the very high absorbance of NADH at 340 nm (the molar absorption coefficient, $\varepsilon = 6.22 \times 10^6 \text{ dm}^3 \text{ mol}^{-1} \text{ cm}^{-1}$), a wavelength where the oxidised form has almost no absorbance at all.

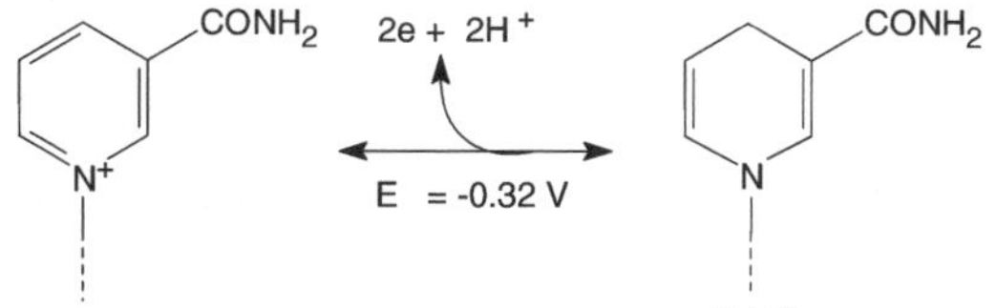

2e + 2H$^+$

E = -0.32 V

(8.19)

* In older literature these were referred to as di- and tri-phosphopyridine nucleotide, DPN and TPN, respectively.

In general the distribution of niacin in foodstuffs is similar to that of the other vitamins that have been discussed so far in this chapter. Although present in small amounts in all biological materials, it is meat (with 5–15 mg per 100 g) that is the richest source in most diets. Dairy products, milk, cheese, and eggs, are surprisingly poor sources of this vitamin, having no more than 0.1 mg per 100 g). Fruit and vegetables, particularly legume seeds, are quite useful sources (0.7–2.0 mg per 100 g). Although many cereals have apparently high niacin contents (*e.g.* whole-wheat flour 5 mg and brown rice 4.7 mg per 100 g), milling can have drastic effects as a very large proportion of the vitamin is located in the germ; white bread and wholemeal bread contain 1.7 and 4.1 mg per 100 g respectively.

To compound this problem much of the niacin in cereals is tightly bound to other macromolecules in the grain. In wheat, rice and maize a small protein has been implicated and in wheat and maize a polysaccharide, probably a hemicellulose, is also involved. The normal digestive processes in the small intestine do not liberate the vitamin and it remains unabsorbed. This is problem is especially important for maize consumers. There is no record of *pellegra*, the disease caused by a deficiency of niacin, before 1753 but from this time it was increasingly found in southern Europe. This increase can now be associated with the replacement of rye with maize as a dietary staple but at the time the connection was not made. It was endemic in the USA, especially the south. Its absence from the Yucatan peninsular, Mexico, where maize was first cultivated, was never noticed and even when the link to maize was spotted it was assumed that a microbial toxin in spoilt maize was to blame. The reason for the Mexican immunity to pellegra is that in Mexico the maize flour that is used for making *tortillas*, which are the dietary equivalent of bread, was traditionally ground from maize kernels that had first been treated with lime water. The apparent effect of the alkali was to weaken the husk of the maize kernel to make milling easier but we now know that under alkaline conditions bound unavailable niacin is liberated by heating! Consumers of 'corn chips' and other Westernised versions of tortillas will be reassured to learn that alkali treatment of maize remains a key step in their manufacture.

A further complication in the assessment of a foodstuff for its niacin content is that the human body has a limited, but significant, ability to synthesise the vitamin for itself. By a rather tortuous metabolic pathway the amino acid tryptophan (*see* Figure 5.2) can be converted into nicotinic acid. The effectiveness of this route is inevitably low owing to the competing demands of protein synthesis. Experiments suggest that about 60 mg of dietary tryptophan is required to replace 1 mg of dietary nicotinic acid. This provides the explanation for the beneficial effect of milk and eggs in cases of pellagra, even though both are poor sources of the actual vitamin. This also illustrates the care needed in the interpretation of food composition tables when diets are being assessed.

Niacin is stable under most cooking and processing conditions, leaching being the only significant cause of losses. During the baking of cereals, especially in products made slightly alkaline by the use of baking powder, the availability of niacin may actually increase.

COBALAMIN (CYANOCOBALAMIN, VITAMIN B₁₂)

The chemical structure of this vitamin is by far the most complex of all the vitamins. As shown in Figure 8.3, its most prominent feature is a 'corrin' ring, which resembles a porphyrin ring system but differs in two respects. The ring substituents are mostly amides, and the coordinated metal atom is cobalt. The fifth coordination position is occupied by a nucleotide-like structure based on 5,6-dimethylbenzimidazole, which is also attached to one of the corrin ring amide groups. The sixth

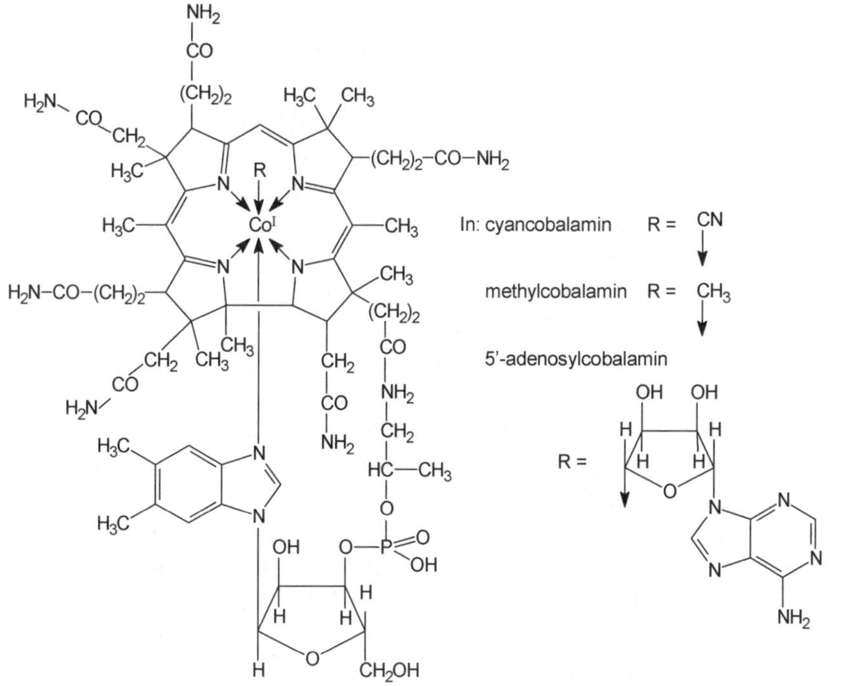

Figure 8.3 *The structure of cobalamin and its derivatives. In the absence of these or other ligands (i.e. R) a water molecule occupies the sixth coordination position.*

coordination position (R in Figure 8.3), on the face of the corrin opposite the fifth, can be occupied by a number of different groups. As normally isolated the vitamin contains a CN group in the sixth position – hence the name cyanocobalamin. In its functional state, when it is the prosthetic group of a number of enzymes, it may have either a methyl group or a 5'-deoxyadenosyl residue in the sixth position. The coordination of the methyl and 5'-deoxyadenosyl residues is unusual in that it is a carbon atom (the 5' in the case of 5'-deoxyadenosyl) that is linked to the cobalt atom.

The distribution of cobalamin in foodstuffs is most unusual. Bacteria are the only organisms capable of its synthesis, and it is therefore never found in higher plants. When traces are detected on vegetables, the source is almost certainly surface contamination with faecal material that has been used as fertiliser. The bacterial flora of the colon secrete about 5 μg daily, but since there is no absorption from the colon this is wastefully excreted. The daily requirement is about 1 μg. The vitamin occurs in useful quantities in most foodstuffs of animal origin [per 100 g: eggs 0.7 μg, milk 0.3 μg, meat (muscle tissue) 1–2 μg, liver 40 μg, kidney 20 μg], but in all these cases the vitamin got there either from micro-organisms or animal tissues consumed in the diet or from synthesis by intestinal micro-organisms high enough in the alimentary tract for absorption to occur.

As a result of the low requirement for this vitamin and its abundance in many foodstuffs, simple deficiency is extremely rare except amongst dietary extremists such as vegans. It has even been suggested that the slowness of the onset of deficiency symptoms in the young children of vegan parents is caused by the lack of toilet hygiene that is inevitable in young children, providing a source of the vitamin from unwashed hands. Unfortunately for this picturesque (?) explanation the observation that it may take as long as 10 years after ceasing consumption of the vitamin for an otherwise healthy individual to show deficiency symptoms, since the vitamin is lost from the body at an extremely slow rate, is probably a better indicator of the truth. The classical disease of apparent cobalamin deficiency is pernicious anaemia. This is caused by a failure to absorb the vitamin rather than a lack of it in the diet. Patients with pernicious anaemia are unable to synthesise a carbohydrate-rich protein known as the Intrinsic Factor. This is normally secreted by the gastric mucosa and complexes the vitamin, facilitating its absorption in the small intestine.

The catalytic activities of cobalamin enzymes are almost all associated with one-carbon units such as methyl groups, an area of metabolism shared in bacteria and animals with folic acid-containing enzymes and taken over by them in plants. The role of cobalamin enzymes in methane synthesis by bacteria has a superficial similarity to that of Grignard reagents. An important example of the synthetic dexterity of cobalamin-containing enzymes is the isomerisation of methylmalonyl coenzyme A to succinyl coenzyme A (8.20).

$$O=C-S-CoA \qquad\qquad O=C-S-CoA$$
$$H-\underset{\underset{COOH}{|}}{\overset{|}{C}}-CH_3 \longrightarrow \underset{\underset{\underset{COOH}{|}}{CH_2}}{\overset{|}{CH_2}}$$

(8.20)

The relationship between the known biochemical functions of cobalamin and the clinical manifestations of its deficiency, a drastically reduced synthesis of the haem moiety of haemoglobin, remains obscure.

In spite of the complexity of its structure cobalamin is fairly stable to food processing and cooking conditions, and what losses do occur are not sufficient to cause concern to nutritionists. As with all the water-soluble vitamins, leaching is the major cause of loss. Microbiological assay (*see* page 270) is the only practical method of determining the vitamin in foodstuffs.

FOLIC ACID (FOLACIN)

Folic acid is the name given to a closely related group of widely distributed compounds whose vitamin activity is related to that of the cobalamins. The structure of folic acid proper is shown in Figure 8.4.

The form that is active as a coenzyme is the tetrahydrofolate with the pteridine nucleus of the molecule in its reduced state and several (*n* from 3 to 7) glutamate residues carried on the p-aminobenzoate moiety. Accurate determinations of the folic acid content of foodstuffs are difficult, even

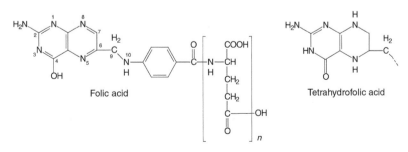

Figure 8.4 *The structures of folic acid and tetrahydrofolic acid.*

using microbiological assays, as it is necessary to remove all but one of the glutamyl residues before consistent results are obtained. This can now be achieved by enzymic methods, but it has meant that many earlier estimates of the folic acid content of foodstuffs were orders of magnitude too low. Only recently have measurements in terms of particular glutamyl/ folic acid conjugates been made. Of animal tissues, where the pentaglutamyl conjugate predominates, liver is a particularly good source (300 μg per 100 g); muscle tissues have much less (3–8 μg per 100 g). Leafy green vegetables, where the heptaglutamyl conjugate predominates, have around 20–80 μg per 100 g) and probably constitute the most important dietary source in modern diets.

As a coenzyme, tetrahydrofolate (FH$_4$) is concerned solely with the metabolism of one-carbon units, which it carries at N-5 or N-10. While bound to the coenzyme, the one-carbon unit may be oxidised or reduced or undergo other changes, giving the structures shown in Figure 8.5. These are intimately involved in the metabolism of many amino acids, purines, and pyrimidines (components of nucleotides and nucleic acids) and, indirectly, haem. In animals the metabolism of methyl groups, which must be supplied in the diet (mostly by choline, *see* page 302), is particularly complex. Both folic acid and cobalamin-requiring reactions are involved, so that it is not surprising that there is some overlap in the symptoms of deficiencies of these two vitamins.

There is growing support for the view that deficiency of folic acid is widespread, even in the comparatively well fed populations of Europe and North America. It is becoming clear that a deficiency of folic acid in the very early stages of pregnancy (technically the 'periconceptual period') leads to neural tube defects such as *spina bifida* in the foetus. Average

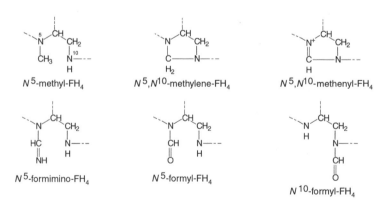

Figure 8.5   *One-carbon units carried by tetrahydrofolate (FH$_4$).*

diets may not supply sufficient folic acid and the UK authorities now recommend that vitamin supplements should be used to ensure that all young women consume 0.4 mg daily to ensure adequate folate reserves in the event of a planned, or unplanned, pregnancy. Intakes ten times this figure are recommended in cases where a neural tube defect occurred in a previous pregnancy. The importance of supplementation is increased by the observation that the bioavailability of naturally occurring folates is actually lower than that of folate supplements.

Routine dietary fortification is not recommended. This is because extra folic acid will alleviate, and therefore mask, the obvious anaemia of cobalamin deficiency. However, it has no effect on the accompanying neurological damage that cobalamin deficiency causes, which, by the time it is detected directly, cannot be reversed.

There is very little useful information on the stability of folic acid and its derivatives. The stability to heat and the acidity vary with both the number of glutamyl residues and the nature of any attached one-carbon units. Oxidation can be a problem, but ascorbic acid has a valuable protective effect. As with the other water-soluble vitamins, most cooking and processing losses are caused by leaching.

The fact that animals depend on their diet for their supply of folic acid makes possible the use of sulfonamides such as sulfanilamide (8.21) as antibacterial drugs. These substances interfere with the formation of the folic acid that the bacteria themselves require without having any great effect on folic acid metabolism in the patient.

$$H_2N-\langle\!\bigcirc\!\rangle-SO_2NH_2$$

(8.21)

BIOTIN AND PANTOTHENIC ACID

These two vitamins have little in common in terms of their structure or function. What brings them together is the total absence of reports of deficiency symptoms in man or animals except under the artificial conditions of the laboratory. Biotin (8.22) occurs as the prosthetic group of many enzymes catalysing carboxylation reactions, such as the conversion of pyruvate into oxaloacetate. Most foodstuffs contain at least a few micrograms per 100 grams (assayed microbiologically), and there appears to be a more than adequate amount in any otherwise satisfactory diet. The only known deficiency of biotin occurs when extremely large quantities of raw egg are eaten. Egg-white contains a protein, avidin, which complexes biotin. Avidin is denatured when eggs are cooked. Biotin is apparently quite stable to cooking and processing procedures.

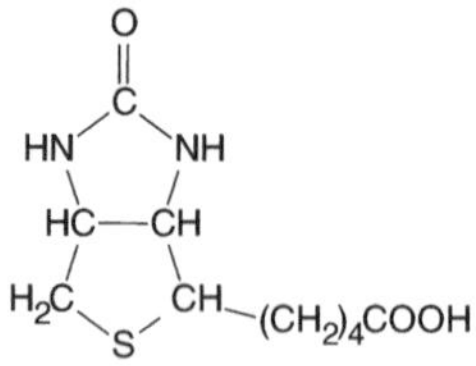

(8.22)

As with biotin, much more is known about the biochemical function of pantothenic acid (Figure 8.6) than about its status as a food component. It occurs in two forms in living systems, as part of the prosthetic group of the acyl carrier protein (ACP) component of the fatty-acid synthetase complex and as part of coenzyme A, the carrier of acetyl (ethanoyl) and

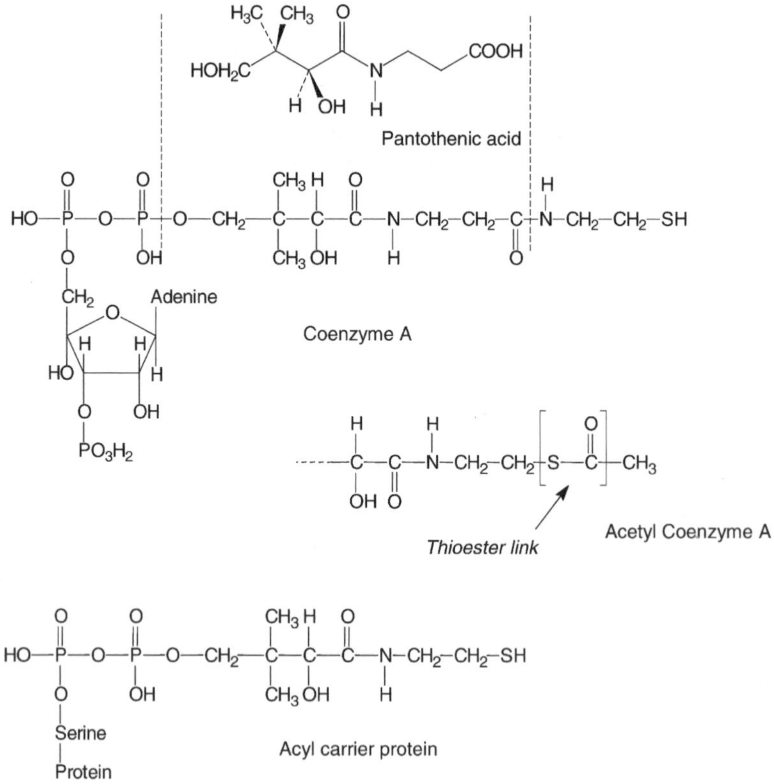

Figure 8.6 *Pantothenic acid and its relationships to coenzyme A and acyl carrier protein.*

other acyl groups in most metabolic systems. In both of these roles the acyl group is attached by a thioester link (*see* Figure 8.6) to the sulfhydryl group which replaces the carboxyl group of the free vitamin. Pantothenic acid is extremely widely distributed, mostly at around 0.2–0.5 mg per 100 g, levels generally assumed to give a more than adequate dietary intake. As with so many vitamins, liver contains especially high levels, in this case about 20 mg per 100 g. Quite high losses of pantothenic acid have been reported to occur during food processing, mostly due to leaching, but some decomposition is known to occur especially during heating under alkaline or acid conditions.

The term 'pro-vitamin' is used to describe substances that are direct precursors of vitamins so that their presence in the diet can substitute for the vitamin itself. The most obvious example is β-carotene, whose relationship to retinol, vitamin A, was described in Chapter 6 and will be given more attention later in this chapter. However, 'pro-vitamin B_5' is used as an ingredient in cosmetics and shampoos. For reasons apparently more connected with marketing than nutrition this name is used for panthenol (8.23) and its ethyl ester. While panthenol can be absorbed by the skin (at least the skin of laboratory rats) and converted into pantothenic acid that then appears in the urine, the relevance of this to the nutrition of the skin or the appearance of one's hair is unclear.

$$HO-CH_2-\overset{\overset{\displaystyle H}{|}}{\underset{\underset{\displaystyle OH}{|}}{C}}-\overset{\overset{\displaystyle O}{\|}}{C}-NH-CH_2-CH_2-CH_2OH$$

(8.23)

ASCORBIC ACID (VITAMIN C)

Ascorbic acid is a rather curious vitamin. It occurs widely in plant tissues and it is also synthesised by almost all mammals, and is not therefore a vitamin for them, the exceptions being the primates, the guinea-pig, and some fruit-eating bats. This latter group's only discernible common feature is a liking for fruit, a rich source of the vitamin. The structures of L-ascorbate (pK_a is 4.04 at 25 °C) and some related substances are shown in Figure 8.7. The oxidised form, dehydro-L-ascorbic acid (DHAA) (an unfortunate, if inevitable, name since the molecule has no ionisable groups), is distributed with the reduced form in small, rarely accurately determined, proportions. DHAA has virtually the same vitamin activity as L-ascorbate. The mirror image isomer, most commonly referred to as erythorbic acid, has no vitamin activity but does behave similarly to

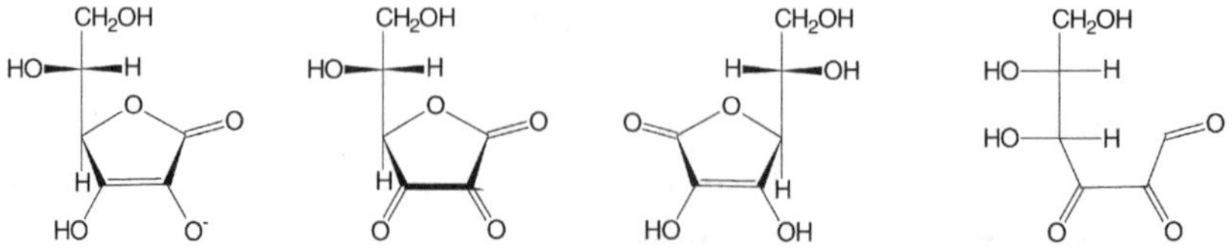

L-Ascorbate	Dehydro-L-ascorbic acid (DHASS)	D-Isoascorbic acid (Erythorbic acid)	2,3-Diketo-L-gulonic acid (DKGA)

Figure 8.7 *The structure of ascorbic acid and related substances.*

ascorbate in many food oxidation/reduction systems and therefore finds some similar applications as a food additive.

Much more is known about the distribution of ascorbic acid in foodstuffs than about any other vitamin. The reason is not that ascorbic acid is inherently more interesting or more important but simply because it is much easier to determine than any of the others, so that when a research worker or student wishes to do a project 'on vitamins' this is the one that usually gets chosen. The richest sources are fruits, but, as the data in Table 8.1 show, wide variations do occur. It has to be remembered that a foodstuff that is eaten in fairly large quantities will be an important dietary source even if it contains only modest levels of the vitamin. The very high levels to be found in parsley are nutritionally insignificant when compared with the quantity in the new potatoes it is used to garnish. Cereals contain only traces, and of the animal tissue foods only liver (30 mg per 100 g) is

Table 8.1 *Typical ascorbic acid contents of some fruit and vegetables. Values for particular samples may vary considerably from the these figures. The content of the edible part when fresh is shown*

	Ascorbic acid mg per 100 g		*Ascorbic acid mg per 100 g*
Plums	3	Carrots	6
Apples	6	Old potatoes	8
Peaches	7	Peas	25
Bananas	11	New potatoes	30
Pineapple	25	Cabbage	40
Tomatoes	25	Cauliflower	60
Citrus fruit	50	Broccoli	110
Strawberries	60	Horse-radish	120
Blackcurrants	200	Sweet red peppers	140
Rose-hips	1000	Parsley	150

a valuable source. Cow's milk and other dairy products contain insignificant amounts of the vitamin.

The role of ascorbic acid in mammalian physiology and biochemistry is
far from fully understood but is quite different from its role in plants. In
plants its function is as an antioxidant (*see* page 98), in particular helping
to protect the photosynthetic machinery of the chloroplasts against the
effects of singlet oxygen and other highly reactive oxygen species (*see*
page 92). Glutathione is also involved in this activity (*see* page 171). In
animals it is involved in hydroxylation reactions. It has been implicated in
a number of reactions, particularly in the formation of collagen, the
connective-tissue protein. Proline (*see* Figure 5.2) is hydroxylated to
hydroxyproline (after it has been incorporated into the collagen polypeptide chain) in a complex reaction which has ascorbic acid as one of the
essential components (8.24):

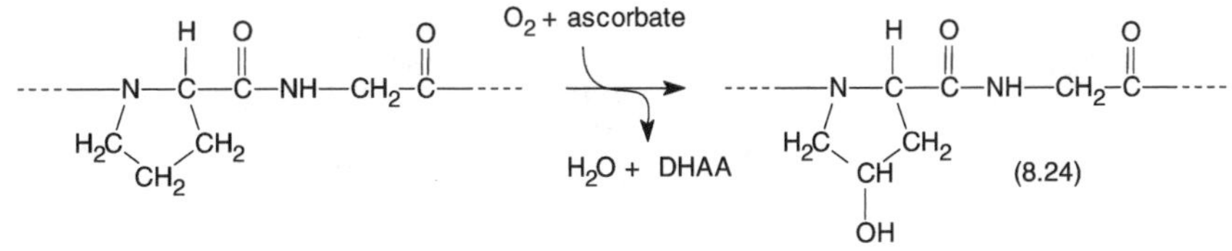

(8.24)

Most of the symptoms of scurvy, the ascorbic acid deficiency disease,
would be expected from a failure of normal connective-tissue formation.
For example, wounds fail to heal, internal bleeding occurs, and the joints
become painful. Another important reaction requiring ascorbate is the
hydroxylation of 3,4-dihydroxyphenylethylamine (dopamine, a derivative
of tyrosine) to noradrenaline (8.25):

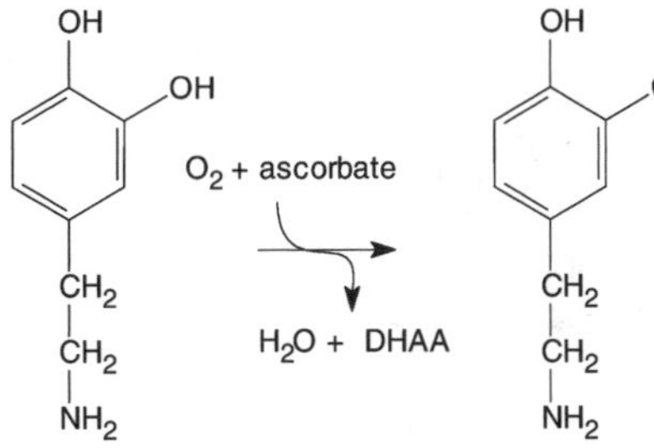

(8.25)

This may be the origin of the disturbances of the nervous system that
accompany scurvy. There is also increasing evidence that the uptake of
non-haem iron (*see* page 381) from the duodenum is enhanced by the
presence of high levels of ascorbic acid. Ascorbic acid reduces iron from

the ferric (Fe^{3+}) state in which it normally occurs in food to its ferrous (Fe^{2+}) state. Compared with ferrous iron very little ferric iron is absorbed as it forms its insoluble hydroxide. Unfortunately the amounts of additional ascorbic acid required to double iron uptake from a typical meal

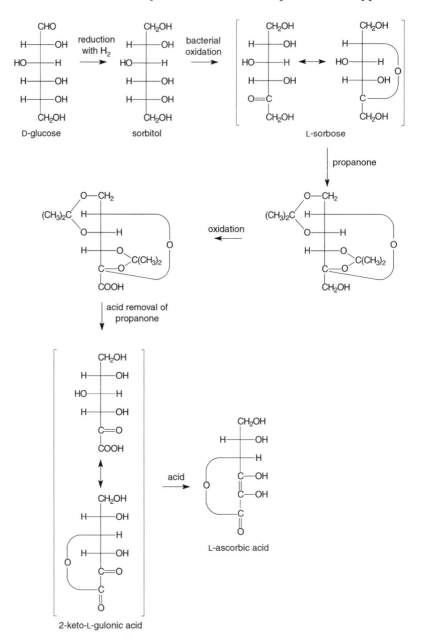

Figure 8.8 *The commercial synthesis of ascorbic acid from glucose.*

are quite high, around 100 mg. Ascorbic acid also has a role in the transport of iron in the blood and its mobilization from reserves.

The diversity of roles for ascorbic acid in human physiology and biochemistry is probably behind its adoption as a 'wonder cure-all' for numerous other conditions and diseases, particularly the common cold. The actual evidence for the value of extra-high doses (up to 5 grams per day, *i.e.* over 100 times the recommended dietary intake) is extremely flimsy, something that cannot be said for the profits to be made in selling ascorbic acid to 'health food' enthusiasts.

There are many different routes available for the chemical synthesis of ascorbic acid, but the most usual commercially adopted route is shown in Figure 8.8. This is an interesting combination of traditional synthetic chemistry with a biotechnology.

Almost all of the numerous methods that have been devised for the determination of ascorbic acid depend on measurements of the reducing property by titration with an oxidising agent, the most popular one being 2,6-dichlorophenolindophenol (DCPIP) (8.26). DCPIP is blue in its oxidised form, colourless when reduced. Highly coloured plant extracts such as blackcurrant juice obviously present special problems because the end-point of the titration is obscured. There are numerous, more or less satisfactory, techniques for circumventing the problem. One example is titration with *N*-bromosuccinimide in the presence of potassium iodide and ether. Excess *N*-bromosuccinimide liberates I_2 from the iodide, and the I_2 dissolves in the ether phase to give it a brown colour when the end-point is reached. Very few of the methods commonly used determine any DHAA present; one that does not discriminate between the two compounds is the fluorimetric method using *o*-phenylenediamine.

(8.26)

The information that has been gained by these studies shows that ascorbic acid is one of the least stable vitamins. Losses during processing, cooking, or storage can occur by a number of different routes, but unfortunately very few experiments have been devoted to discovering which routes are involved in particular foodstuffs. Leaching is obviously important in vegetable preparation when the cooling or processing water is discarded. Green vegetables may lose more than 50% of the vitamin in this

Box 8.1   Further Details of the Role of Ascorbic Acid

While a full treatment of the reactions in which ascorbic acid participates are the proper concern of biochemists it is instructive for food chemists to have a systematic overview of the full range of this vitamin's potential activities.

Ascorbic acid is an effective scavenger of the various free radical forms of oxygen that arise in the tissues, notably as by-products of oxygen metabolism or the autoxidation of unsaturated fatty acids (*see* Chapter 4). As shown in Figure 8.9 the free radicals' unpaired electrons are, in effect, transferred to ascorbic acid. Ascorbic acid is then regenerated in reactions with other reducing agents such as glutathione or NADH (*see* page 172).

Ascorbic acid is an important cofactor in many reactions catalysed by

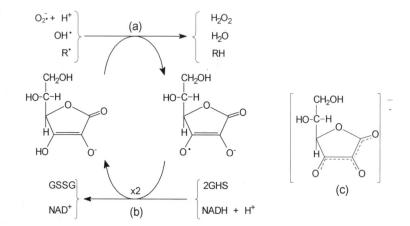

Figure 8.9   *The reactions of ascorbic acid with free radicals (a) and regeneration by reducing agents such as reduced glutathione (GSH) or reduced nicotinamide dinucleotide (NADH + H⁺) (b). The stability of the ascorbic acid free radical is the result of extensive delocalisation as indicated in the alternative structure shown (c).*

way if they are boiled for prolonged periods. Losses from root vegetables are usually much smaller simply because they present a smaller surface area in relation to their weight.

In the presence of air most actual degradation of the vitamin occurs through the formation of the less stable DHAA. Once formed, DHAA rapidly undergoes an irreversible ring-opening reaction to form 2,3-diketo-L-gulonic acid (DKGA), which has no vitamin activity. The oxidation to DHAA can occur by various mechanisms. In plant tissues stored without blanching, especially if they have been sliced, peeled, or otherwise *oxygenases*, *i.e.* reactions where an oxygen atom, in the form of a hydroxyl group, is incorporated into a substrate. In monooxygenase reactions, shown in Figure 8.10, one of the two oxygen atoms emerges from the reaction as water. However, in dioxygenase reactions the second oxygen atom is utilised in the oxidation of the auxiliary substrate, α-ketoglutarate, to succinate:

$$R\text{–}H + O_2 + \alpha\text{-ketoglutarate} + \text{ascorbate} \rightarrow R\text{–}OH + \text{succinate} + \text{DHAA}$$

The hydroxylation of proline side chains in collagen (8.24) is a dioxygenase reaction whereas the conversion of dopamine into noradrenaline (8.25) is a monooxygenase reaction. Two essential stages in the biosynthesis of carnitine (8.27) are catalysed by dioxygenases. Carnitine is an essential component of the system for transporting long chain fatty acids across mitochondrial membranes and its depletion in vitamin C deficiency may well be the basis of some of the symptoms of scurvy.

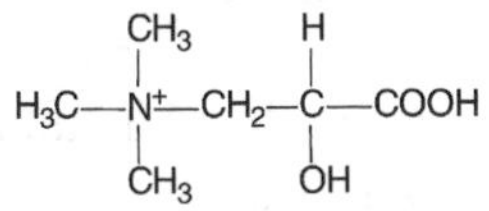

$$H_3C\text{—}\overset{\overset{\displaystyle CH_3}{|}}{\underset{\underset{\displaystyle CH_3}{|}}{N^+}}\text{—}CH_2\text{—}\overset{\overset{\displaystyle H}{|}}{\underset{\underset{\displaystyle OH}{|}}{C}}\text{—}COOH$$

(8.27)

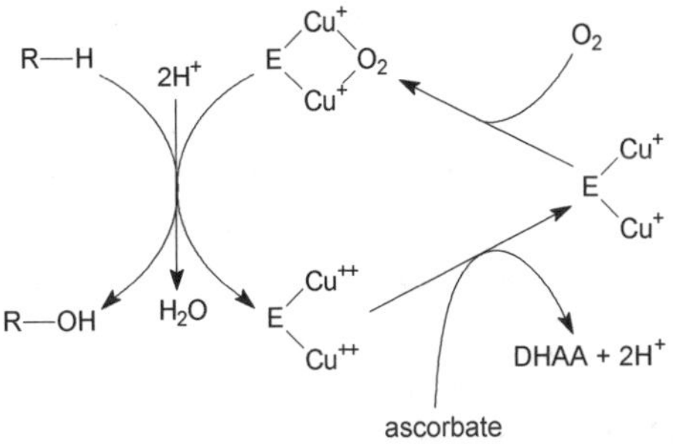

Figure 8.10 *The role of ascorbic acid and copper in monooxygenase reactions.*

damaged, the enzyme ascorbic acid oxidase will be very active, catalysing the reaction:

$$\text{ascorbate} + \tfrac{1}{2}O_2 \rightarrow \text{DHAA} + H_2O_2$$

The enzyme phenolase is also responsible for ascorbic acid losses when, as was discussed in Chapter 6, ascorbic acid reduces *o*-quinones back to the original *o*-diphenols. Metal cations, particularly Fe^{3+} and Cu^{2+}, will catalyse the oxidation and may cause serious losses in food products. Even in the absence of a catalyst the oxidation occurs quite rapidly at elevated temperatures. Formation of DKGA is virtually instantaneous at alkaline pH values, rapid around neutrality, and slow under acid conditions. The reason for the undesirability of adding sodium bicarbonate to green-vegetable cooking water to preserve the colour now becomes clear.

The vitamin is quite stable in the acid environment of fruit juice processing. The stability is enhanced by the presence of citrate and flavanoids, which both complex metal cations, but it is still necessary to keep juices de-aerated as much as possible. At elevated temperatures under anaerobic conditions ascorbic acid will also undergo breakdown along similar lines to other sugars (*see* Chapter 2) except that CO_2 is evolved. However, other details of the pathway remain the subject of speculation.

The low cost of ascorbic acid and its obvious acceptability as a nutrient make it a valued food additive for technological rather than nutritional purposes. In fruit and vegetable products, notably dehydrated potato, it is used as an antioxidant to prevent the browning reactions that would be catalysed by phenolase. In cured-meat products it is used as a reducing agent to lower the concentration of nitrite needed for a good pink colour (*see* page 158), and in modern breadmaking it is used as a flour improver (*see* page 171).

At the beginning of this chapter the author disclaimed responsibility for describing how each of the vitamins was discovered. In spite of this disclaimer no apology is offered for the inclusion of the following few paragraphs which are a useful reminder that scientists have not always taken themselves so seriously as they do today.

Chemists who have, in exasperation, referred to an unidentifiable sugar on their chromatograms as '@*!-knows' may not realise that this terminology has an honourable history. In 1928 Albert Szent-Györgyi set out to publish the details of the sugar acid with strong reducing properties that he had isolated, initially from the adrenal cortex of cattle and later from orange juice and cabbage water. In the draft of his paper he suggested the name 'ignose' (Latin *ignorare*, not to know). This and a second suggestion of 'Godnose' were both rejected by the editor of *Biochemical Journal* on the grounds that they didn't publish jokes. Szent-Györgyi had to settle for 'hexuronic acid', a very poor third choice. His paper did hint at the possibility that this substance might be the antiscorbutic (*i.e.* scurvy preventing) factor already designated 'vitamin C', but this was not confirmed until 1932.

When he first suspected that it might be a vitamin Szent-Györgyi proposed the designation 'vitamin P' on the simple grounds that vitamins had not got up to 'P' at that time so that if he was wrong its elimination from the sequence would not be too disruptive.*

Once the structure was firmly established by Haworth in Birmingham, using the kilogram quantities Szent-Györgyi was able to isolate from a local vegetable, red pepper, the name 'ascorbic acid' was adopted in recognition of its antiscorbutic properties. Food chemists should forgive Szent-Györgyi's later remark that '... *vitamins were, to my mind, theoretically uninteresting. "Vitamin" means that one has to eat it. What one has to eat is the first concern of the chef, not the scientist.*'

RETINOL (VITAMIN A)

This vitamin is the first of the 'fat-soluble' vitamins to be considered in this chapter. The form that is active in mammalian tissues is the alcohol retinol (8.28), but in the diet most of the vitamin occurs as precursors, of which there are several.

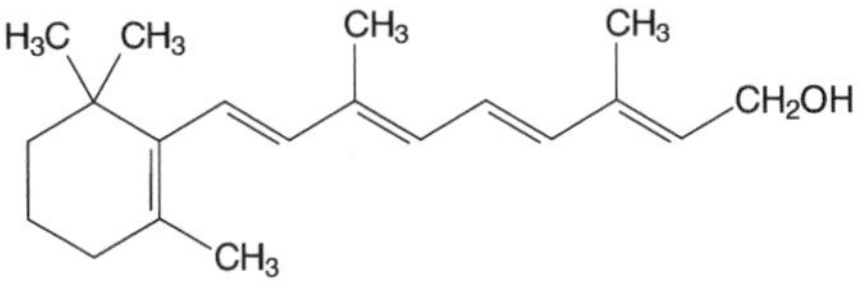

(8.28)

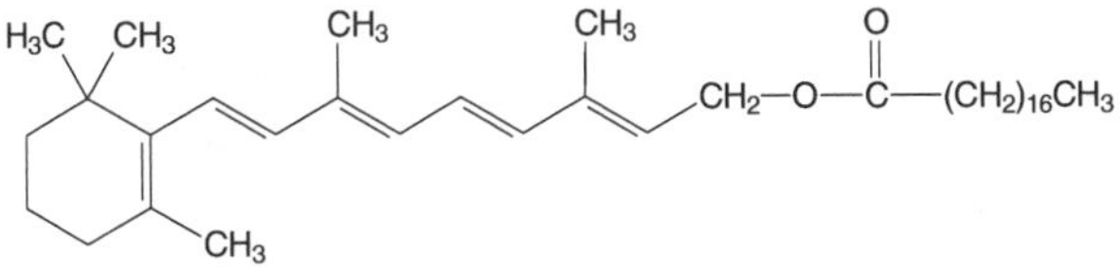

(8.29)

Plant tissues provide the vitamin in the form of carotenoids. As was indicated in Chapter 6, any carotenoid that contains the β-ionone ring system, as do β-carotene (at both ends of the molecule), α- and γ-carotene and β-apo-8′-carotenal (at one end only), is converted by enzymes in the mucosa of the small intestine into retinol. The absorption of carotenoids and their conversion into retinol do not occur with total or uniform efficiency, so that estimation of the vitamin activity of different foodstuffs is rather complicated. At the present time the equivalence figures recommended by the FAO/WHO[†] in 1967 are in general use, *i.e.* 6 μg of

* A. Szent-Györgyi, in 'Annual Reviews of Biochemistry', eds. E. E. Snell. J. M. Luck, P. D. Boyer and G. Mackinney, Annual Reviews Inc., Palo Alto, 1963.
[†] FAO = Food and Agriculture Organisation (of the United Nations), WHO = World Health Organisation.

β-carotene or 12 μg of other active carotenoids being regarded as the equivalent of 1 μg of retinol. It is increasingly recognised that this approach is over-simplified, firstly because the efficiency of conversion is inversely related to the level of β-carotene intake, giving a range of 4 to 10 μg equivalent to 1 μg of retinol. Allowance should also be made for the source of dietary carotenoids since less than 10% may be absorbed from raw vegetables whereas around 50% is absorbed from cooked vegetables.

Leafy green vegetables contain between 1000 and 3000 retinol equivalents (*i.e.* an amount equivalent to 1 μg of retinol per 100 g), mostly as β-carotene. Carrots, the classic source, contain about 2000 retinol equivalents per 100 g. The only other major plant source is red palm oil, whose β-carotene content is sufficient to give some 20 000 retinol equivalents (per 100 g).

In animal tissues the vitamin (always originally derived from plant sources) is stored and transported as the retinyl ester of long-chain fatty acids, mostly palmitate (8.29) and stearate. Dietary retinyl esters are first hydrolysed and then re-esterified by the intestinal mucosal cells during the process of absorption. A specific protein binds the retinyl esters in the liver with the result that liver and liver oils are particularly good sources in the diet. Cod liver oil and halibut liver oil are well known for their vitamin A contents, around 10^5 and 10^7 retinol equivalents per 100 g, respectively. The vitamin A content of mammalian liver varies widely, from 3000 in pig's liver and 17 000 in sheep's liver to 600 000 retinol equivalents per 100 g in polar bear liver! The relationship between the diet of these animals and the vitamin A content of their livers is obvious. Meat contains very little of this vitamin. Although milk itself has only a low concentration of the vitamin, dairy products in which the lipid phase of the milk is concentrated, such as butter and cheese, contain 500–1000 retinol equivalents per 100 g. Margarine sold in Britain is required by law to be supplemented with vitamin A, usually in the form of synthetic retinyl acetate or red palm oil, to give a level of 800–1000 retinol equivalents per 100 g, comparable with butter.

The chemical synthesis of vitamin A was first achieved in 1947. In the form of retinyl acetate it is now produced commercially on a large scale in a simple process using β-ionone as the starting material. As one might expect from its chemical structure, ensuring the resistance of vitamin A preparations to autoxidation, especially in the presence of trace metals, is a major problem.

The involvement of vitamin A in the visual process is well known and the sequence of events in the 'visual cycle' of the rod cells in the retina is shown in Figure 8.11. The inevitable inefficiency of the cycle results in

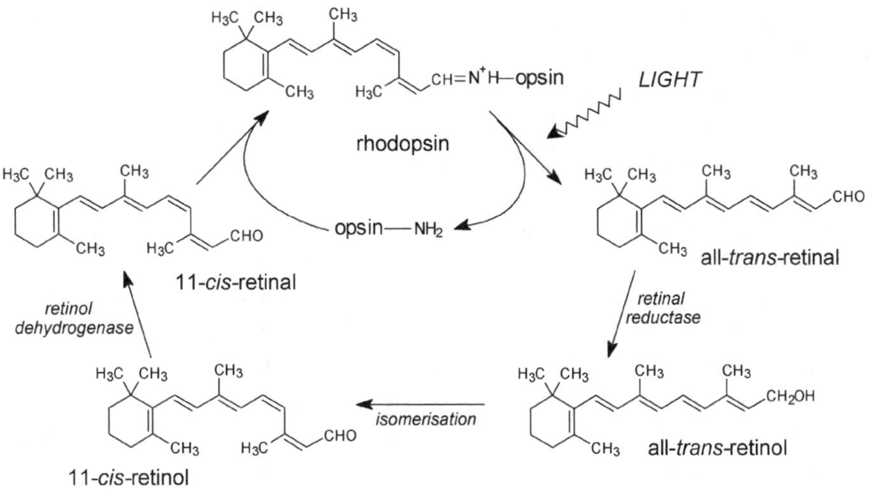

Figure 8.11 *Vitamin A in the visual cycle. In the dark 11-cis-retinal combines with the protein opsin to form rhodopsin. When this absorbs light (λ_{max} 500 nm) the retinal isomerises to the all-trans configuration and detaches from the protein. The 11-cis-retinal is then regenerated to recombine with opsin in darkness. One stage in the complex process by which rhodopsin is converted into opsin (not shown in detail here) is linked to the initiation of a nerve impulse.*

losses and the supply of all-*trans*-retinol is replenished from the circulation. The absence of adequate supplies results in loss of vision in dim light, so-called 'night-blindness'. Night blindness was linked to a deficiency of this vitamin as early as 1925.

Deficiency of the vitamin leads ultimately to death due to effects quite unconnected with night blindness. These include xerophthalmia, a drying and degenerative condition of the cornea of the eye which, untreated, will cause total blindness, abnormal bone development, and disorders of the reproductive system. All these effects concern epithelial cells but the mechanism's underlying vitamin A's involvement is unclear. There are indications of a role in the synthesis of the glycoproteins that occur in mucous membranes; deficiency is known to lead to keratinisation of these membranes. This can lead to enhanced susceptibility to infections following the degeneration of the epithelia that line the digestive and respiratory tracts.

Only since the 1960s has it been realised that vitamin A deficiency afflicts millions of people in the developing countries of Asia. In these areas neither dairy products nor green vegetables feature much in the diet of the poor, and a variety of methods are being used to supplement diets with synthetic retinyl acetate.

Considerable attention has been focused in recent years on the possibility that vitamin A has an anticancer function. Laboratory experiments with unnaturally high doses of the vitamin have shown some promising effects, but it has not proved possible to put these to therapeutic effect. Large doses simply lead to increases in the liver reserves and to the dangers of hypervitaminosis but no increases in the target tissues. The beneficial effect of vitamin A in this context is linked to its antioxidant activity. Both retinol and β-carotene are weak antioxidants but β-carotene also has a specific reaction with singlet oxygen. This quenching reaction may help to eliminate this dangerous by-product of some of the body's reaction with oxygen. There is growing support for the view that vitamin A, vitamin E (*see* page 296), selenium (as a component of the enzyme glutathione peroxidase), and ascorbic acid function collectively as the body's antioxidant system and should therefore be considered collectively when the relationship between health and the vitamin content of diet is being examined.

An unusual feature of retinol is its toxicity in excess. Most victims of this hypervitaminosis have been 'health food' enthusiasts, but it has been reported that Polar explorers have also suffered following the ill-advised consumption of a polar bear. As the recommended daily intake for the vitamin is only about 750 retinol equivalents, one modest portion of the polar bear liver is equivalent to over two years' supply of retinol. Fortunately there is no risk of hypervitaminosis from excess β-carotene consumption: as the level in the diet increases, the efficiency of conversion into retinol falls.

Both retinyl esters and β-carotene are fairly stable in food products. Most information on stability concerns β-carotene. High temperatures in the absence of air, as in canning, for example, can cause isomerisation to neocarotenes (*see* Chapter 6), which only have vitamin activity if one end of the molecule remains unaffected. In the presence of oxygen, breakdown can be rapid, especially in dehydrated foods of which a large surface area is exposed to the atmosphere. The hydroperoxides that result from the autoxidation reactions of polyunsaturated fatty acids will bleach carotenoids and can be presumed to have a similar destructive effect on retinyl esters.

The determination of the amount of this vitamin in a foodstuff is complicated by the diversity of forms in which it occurs. An indication can be gained from the absorbance of a solvent extract at 328 nm, the absorption maximum of retinol. Elaborate procedures are available to compensate for the interfering absorbance of other substances at this wavelength, but nowadays this problem is usually avoided by preliminary column chromatography to separate and identify the different carotenoids and retinyl esters.

CHOLECALCIFEROL (VITAMIN D, CALCIFEROL)

Only one form of this vitamin, known either as cholecalciferol (8.30) or vitamin D$_3$, occurs naturally in the diet. Another form, ergocalciferol (8.31) (vitamin D$_2$), is a form of the vitamin used for pharmaceutical applications and for food supplementation since it can be readily synthesised on an industrial scale from ergosterol (8.32), a cheap starting material.

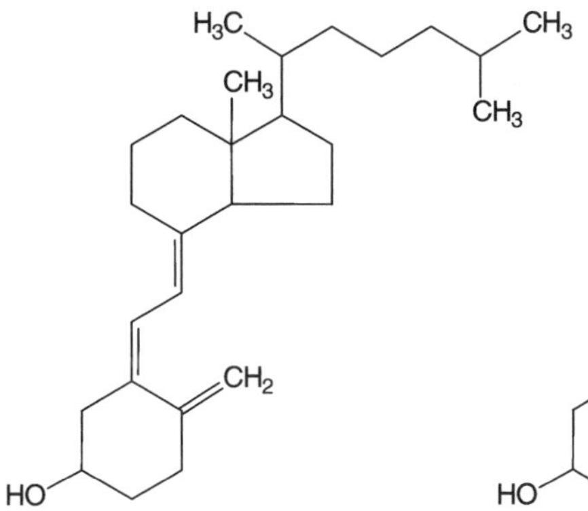

(8.30) 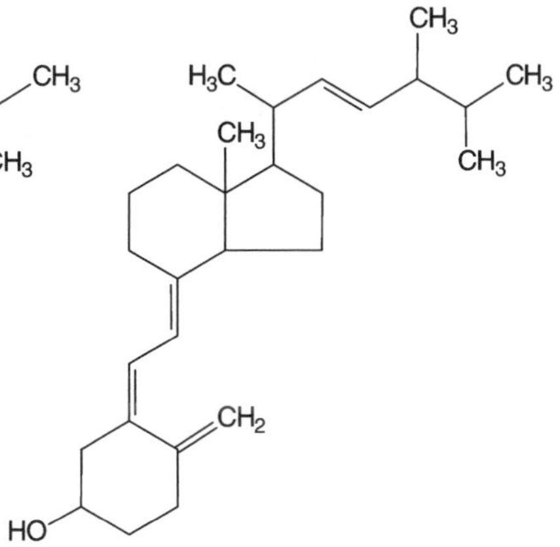 (8.31)

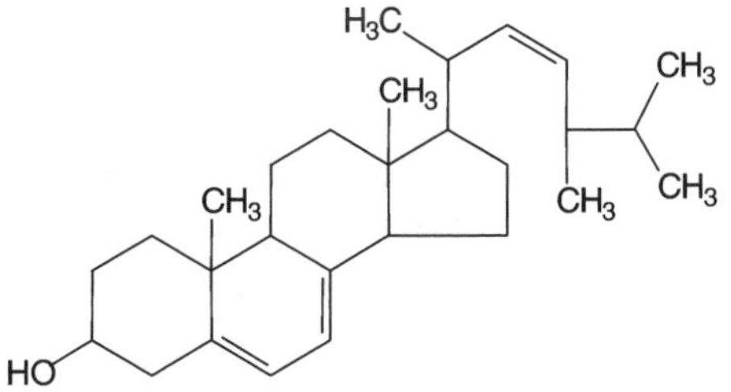

(8.32)

There is an increasing tendency for physiologists, but not nutritionists, to regard cholecalciferol as a hormone rather than a vitamin. This is because, as we shall see, its function in the body is that of a hormone rather than an enzyme prosthetic group like most other vitamins.

Furthermore, man and other mammals have, under the right circumstances, a more than adequate capacity for synthesising it for themselves.

Animals are able to synthesise cholesterol and other steroids for themselves, one intermediate in the pathway being 7-dehydrocholesterol. The epidermal cells of the skin contain 7-dehydrocholesterol, which is converted into cholecalciferol by the action of the UV component of sunlight. The course of the reaction is shown in Figure 8.12.

When human skin is adequately exposed to sunlight, the body is able to provide sufficient vitamin D for its needs (2.5–10 μg per day, infants, children, and pregnant or lactating women having the highest needs). The body is able to maintain useful stores of the vitamin but unlike the other fat soluble vitamins it is not stored in only in the liver but is found in the lipid elements of most tissues, particularly adipose tissue. These reserves provide some protection against reduced formation in the skin during the winter months but without dietary supplies many people would still have a sub-optimal vitamin D status for part of the year. The supply in the diet becomes important in parts of the world that do not get much sunshine, especially since the UV content of sunlight is much lower at the poles than it is at the equator and such regions are often too cold to encourage sunbathing. The heavy skin pigmentation associated with ethnic groups

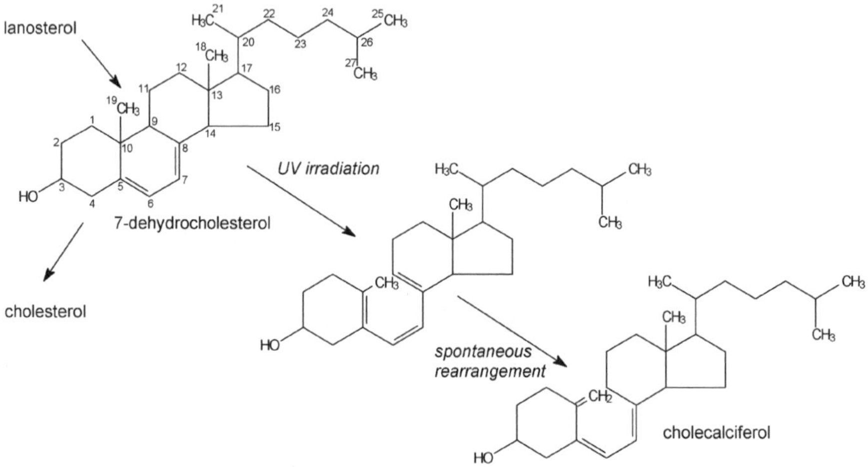

Figure 8.12   *The formation of cholecalciferol in the skin. The standard numbering for the carbon atoms of steroids is shown for 7-dehydrocholesterol, the immediate biosynthetic precursor of cholesterol. The optimum wavelength for the first reaction is 300 nm. The structure of cholecalciferol here differs from that shown in the text (8.30) only in that the single bond between carbons 6 and 7 has been rotated through 180° to show the relationship between cholecalciferol and its parent compounds more clearly.*

originating in tropical regions prevents a high proportion of the UV light from reaching the epidermal cells where cholecalciferol is formed. It may well be that the pale complexions of the northern European races are an adaptation to ensure the most efficient use of the little sunlight available.

The pattern of distribution of cholecalciferol in foodstuffs is strikingly similar to that of retinol. As with retinol, the liver oils of fish have spectacular levels, *e.g.* mackerel liver oil has 1.5 mg per 100 g. The muscle tissues of fatty fish such as salmon, herring, or mackerel are valuable dietary sources (5–45 μg per 100 g), but mammalian tissues, including liver, have less than 1 μg per 100 g. Milk contains only about 0.1 μg per 100 g, but fatty dairy products such as butter and cream do contain useful amounts (1–2 μg per 100 g); egg yolk has about 6 μg per 100 g. An unfortunate contrast with retinol is that plant materials contain no useful amounts of the vitamin.

It is nowadays common practice to enhance vitamin D contents of certain foods including breakfast cereals, milk (in the USA), and margarine. In the UK there is a legal requirement for the vitamin D content of margarine to lie between 7.05[*] and 8.82 μg per 100 g, a level some ten times that found in butter. These levels are normally expressed in terms of International Units; 1 IU of vitamin D equals 0.025 μg of cholecalciferol or ergocalciferol. The growing unpopularity of fatty dairy foods (*see* Chapter 4) may lead to a requirement for vitamin D supplementation of milk in the UK.

Rickets, the childhood disease that is most closely linked to vitamin D deficiency, is a failure of proper bone development. First described in 1645 it was always regarded as a disease of the urban poor. By the late 19th century 80% of London children showed symptoms. Although some authorities guessed, correctly, that it was a 'sunlight deficiency disease' syphilis and heredity were also blamed. The fact that poverty and nutritional deprivation could be 'inherited' escaped attention. Cod liver oil was being used as cure as early as 1848 (at Manchester Infirmary) but most of the medical profession remained sceptical until the 1920s.

Whether from the diet or formed in the skin, cholecalciferol is converted in the liver into the physiologically active compound 1,25-dihydroxycholecalciferol (calcifetriol) (8.33). This is one of the three hormones (the others are calcitonin and parathormone) that together control calcium metabolism. Calcifetriol promotes the synthesis of the proteins that transport calcium and phosphate ions through cell membranes. A lack of

---

[*] The curious degree of precision has no nutritional significance; it is merely the side effect of ounces to grams conversion.

calcifetriol prevents the uptake of calcium from the intestine, and the resulting shortage of calcium for bone growth is manifested as rickets.

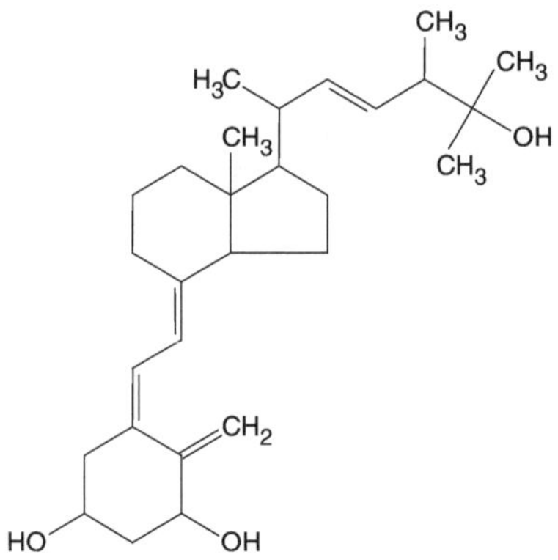

(8.33)

Another feature of this vitamin that it has in common with retinol is toxicity in excess. Babies and young children are the usual victims, the result of confusion in the dosages required of halibut liver oil and cod liver oil (3000 μg and 250 μg per 100 g, respectively). Massive overdoses cause calcification of soft tissues such as the lungs and kidneys, but in babies even modest overdoses will cause intestinal disorders, weight loss, and other symptoms.

At the levels occurring in most foodstuffs the determination of cholecalciferol is extremely difficult. Biological assays using rats reared on rickets-inducing diets were essential until the development of gas chromatography (GC) and high-performance liquid chromatography (HPLC). As a result very little work has been done on the behaviour of cholecalciferol in food. In general it is believed to be fairly stable to heat processing and only subject to oxidative breakdown in dried foods such as breakfast cereals.

VITAMIN E (α-TOCOPHEROL)

Of all the vitamins this one and also the one that concludes this chapter, vitamin K, have defeated efforts to replace their old alphabetical designations with a more scientific terminology. In both cases this is because no single chemical name is both sufficiently comprehensive and sufficiently exclusive. The vitamins E are a group of derivatives of 6-hydroxychroman carrying a phytyl side chain. As shown in Figure 8.13, tocopherols have a fully saturated side chain whereas in tocotrienols the side chain is

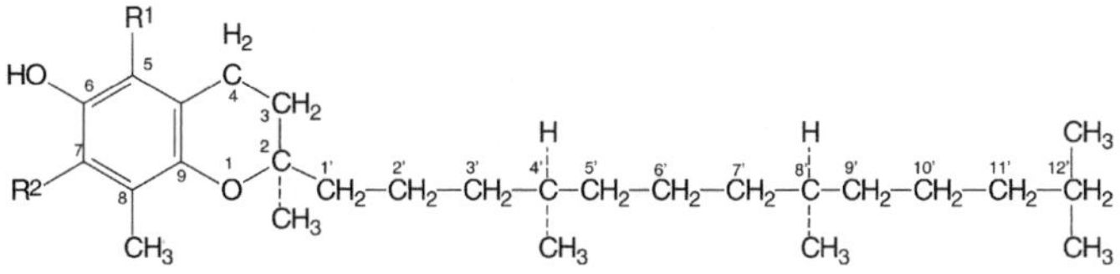

General tocopherol structure

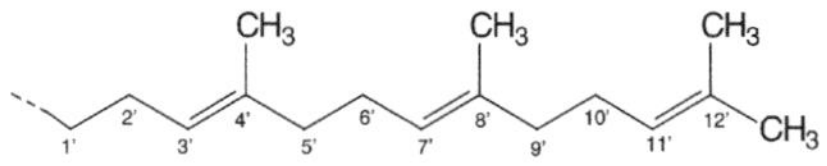

Tocotrienol side chain

Figure 8.13 *Vitamin E structures. The naturally occurring tocopherols and tocotrienols all have the configurations of their asymmetric centres (2, 4' and 8') as shown here and are often distinguished by the prefix 'd' included in the name, e.g. α-(d)-tocopherol. In the absence of 'd' or 'dl', 'd' is assumed. Synthetic tocopherols are mixtures of all eight possible isomers and the prefix 'dl' is used instead.*

unsaturated. Variations in the degree of methylation of the chroman nucleus give the α-, β-, γ-, and δ-members of each series. All eight of the compounds shown in Figure 8.13 are found in nature, but only α-, β- and γ-tocopherols and α- and β-tocotrienols are widespread.

The most important sources of vitamin E in the diet are the plant seed oils (50–200 mg per 100 g). Most other plant tissues contain less than 0.5 mg per 100 g. Animal tissues, including liver, milk, and eggs, have similarly low levels. Unlike other fat-soluble vitamins, this one is not found in particularly large amounts in fish liver oils; cod liver oil has $\sim$25 μg per 100 g. The evaluation of data such as these is complicated by the variations in vitamin activity between the different tocopherols and tocotrienols. Relative to α-(d)-tocopherol (1.0), the vitamin activities of the β-, γ-, and γ-(d)- forms are 0.27, 0.13, and 0.01, respectively. The *dl*-racemic mixtures have about three-quarters the activity of the corresponding pure *d*-isomers. Of the tocotrienols only the α-form has significant vitamin activity (about 0.3). Data like this make any statement of 'total tocopherol content' in milligrams in a given weight of the foodstuff nutritionally meaningless. In situations like this we have to resort to the International Unit (IU). In the case of vitamin E this is defined as the average minimum amount of the vitamin necessary daily to prevent sterility in a female rat, which actually turns out to be an amount close to 1 mg of α-tocopherol. It is common, as for example in Table 8.2, to use the term 'as α-tocopherol' rather than IU. Although α-tocopherol is the

Table 8.2   *The vitamin E, tocopherol and tocotrienols content of vegetable oils. These figures are based on data from 'McCance and Widdowson' (see Appendix II) and 'Handbook of vitamins', 2nd edn, ed. L. J. Machlin, Dekker, New York,1990. They should be regarded as typical values for the particular oil rather than absolute figures which apply to any sample of the oil*

	Vitamin E as α-(d)-tocopherol	d-Tocopherols				Total tocotrienols
		α	β	γ	δ	
	mg per 100 g					
Cottonseed oil	43	39	–	39	–	–
Olive oil	5	5	–	trace	–	–
Palm oil	33	26	–	32	7	46
Soybean oil	16	8	–	59	26	–
Wheatgerm oil	137	119	71	26	27	21
Maize (corn) oil	17	11	5	60	2	–

most potent, these data show that the less potent but often more abundant γ-tocopherol is at least as important in nutritional terms.

The tocopherols are the natural antioxidants of animal (and possibly plant) tissues. Vegetable oils, such as those listed in Table 8.2, are also protected by their tocopherol content. Antioxidants block the free-radical chain reactions of lipid peroxidation (*see* Chapter 4). The tocopherols have a special affinity for the membrane lipids of the mitochondria and endoplasmic reticulum of animal cells. It has been suggested that the shape of the tocopherol side chain will permit the formation of a complex with the arachidonic acid component of the phospholipids of these membranes. The mitochondrial membranes from animals reared on vitamin E-deficient diets are exceptionally prone to peroxidation. In view of the antioxidant role of vitamin E it would be expected that dietary supplements would be beneficial in reducing the likelihood of cardio-vascular disease. Unfortunately this attractive hypothesis has not been borne out in recent trials although the possibility that supplementation would need to be carried out for extended periods, up to 10 years, for effects to be observed remains open.

A deficiency of dietary selenium can cause similar symptoms in farm and laboratory animals to those invoked by vitamin E deficiency. The only known physiological role of selenium is in the active site of the enzyme glutathione peroxidase, the enzyme responsible for the breakdown of hydrogen peroxide and organic peroxides and so constituting a second defence mechanism against the consequences of lipid autoxidation.

In spite of health food enthusiasts' tendency to regard vitamin E as beneficial to all aspects of human sexual behaviour,* no deficiency disease or condition has ever been identified in man under otherwise normal circumstances. It is now well established that in order to maintain what are thought to be satisfactory tissue levels of the vitamin the minimum intake is related to the quantity of polyunsaturated fatty acids in the diet. A diet high in vitamin E-rich vegetable oils will, unfortunately, also be rich in the polyunsaturated fatty acids that these oils contain (*see* Chapter 4).

Determination of vitamin E in foodstuffs is complicated by the diversity of forms with different vitamin activity. Straightforward chemical tests for total tocopherols can be carried out, but they are clearly valueless. Chromatographic methods, GC and HPLC systems that can provide a distinction between the different forms are now well established.

Little is known of the stability of vitamin E in the foodstuffs that contain only modest proportions. In extracted vegetable oils it is quite stable unless conditions allowing autoxidation of the polyunsaturated fatty acids arise. Its presence in the oil will delay the onset of rancidity in the oil, but inevitably the buildup of peroxides in the oil will eventually overcome the tocopherols and cause their oxidation to compounds lacking antioxidant, and vitamin, activity. Vitamin activity is rapidly lost from the oil content of commercially deep-fat-fried frozen products such as potato chips. Low-temperature storage *does not* prevent the loss of most of the vitamin activity.

VITAMIN K (PHYLLOQUINONE, MENAQUINONES)

This is the last of the fat-soluble vitamins to be considered. As has already been mentioned, the alphabetical designation has been maintained in view of the lack of suitably concise chemical names for the various substances with vitamin K activity. All the vitamins K are derivatives of menadione (2-methyl-1,4-naphthoquinone) (8.34) or menadione itself. Phylloquinone (8.35), referred to as a vitamin K_1, has a phytyl side chain. The vitamin K_2 series, the menaquinones, have side chains of varying lengths, up to 13 isoprenoid units, but those having between 4 and 10 are most commonly encountered, particularly the compound referred to as 'vitamin K_2', which has seven (8.36).

---

* The most studied symptom of vitamin E deficiency in laboratory animals is reproductive failure caused by foetal resorption or testicular degeneration. The name 'tocopherol' was derived from the Greek, τοκος offspring, and φερος to bear.

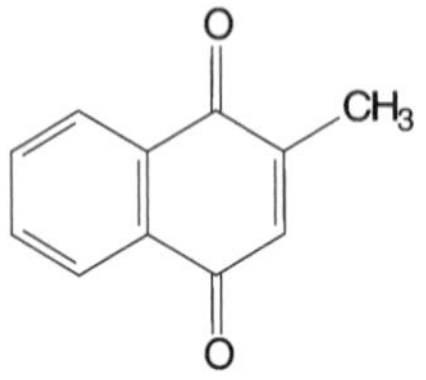

(8.34) 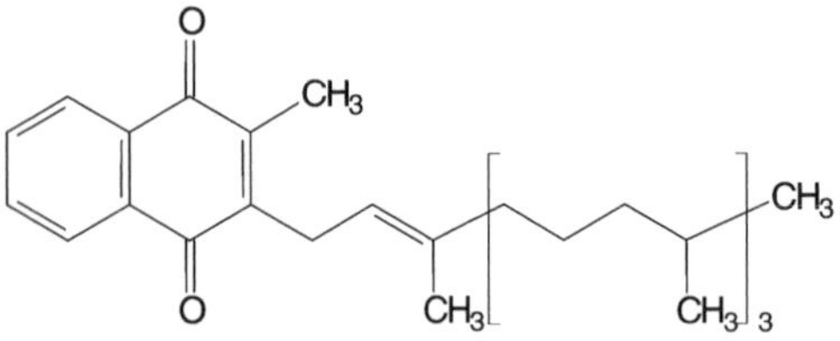   (8.35)

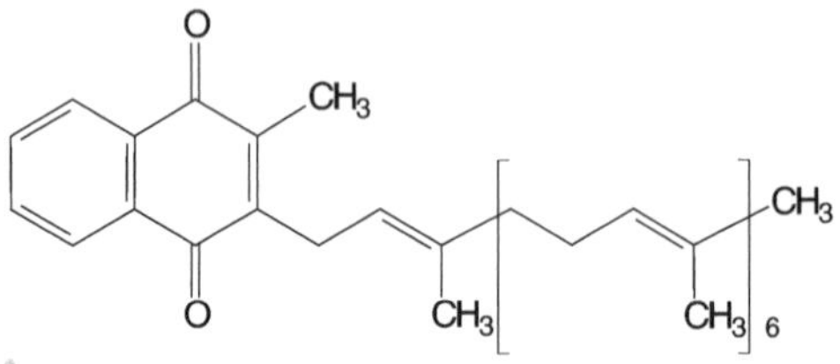

(8.36)

Vitamins K are widely distributed in biological materials in small amounts but reliable data is not abundant. Only leafy green vegetables such as spinach or cabbage are particularly rich in the vitamin (>100 μg per 100 g). Most other vegetables such as peas or tomatoes contain 50–100 μg per 100 g. Animal tissues, including liver, contain similar levels. Potatoes, fruit and cereals contain less than 10 μg per 100 g. Cow's milk and human milk contain around 6 and 1.5 μg per 100 g, respectively. Although animals are unable to synthesise this vitamin, humans are not particularly dependent on dietary supplies. This is because considerable quantities are synthesised by the bacteria of the large intestine. Vitamin K_2 was first isolated from putrefying fish meal. Deficiency of vitamin K is an important problem in poultry farming but is not encountered in otherwise healthy adult humans. Very occasionally, new-born infants do show deficiency symptoms.

The only physiological role for vitamin K that has been clearly identified is in blood clotting. The process of clot formation (the details of which are outside the scope of this book) depends upon the conversion of a number of different protein factors from inactive into active forms. The best understood of these is the conversion of the inactive prothrombin into the active proteolytic enzyme thrombin. This conversion requires the carboxylation of several glutamate residues (8.37). A vitamin K molecule in its reduced form, *i.e.* a hydroquinone (8.38), is an essential component of the carboxylation reaction system.

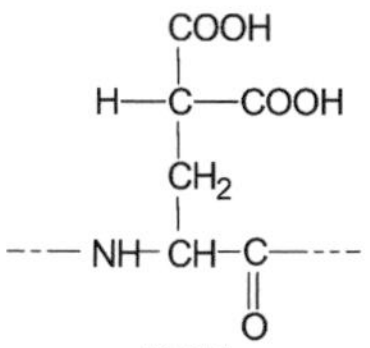

(8.37)

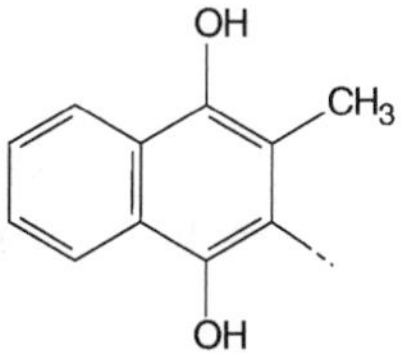

(8.38)

Vitamin K deficiency therefore results in a failure of the bloodclotting mechanism. Many substances which antagonise the activity of vitamin K are now known. Most are derivatives of coumarin such as dicoumarol, 3,3′-methylenebis-4-hydroxycoumarin (8.39) and warfarin, 3-(α-acetonyl-benzyl)-4-hydroxycoumarin (8.40). Dicoumarol causes disease in cattle and is produced when clover for animal feeding is spoiled by fermentation. Warfarin is used as a rat poison.

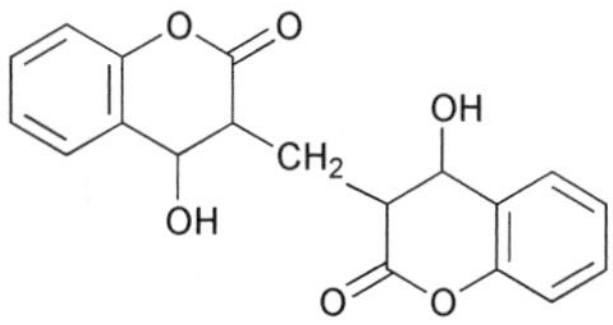

(8.39)

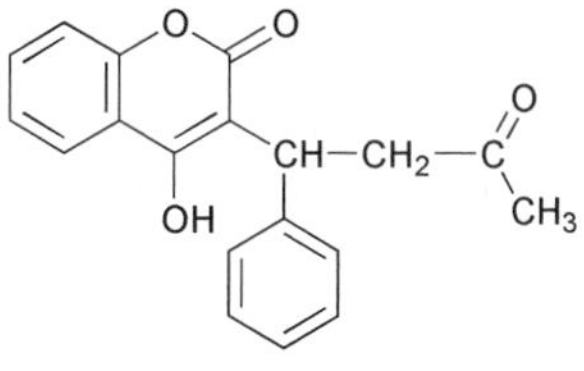

(8.40)

NON-VITAMINS

A number of other substances, not mentioned so far in this chapter, are often described as vitamins or listed with them. As in the case of vitamin F, mentioned on page 264, these are sometimes terminological hangovers from the 1930s but others have even less credibility.

Carnitine (8.26) is required for the growth of mealworm (*Tenebrio molitor*) larvae and *p*-aminobenzoic acid (PABA) (8.41) is a growth factor for a variety of micro-organisms but there is no evidence to suggest that they are required by humans. *myo*-Inositol (8.42), the only one of nine isomers of inositol that occurs widely in nature, has been shown to be a necessary growth factor for a few types of human cells growing in culture but there is no evidence of a requirement by an intact human. There is therefore no reason to regard any of these three substances as vitamins or seek them out as dietary supplements.

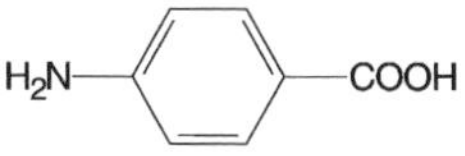

(8.41)

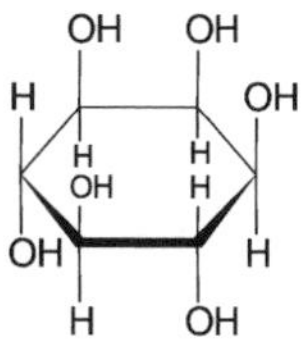

(8.42)

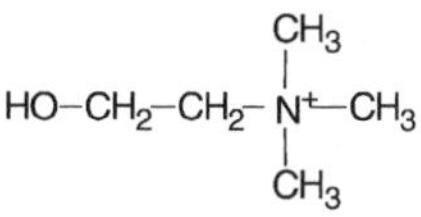

(8.43)

Choline (8.43) has been listed as a vitamin from time to time but only recently has its status been clarified. The body uses considerable quantities of choline, in the form of phosphatidyl choline or lecithin as a component of cell membranes. Although humans can synthesise choline, by the addition of methyl groups to phosphatidyl ethanolamine, this process still depends on a source of methyl groups. These can be supplied by the essential amino acid methionine but a normal diet is unlikely to provide sufficient surplus of methionine beyond that required for synthesising protein. We therefore require, as adults, around half a gram of choline per day. Growing infants and children require somewhat higher levels, in relation to body weight. Fortunately any normal diet supplies at least one gram of choline per day. The amounts required and its function as a tissue component mean that it should be considered as an essential nutrient alongside the essential fatty acids (Chapter 4) and essential amino acids (Chapter 5) and not a vitamin.

Three other substances are frequently referred to as vitamins beyond the realms of conventional nutrition. 'Vitamin B_{13}' is actually orotic acid (8.44), a normal intermediate in the biosynthesis of pyrimidines. 'Vitamin B_{15}' is otherwise known as pangamic acid but both names have been applied to a number of different substances, notably D-gluconodimethyl aminoacetic acid. (8.45). 'Vitamin B_{17}' was encountered in Chapter 2 under the name amygdalin. Under the name *laetrile* it achieved notoriety as a supposed cancer cure. The theory, unsupported by experimental evidence, was that cancer cells are especially rich in the β-glucosidases that break it down, ultimately releasing cyanide (8.46) which selectively kills the cancer cell. These levels of cyanide released would normally be detoxified by conversion into thiocyanate by the enzyme rhodanase.

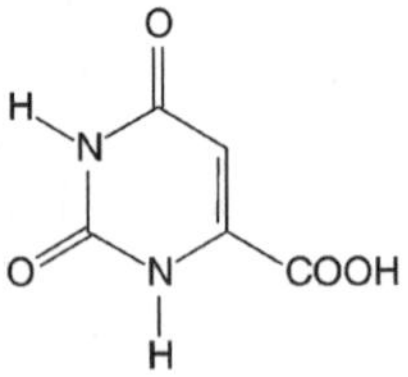

(8.44)

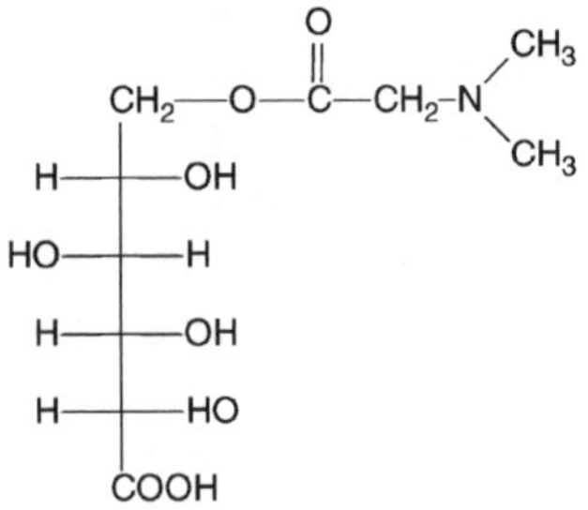

(8.45)

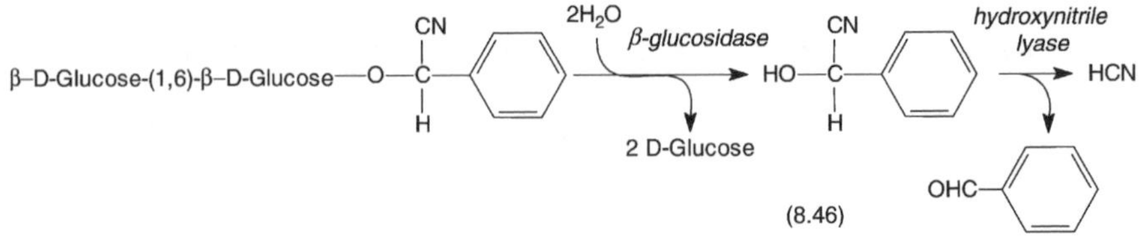

(8.46)

FURTHER READING

M. B. Davies, J. Austin and D. A. Partridge, 'Vitamin C: its Chemistry and Biochemistry', The Royal Society of Chemistry, Cambridge, 1991.

A. E. Bender, 'Food Processing and Nutrition', Academic Press, London, 1978.

G. F. Combs, 'The Vitamins: Fundamental Aspects in Nutrition and Health', Academic Press, San Diego, 1992.

'Fat-soluble Vitamins: their Biochemistry and Applications', ed. A. T. Diplock, Heinemann, London, 1985.

T. K. Basu and J. W. Dickerson, 'Vitamins in Health and Disease', CAB International, Wallingford, Oxon., 1996.

'The Technology of Vitamins in Food', ed. P. B. Ottaway, Blackie, Glasgow, 1993.

'Nutritional Aspects of Food Processing and Ingredients', eds. C. J. K. Henry and N. J. Heppell, Aspen, Gaithersburg, Md., 1998.

'Dietary Reference Intakes for Thiamine, Riboflavin, Niacin, Vitamin B_6, Folate, Vitamin B_{12}, Pantothenic Acid, Biotin and Choline', Institute of Medicine Standing Committee on the Scientific Evaluation of Dietary Reference Intakes, National Academy of Sciences, Washington D.C., 1998.

'Dietary Reference Intakes. Calcium, Phosphorus, Magnesium, Vitamin D, and Fluoride.', Institute of Medicine Standing Committee on the Scientific Evaluation of Dietary Reference Intakes, National Academy of Sciences, Washington D.C., 1997.

'Dietary Reference Intakes for Vitamin C, Vitamin E, Selenium, and Carotenoids', Institute of Medicine Food and Nutrition Board, National Academy of Sciences, Washington D.C., 2000.

Chapter 9

Preservatives

Micro-organisms, particularly bacteria, yeast, and moulds, have nutritional requirements remarkably similar to our own. Unless they use photosynthesis (as some bacteria do), energy is obtained by the oxidation or fermentation of organic compounds.* Similarly, organic compounds are the source of carbon, nitrogen and other elements for the biosynthesis of cell material. It is hardly surprising therefore that few types of micro-organisms will decline to utilise the same nutrients that make up a typical human diet – given the opportunity. The popularity of human foods with micro-organisms is enhanced by their tendency to be at moderate pH values and mild temperatures and, in one sort of food or another, to offer a wide enough range of oxygen tensions to tempt obligate anaerobes and obligate aerobes as well as less fastidious types.

Thus it is to be expected that a major goal of food technology has always been the control of food-borne micro-organisms. The word 'control' is stressed simply because the elimination of micro-organisms from food is frequently not even attempted and in some foodstuffs will be totally undesirable. In fact there are four quite distinct aspects of food science to which microbiologists contribute. The most important is that of safety. Foodstuffs are inevitably ideal carriers of pathogenic bacteria, *e.g.*, *Salmonella sp.*, as well as being ideal substrates for the growth of bacteria and fungi which secrete toxins, *e.g.*, *Clostridium botulinum* and *Staphylococcus aureus*.

A second aspect of food microbiology is that of biodeterioration. If food materials were not vulnerable to degradation by the extracellular enzymes secreted by invading bacteria and moulds, there would be little prospect of success for the parallel processes of digestion that occur in our own alimentary canal.

---

* There are some bacteria, irrelevant to human food that obtain their energy by the oxidation of inorganic compounds.

The second two aspects of food microbiology are less negative in outlook, *i.e.* where microbial activity is utilised in food production. Many processes, often under the blanket term 'fermentation', have been practised for thousands of years. For example, cheese-making and pickling are both exploitations of 'natural' processes of biodeterioration. Microbial action reduces the raw material to a more stable product in which there is much less scope left for further breakdown. This is usually because microbial activity has been accompanied by loss of water and/or the accumulation of ethanol, lactic acid, or acetic acid (ethanoic acid).

The fourth aspect of food microbiology, which attracts less interest than it did a few years ago, is the use of the micro-organisms themselves as food. While we have always eaten fungi such as mushrooms, scientists have explored the use of single-celled micro-organisms as food. These could be utilised directly for human consumption, but with one notable exception the plan was for the protein to extracted and used as a food supplement for livestock. Strains of micro-organisms were developed which grew readily on methanol and other by-products of the oil industry. This almost all came to a halt when rises in world oil prices ruined the economics of SCP ('single-cell protein') production. The exception is the mycoprotein known commercially as Quorn™ (*see* page 163 and R. Angold *et al.* 'Food Biotechnology' *in* Further Reading). The carbon source for the production of Quorn is the starch that is an inevitable by-product from the commercial production of wheat gluten.

Of these various aspects of food microbiology the use of chemical substances to deter unwanted microbial activity is of greatest interest to chemists. Even with regard to unwanted micro-organisms, the inhibition of growth is often the most one may hope to attain. As we have seen elswhere in this book, the heat-processing methods required to obtain sterility, *i.e.* the absence of viable vegetative cells or spores, cannot be regarded as compatible with maximum nutritive value. Furthermore the essential similarity between the cellular components and metabolic processes of micro-organisms and those of man will ensure that chemical sterilants (*e.g.*, chlorine or phenol in high concentrations) will hardly be acceptable as food ingredients. The advantage of a chemical-additive approach to the reduction of microbial activity is that many products will need to remain stable for some time after the package has been opened. A jar of jam or a bottle of tomato ketchup will be subjected to repeated and massive recontamination once it comes under the influence of the junior members of a household.

It is actually quite unusual for a single antimicrobial procedure to be used alone to protect a food product. Wherever possible, several relatively mild procedures are combined to maximise the inhibition of microbial

Table 9.1 *Anti-microbial preservatives permitted in Europe and mentioned in the text. There are considerable restrictions on the range of foods and concentrations in which particular preservatives may be used. Sodium chloride, alcohol and sugars, which exert their anti-microbial action by reducing the availability of water, are not treated as food additives in the context of regulation.*

Preservative	E number
Sorbic acid and sorbates	E200–E203
Benzoic acid and benzoates	E210–E213
Esters of *p*-hydroxybenzoate	E209, E214–E219
Sulfur dioxide, sulfites, hydrogen sulfite and metabisulfite	E220–E228
Nisin	E234
Pimaricin (*Natamycin*)	E235
Nitrites	E249, E250
Nitrates	E251, E252
Acetic acid and acetates	E260–E266
Lactic acid	E270
Propionic acid and propionates	E280–E283

activity while minimising adverse effects on nutritional value or acceptability. For example, the long-term storage of meat, fish, and vegetables at ambient temperatures demands the use of canning with the temperature at the centre of the can being held above 115 °C for over one hour. Otherwise the surviving spores of *C. botulinum* would find the anaerobic, nutrient-rich, neutral pH environment ideal for germination, growth, and toxin production. However, fruit's natural acidity eliminates the risk of botulism, and the more modest heat treatments we associate with home bottling are adequate. As we shall see in the accounts of various food preservatives that follow, they are very often combined in use with other anti-microbial procedures. For example, the safety of cooked ham depends on (i) salt content to maintain a low water activity, (ii) cooking to destroy most vegetative bacterial cells and some spores, and (iii) nitrite content to prevent spore germination and bacterial growth.

Other than sodium chloride and smoke, preservatives are recognised as food additives and in Europe (as elsewhere) their use is covered by legislation. The principal preservatives approved for use in Europe are listed in Table 9.1.

SODIUM CHLORIDE

Common salt, sodium chloride, was undoubtedly the first antimicrobial substance to be used. One can be confident that it was its use as a preservative rather than as a flavouring that gave it its value to early civilisations. Salting is the traditional method of preserving meat, often in combination with smoking and drying. Modern technology has provided more rapid methods of getting the salt into the meat, but the essentials have remained unchanged for centuries. Salt solutions containing 15–25% salt are used to bring the water activity, a_W,* down to about 0.96. This has the effect of retarding the growth of most micro-organisms, including the majority of those responsible for meat spoilage. With the advent of other preservative methods, notably canning and refrigeration, the importance of salting has diminished, but not without some salt-preserved meat products such as ham and bacon becoming firmly established in our diet. As the necessity for high salt content for preservation has lessened, we have adapted our taste to less salty and less dry bacon and ham.

NITRITES

Nowadays very little meat is preserved by the use of common salt alone. At some unknown point in history it was realised that it was the unintended presence of saltpetre (sodium nitrate) as an impurity in the crude salt that resulted in an attractive red or pink colour in preserved meat. Subsequently nitrates and/or nitrites have become an almost indispensible component of the salt mixtures (known as 'pickles') used for curing bacon and ham. In spite of the antiquity of the process, it is only very recently that the special antimicrobial properties of the nitrite in ham have been recognised. There are many subtle variations in curing procedures, but the well-known 'Wiltshire cure' is typical. Sides of pork are injected with about 5% of their own weight of a pickling brine containing 25–30% sodium chloride, 2.5–4% sodium or potassium nitrate, and sometimes a little sugar. The sides are then submerged in a similar solution for a few days. After removal they are stored for 1–2 weeks for 'maturation' to occur. If prolonged storage is then called for, a second preservative, smoke, is applied.

By the time the maturation stage is completed, the salts will be evenly distributed throughout the muscle tissue and a complex series of reactions will have given rise to the characteristic red colour of uncooked bacon and ham. The essential features of these reactions can be summarised as follows.

---

* The water activity of an aqueous solution is equal to its vapour pressure divided by the vapour pressure of pure water at the same temperature. This concept is discussed in much greater detail in Chapter 12.

(i) Some of the nitrate present is reduced to nitrite:

$$NO_3^- + 2[H] \rightarrow NO_2^- + H_2O$$

either by salt-tolerant micro-organisms in the brine or by the respiratory enzymes of the muscle tissue.

(ii) The nitrite oxidises the iron of the muscle myoglobin to the iron(III) state (*see* Chapter 5):

$$Fe^{II} + NO_2^- + H^+ \rightarrow Fe^{III} + NO + OH^-$$

i.e. myoglobin (Mb) is converted into metmyoglobin (MMb) and nitrogen oxide is formed.

(iii) The resulting nitrogen oxide reacts with the iron MMb to form nitrosyl metmyoglobin (MMbNO).

(iv) The MMbNO is immediately reduced by the respiratory systems of the muscle tissue to nitrosyl myoglobin, MbNO, the red pigment of uncooked bacon and ham.

The distribution of electrons around the iron of MbNO is similar to that in oxymyoglobin (MbO$_2$), hence the similarity in colour. When bacon is grilled or fried and when ham is boiled, the nitrosyl myoglobin is denatured and a bright pink pigment, often referred to as nitrosylhaemo-chromogen, is formed. There is no certainty as to its structure, but it is believed that the denaturation of the globin allows a second nitrogen oxide molecule to bind to the iron in the place of the histidine residue.

In recent years the use of nitrates in food has been regarded with increasing suspicion owing to the risk of nitrosamine formation, nitrosa-mines being potent carcinogens. Secondary amines react readily with nitrous acid to form stable *N*-nitroso compounds (9.1):

$$\begin{array}{c} R' \\ {\Large\diagdown} \\ N-H \\ {\Large\diagup} \\ R'' \end{array} + NO_2^- \longrightarrow \begin{array}{c} R' \\ {\Large\diagdown} \\ N-N{=}O \\ {\Large\diagup} \\ R'' \end{array} + OH^- \qquad (9.1)$$

The reaction with primary amines, which are of course abundant in meat as free amino acids, leads simply to deamination:

$$RNH_2 + NO_2^- \rightarrow N_2 + OH^-$$

The reaction with amino acids is important in that it will lead to some elimination of the excess nitrite in cured meat. Secondary amines are much less abundant in meat but are assumed to arise as a result of microbial action, especially by anaerobes. The decarboxylation of proline, which occurs spontaneously at the high temperatures involved in frying, leads to the formation of the very important nitrosamine N-nitrosopyrroli-dine (NOPyr) (9.2):

$$
\begin{array}{ccccc}
\text{H}_2\text{C}-\text{CH}_2 & & \text{H}_2\text{C}-\text{CH}_2 & & \text{H}_2\text{C}-\text{CH}_2 \\
| \quad | & \xrightarrow{-\text{CO}_2} & | \quad | & \xrightarrow{+\text{NO}_2^-} & | \quad | \\
\text{H}_2\text{C} \quad \text{C}-\text{COOH} & & \text{H}_2\text{C} \quad \text{CH}_2 & & \text{H}_2\text{C} \quad \text{CH}_2 \\
\text{N} & & \text{N} & & \text{N} \\
\text{H} \quad \text{H} & & \text{H} & & \text{NO}
\end{array}
\quad (9.2)
$$

Although nitrosamines have been detected in many cured-meat products such as salami and frankfurters, it is clear that they are most significant in cured meat that has been cooked at high temperatures, such as fried bacon. Levels of NOPyr in fried bacon are consistently found to be around 100 μg kg^{-1} with rather lower levels of N-nitrosodimethylamine (NDMA). Although there is no doubt as to the carcinogenicity of these volatile nitrosamines when tested in laboratory animals, there is little evidence that the consumption of cured meats has actually been responsible for disease in man. However, a recently published study from Finland has shown that over a long period there is a positive correlation between NDMA intake and the incidence of colo-rectal cancer. In this study a high intake of smoked or salted fish was found to be much more important whereas there was no correlation with cured meat consumption. The potential hazard is sufficient to ensure that steps continue to taken to reduce the consumption of N-nitrosamines. Beer is another product whose nitrosamine content has come under attention. Studies in the late 1970s showed that levels of NDMA around 2–3 ppb were commonplace. This arose when the malted barley was kilned, *i.e.* heated and dried to prevent continued germination once the starch degrading α- and β-amylases had been synthesised in the grain. Nitrogen oxides were generated in the flames of the gas, oil or solid fuel burners and these were transferred to the germinated barley by the direct exposure to the flue gases. It was usual to expose the malt directly to the flue gases from an oil or solid fuel fired furnace. It was discovered that the malt kilning was to blame. Changes in the construction of the dryers, and incorporation of sulfur dioxide, which inhibits nitrosation, into the brewing process have effectively eliminated the NDMA problem.

Improvements in processing controls are making possible a steady reduction in residual nitrite levels, but the most important measure is the inclusion of ascorbic acid in curing-salt mixtures. Ascorbic acid is beneficial in two ways. Firstly, being a reducing agent, it enhances, directly or indirectly, the rates of the key reducing reactions in the formation of

MbNO, thereby allowing lower levels of nitrites or nitrates to be used in the pickles. Secondly, it actually inhibits the nitrosation reaction.

Even the remote possibility of a hazard in the use of nitrite for curing would justify its prohibition, were its sole value as a colouring agent. (The question of whether nitrite actually contributes to the flavour of cured meat remains unresolved.) However, its antimicrobial properties would justify its inclusion even if it had no effect on colour. It has been known for some years that the growth of many types of anaerobic bacteria, including the causative organism of botulism, *C. botulinum*, is prevented in cured meat by an unidentified product of the interaction of the residual nitrite with the meat that arises when the meat is cooked. If a ham were given sufficient heat treatment to ensure that all spores of *C. botulinum* had been killed, it would be unacceptably overcooked. Canned stewing beef, which is given such a heat treatment, provides a good illustration of the sort of texture the ham would have. This effect of the residual nitrite in ham was termed the 'Perigo effect' after its discoverer, and a great deal of work was done to identify the 'Perigo inhibitor'. It is now well established that during cooking much residual nitrite is broken down to nitrogen oxide. This is not liberated from the meat but becomes loosely associated with some of the exposed amino acid side chains and iron atoms of the denatured meat proteins. From this reservoir nitrogen oxide is available for the inhibition of sensitive bacteria. It appears that nitrogen oxide is a specific, and potent, inhibitor of at least one enzyme (namely pyruvate:ferredoxin oxidoreductase) that has an essential role in the energy metabolism, and therefore growth and toxin production, of anaerobes such as *C. botulinum*.

The use of nitrite therefore presents those concerned with food safety with a paradox. On the one hand we have nitrite the preservative, which has been used for centuries and is now known to be essential for microbiological safety, and on the other hand we have nitrite the colouring agent, suspected of giving rise to carcinogens. Apart from the obvious but unacceptable solution of eliminating cured meats from our diet, the most we can do is to ensure that residual nitrite levels are never much more than the minimum, about 50 μg g^{-1}, needed to prevent toxin production. It is also worth remembering that nitrate naturally present in other foods, particularly vegetables, and in drinking water can be reduced to nitrite by our intestinal bacteria and therefore has the potential to cause problems even for those who never eat cured meat (*see* Chapter 10).

SMOKE

Smoke is the other preservative traditionally associated with meat and fish. We can be fairly sure that the flavouring action of wood smoke was initially regarded only as a valuable side effect of drying out over a wood fire. The preservative action of smoke most certainly went unnoticed. Nowadays meat and fish are rarely preserved by drying, and refrigeration has made the preservative action of smoke less important than its flavour.

Smoke consists of two phases, a disperse phase of liquid droplets and a continuous gas phase. In smoking, the absorption of gas-phase components by the food surface is considered to be much more important than the actual deposition of smoke droplets. The gas phase of wood smoke has been shown to include over 200 different compounds, including formalde-hyde (methanal, HCHO), formic acid (methanoic acid, HCOOH), short-chain fatty acids, vanillic (9.3) and syringic (9.4) acids, furfural (9.5), methanol, ethanol, acetaldehyde (ethanal), diacetyl (butanedione) (2.43), acetone (propanone), and 3,4-benzpyrene (9.6). Of these the most important antimicrobial compound is almost certainly formaldehyde.

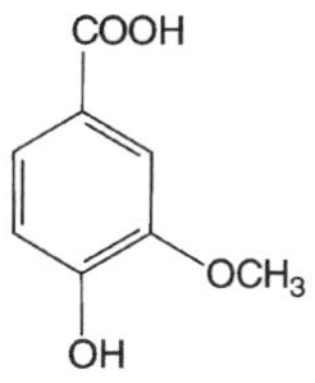

(9.3) 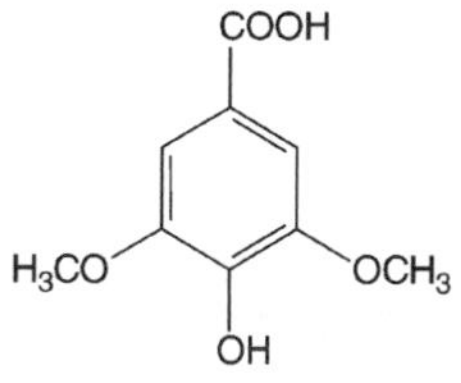   (9.4) 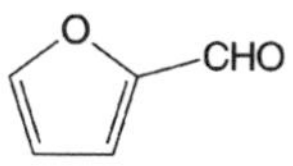   (9.5) 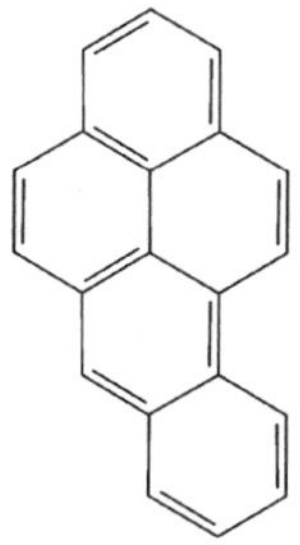   (9.6)

The detection of known carcinogens such as 3,4-benzpyrene and other polynuclear aromatic compounds in wood smoke has led to concern over the safety of smoked foods. It has been suggested that the high incidence of stomach cancer in Iceland and the Scandinavian countries is due to the large amounts of smoked fish consumed there. Although there is no suggestion that the amount of smoked food consumed elsewhere in the world is sufficient to cause similar problems, liquid-smoke preparations which do not contain the polynuclear aromatics are being adopted to an increasing extent. Liquid smokes are prepared from wood smoke con-densates by fractional distillation and water extraction. The undesirable polynuclear hydrocarbons are not soluble in water.

It is sad that so many consumers overlook the fact that these modern developments in curing and smoking have been introduced to reduce the small, but very real, risks involved in traditional processes and that with both smoke and nitrite it is the preservative effect rather than flavour or colour that commends them to the food processor.

SULFUR DIOXIDE

Sulfur dioxide (SO_2) has been used in wine-making for hundreds of years to control the growth of unwanted micro-organisms. It was originally obtained by the rather haphazard process of exposing the 'must', *i.e.* the unfermented grape juice, to the fumes of burning sulfur. Nowadays the free gas is rarely used, being replaced by a number of SO_2-generating compounds, particularly sodium sulfite (Na_2SO_3), sodium hydrogen sulfite ($NaHSO_3$), and sodium metabisulfite (sodium disulfite, $Na_2S_2O_5$). The relationships between these, sulfur dioxide, and sulfurous acid are outlined in Figure 9.1. The antimicrobial activity of these compounds increases dramatically as the pH falls, and it is therefore assumed that it is the undissociated sulfurous acid that has the antimicrobial activity.

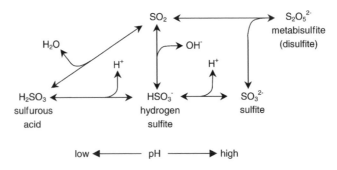

Figure 9.1 *Structural relationships of the sulfur dioxide-generating compounds.*

Total sulfur dioxide levels of around 100 p.p.m. are added to musts to achieve a differential effect: the desirable wine yeast *Saccharomyces cerevisiae* is able to grow and ferment the sucrose to ethanol, but some undesirable yeast species such as *Kloeckera apiculata*, as well as lactic acid-producing bacteria, are suppressed. Sulfur dioxide inhibits many of the NAD^+ dependant dehydrogenases of yeast and bacteria but is a particular inhibitor of the glyceraldehyde dehydrogenase of yeast and the malate dehydrogenase of *Escherichia coli*. Both these enzymes play central roles in energy metabolism. About two-thirds of the total sulfur dioxide in a must or wine is bound to anthocyanins and other flavonoids (*see* Chapter 6), sugars (9.7):

and other aldehydes. Besides the sulfur dioxide that is added, many strains of wine yeasts actually produce sulfur dioxide themselves by the reduction of sulfate present in the grape juice. When the wine is bottled, further sulfur dioxide is added to prevent secondary fermentation, *i.e.* fermentation of residual sugar in the bottle; concentrations of total sulfur dioxide of up to around 500 p.p.m. are commonly used.

A wide range of other foodstuffs, particularly fruit or vegetable-based products, have sulfite added as a preservative or as a residue from processing operations. The use of sulfite as an antioxidant has already been examined (Chapter 6).

The use of sulfur dioxide has generally been regarded as being without any toxicity hazard at the usual levels. Of all the preservatives sulfur dioxide and its relatives are consumed in the largest amounts. Although there is no more recent data than 1986 for the UK consumption cannot be expected to have changed significantly since then. At that time average consumption per person was estimated to be 18.2 mg day^{-1} with alcoholic drinks, sausages, hamburgers and dried fruit being the major sources. The margin between this level of intake and the ADI (Acceptable Daily Intake, *see* page 353) is not large, 3.5 mg kg^{-1} body weight day^{-1} (*e.g.*, 245 mg day^{-1} for a 70 kg person), especially when the intake at the top of the range (the 97.5th percentile), 64.3 mg day^{-1}, is considered. Nevertheless the UK authorities do not consider sulfur dioxide intake levels to be a cause for concern.

Recently levels at the top end of the usual range for food use have been shown to cause symptoms similar to a severe allergic response in a very small proportion of asthmatics. It remains to be seen whether the extent of this problem will prove sufficient to justify a reappraisal of the GRAS (*i.e.* 'Generally Recognised As Safe') status of sulfur dioxide. There is a quite different nutritional problem with the use of sulfur dioxide. The bisulfite ion reacts readily, and destructively, with the vitamin thiamin (*see* Chapter 8). Vegetables, such as potatoes, that are often stored in sulfite solutions at intermediate stages of processing, will lose considerable proportions of their thiamin content. It is for this reason that many countries prohibit the use of sulfur dioxide in foodstuffs that are important sources of thiamin in that country's diet.

A further drawback to the use of sulfite/sulfur dioxide is the taste. Above 500 p.p.m. most people are aware of its disagreeable flavour, and some can detect it at much lower levels. Some white wines are actually characterised by their slight sulfur dioxide flavour, notably *Gewurztraminer* from the Alsace region of France.

Sulfur dioxide is one of the easier preservatives of which to measure the concentration in foodstuffs and drinks and one that has not yet been taken over by the chromatographers. In the most popular version of the Monier–Williams method the sample is placed in a distilling flask and acidified in order to convert bisulfites *etc.* sulfite/SO_2 and liberate the sulfur dioxide from any complexes with aldehydes, *etc.* The volatile sulfur dioxide is then removed by distillation through a reflux condenser, which holds back any volatile organic acids. The sulfur dioxide that comes over is trapped in a hydrogen peroxide solution:

$$SO_2 + H_2O_2 \rightarrow SO_4^{2-} + 2H^+$$

facilitating titrimetric estimation. Sulfur dioxide is also conveniently measured enzymically using a combination of the two enzymes, sulfite oxidase and NADH peroxidase:

$$SO_3^{2-} + O_2 + H_2O \rightarrow H_2O_2 + SO_4^{2-}$$
sulfite oxidase

$$H_2O_2 + NADH + H^+ \rightarrow 2H_2O + NAD^+$$
NADH peroxidase

A concentration of 0.10 mM sulfite gives a fall in the absorbance at 340 nm (due to NADH, *see* page xx) of 0.622, providing ample sensitivity for food analysis purposes.

BENZOATES

Benzoic acid (9.8) occurs naturally in small amounts in some edible plants, notably the cloudberry (at about 0.8%) but whether this means that the synthetic product could be described as 'natural' is another matter[*]. The sodium salt, being much more soluble than the free acid, is the form that is most usually added to food as a preservative. However, it is the undissociated acid that is active against micro-organisms, particularly yeasts and bacteria, through its inhibitory effect on the activities of the

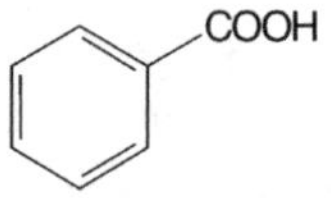

(9.8)

---

[*] It is cases such as this that emphasise the futility of the 'naturalness equals wholesomeness' debate.

tricarboxylic acid cycle (*i.e.* the Krebs or citric acid cycle) enzymes, α-ketoglutarate dehydrogenase and succinate dehydrogenase. Since it is the protonated form that is active, benzoic acid use is restricted to acid foods (pH 2.5–4.0) such as fruit juices and other beverages. At the levels used, 0.05–0.1%, no deleterious effects on humans have been detected. Benzoate does not accumulate in the body but is converted, by condensation with glycine into hippuric acid (*N*-benzoylglycine, 9.9), which is excreted in the urine:

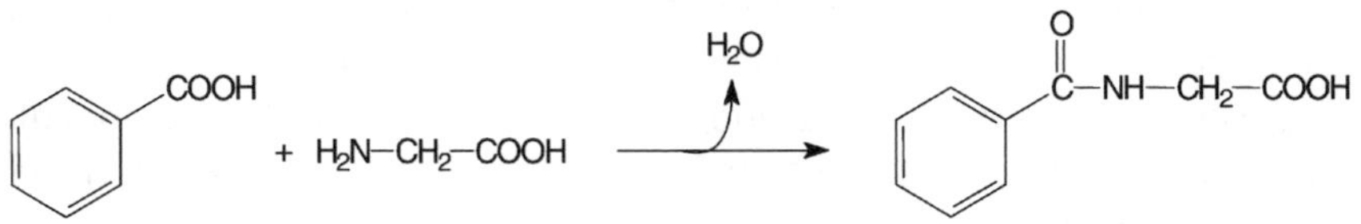

(9.9)

Esters of *p*-hydroxybenzoic acid with methanol, propanol (9.10), and other alcohols, known collectively as 'parabens', are also commonly used in most of the same situations as benzoic acid and present similarly little problem of toxicity.

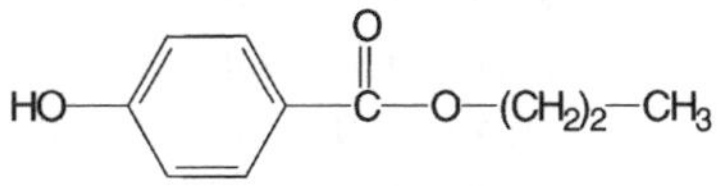

(9.10)

OTHER ORGANIC ACIDS

There are a number of organic acids that are useful preservatives in spite of having rather innocuous-looking chemical structures. Sorbic acid (9.11) is a particularly effective inhibitor of mould growth, inhibiting a wide range of enzymes important in intermediary metabolism including enolase, and the tricarboxylic acid cycle dehydrogenases. As with benzoate it is the undissociated form rather than the anion that is the effective agent. However, sorbic acid's higher pK_a value, 4.8, compared with benzoic acid's 4.2, makes it useful at pH values up to 6.5. This means that it can be used in a wider range of food products including processed cheese and flour confectionery (*e.g.* cakes and pastries). It cannot be used in bread because of its inhibitory effect on the yeast. As it does show some activity against bacteria, there is growing interest in the possibility of using it to replace all or some of the nitrite in cured-meat products.

At the levels normally used (up to 0.3%) no toxic effects have ever been detected. It has been suggested that it is readily metabolised by mammals using much the same route as naturally occurring unsaturated fatty acids.

$$CH_3-CH=CH-CH=CH-COOH$$
(9.11)

It might be assumed that the association of acetic acid with preserved foods such as pickles was owed to its acidifying property, but in fact it is an effective inhibitor of many types of spoilage bacteria and fungi at concentrations too low (0.1–0.3% of the undissociated form) for it to have much effect on the pH of a foodstuff. Aside from its inevitable presence in vinegar-based food products, acetic acid is becoming increasingly popular as an inhibitor of moulds and 'rope'* in bread. One of its special virtues, and the one that has led to it superseding propionic acid and its salts in this role, is that it can be added as vinegar, which can be listed as a simple ingredient on the packaging and not described as an 'added chemical'.

Like acetic acid, propionic acid ($CH_3CH_2CH_2COOH$) is naturally found in many foodstuffs. Some Swiss cheeses contain as much as 1%. It has a similar spectrum of antimicrobial activity to that of acetic acid and shares many applications with sorbic and acetic acids.

There is a surprising lack of information about the mechanisms by which these organic acid preservatives actually inhibit microbial growth. The most probable site of action is the cell membrane, where it may be imagined that molecules would be able to form close associations with the polar membrane lipids. In doing so, it is likely that they would disrupt the normal processes of active transport into the cell.

NISIN AND NATAMYCIN

These two antibiotics are unlikely preservatives. The better known of the two, nisin, is a polypeptide consisting of 34 amino acids in a chain having a number of small loops formed by sulfhydryl bridges involving unusual amino acids such as lanthionine. Though fascinating to biochemists, the details of nisin's structure, beyond the simplified diagram shown in Figure 9.2, need not detain us here.

Nisin was first identified, in the 1930s, as a problematical inhibitory substance that occurred naturally in many dairy products. It was found to be produced by many strains of the bacterium *Streptococcus lactis*. This organism commonly occurs in milk, and it was found to inhibit the growth

---

* Rope is the disorder of bread that occurs when spores of *Bacillus mentericus* survive the baking temperatures in the centre of a loaf and then germinate when the loaf cools. The bacteria then attack the starch to produce a foul smelling mass of sticky dextrins down the centre of the loaf – the 'rope'.

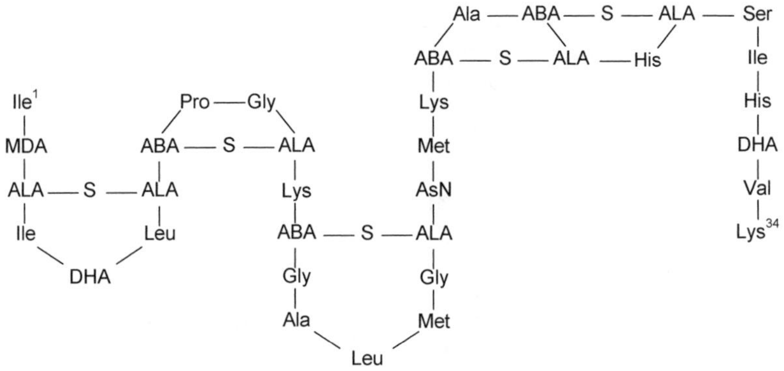

MDA = β-methyldehydroalanine, $CH_3CH=CNH_2COOH$
DHA = dehydroalanine, $CH_2=CNH_2COOH$
ALA-S-ALA = lanthionine, $S(CH_3CH=CNH_2COOH)_2$
ABA-S-ALA = β-methyl-lanthionine, $HOOCCH(NH_2)CH_2SCH(CH_3)CH(NH_2)COOH$

Figure 9.2 *The amino sequence of nisin. The amino acids are numbered from the NH₂ terminal in the conventional way. The usual abbreviations for amino acids are used, plus those shown in the figure. The first 21 amino acids are predominantly hydrophobic and could form a region of the molecule that is able to bind with the membrane lipids of susceptible micro-organisms.*

of other bacteria (but not fungi or yeasts) important in cheese-making, including both desirable lactobacilli and undesirable clostridia. Nisin's origins and activity mean that it is correctly defined as an antibiotic, but this term is not normally be applied to nisin as it has never been used in the treatment of disease.

Nowadays nisin is manufactured by growing selected strains of *S. lactis* and isolating the preservative from the culture medium. It is likely to prove valuable in a wide range of foodstuffs where spoilage is caused by Gram-positive bacteria, but at present it is mainly being used in processed cheese. Here it will prevent the growth of bacteria such as *Clostridium butyricum* that have spores that can survive the 85–105 °C melting temperatures involved when the cheese is processed. As with antibiotics and enzyme preparations, it is usual to describe amounts of nisin in terms of units of activity rather than absolute units of weight: 1 gram of pure nisin is actually equivalent to 4×10^7 international units (IU). Levels of nisin giving between 250 and 500 IU g^{-1} are required for most applications.

Nisin functions by binding to a specific protein in the cell membrane of susceptible bacteria and in doing so forming a pore in the membrane, allowing ions to leak across. Nisin's highly specific requirements for binding to target cells is clearly the reason why there has never been any evidence of unwanted effects on humans. It has a very long history of natural occurrence in dairy products.

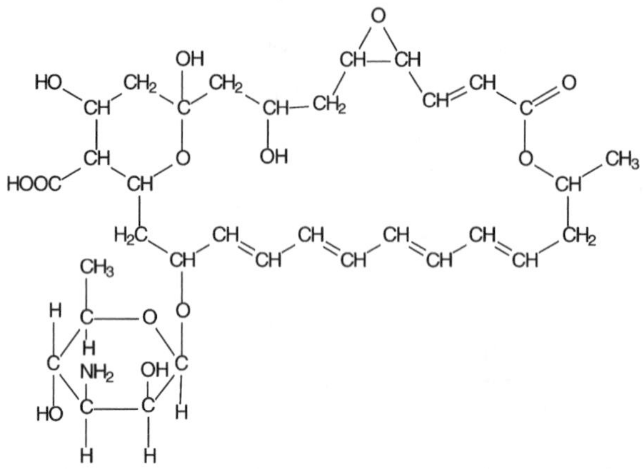

(9.12)

Natamycin (9.12) is an antifungal antibiotic (classed as a polyene) which has a limited clinical application in ophthalmics besides its equally restricted application in food preservation. It is permitted for use as a surface treatment for cheese to deter unwanted mould growth. It is produced by a number of *Streptomyces* species including the splendidly named *S. chattanoogensis*.

FOOD IRRADIATION

It could be argued that the use of ionising radiation to reduce or eliminate the microbial burden of foodstuffs belongs in textbooks of microbiology or physics. However, chemists are concerned to understand what effects irradiation may have on the nutrients and other components of food and must therefore have some appreciation of the process. The side effects of the other physical method of food preservation, heating, have been one subject of all the preceding chapters of this book.

The possibility that food may be preserved by irradiation was first recognised in the early 1950s, but in spite of varying degrees of acceptance by the authorities in many countries it remains a potential rather than a popular process. Irradiation was seen as a method of killing contaminating pests and micro-organisms without damaging a food's physical structure, flavour, or nutritive value.

A number of different levels of irradiation treatment can be used to achieve different process objectives. Novel names have been given to some of these processes, as shown in Table 9.2, but for the sake of the English language we must hope that they do not become widely adopted.

Although X-rays or electron beams from accelerators could provide the necessary radiation, the ideal penetration into food material is given by the γ-rays emitted by the radioactive decay of either ^{60}Co (1.17 and 1.33 MeV) or ^{137}Cs (0.66 MeV). The mechanism by which irradiation kills micro-organisms or induces other biological effects depends in the first instance on the absorption of energy by the atoms of the irradiated material. In the case of food materials the vast majority of the atoms present are of low atomic number. In a typically 'wet' biological tissue hydrogen ($\sim$10%), carbon ($\sim$12%), nitrogen ($\sim$4%), and oxygen ($\sim$73%) account for all but about 1% of the elements present. None of these elements is convertible into unstable, radioactive isotopes by radiation at the energy levels used. It follows that induced radioactivity will be at a very low level, in fact much lower than that from naturally occurring radioisotopes of heavier elements.

In the majority of foods the energy of irradiation is absorbed by the water present according to the Compton effect. The incident photon of the γ-ray transfers some of its energy to an orbital electron that is ejected from the absorbing atom. The residual energy is emitted as a photon of reduced energy. When water is the absorbing substance the loss of an electron leads to a range of free radicals, ions, and other products, including $\cdot$OH, H$\cdot$, H_2, H_2O_2, and H_3O^+. The ejected electron is immediately hydrated (indicated by e_{aq}^-) and gives rise to further free radicals. It is these various free radicals that are responsible for the observed chemical and biological effects of irradiation. The distinction between irradiation and other processes, such as heating, for delivering

Table 9.2 *Levels of irradiation treatment.*

Process	Treatment/ kGy^a	Effect
Inhibition	0.02–3	Delayed sprouting, ripening, *etc.* of vegetables and fruit
Disinfestation	0.1–1	Elimination of insect pests from stored food materials
Radurisation	1–5	Reduction of numbers of major spoilage bacteria and fungi
Radicidation	2–8	All vegetative microbial cells (but not spores) eliminated
Decontamination (hygenisation)	3–30	Reduction of total microbial load in food ingredients; diet sterilisation for specialised clinical applications
Radapperisation	$\sim$50	Effective sterilisation

^{a} kGy = kilogray (the gray is the SI unit of radiation dose: 1 Gy 1 Jkg^{-1} = 100 rad).

energy to food materials is that irradiation focuses a high proportion of the energy into a small number of chemically reactive molecules. By contrast, heat energy is dispersed throughout all the molecular compounds of the material and selectivity is lost.

The lethal effect of irradiation on the reproductive capabilities of micro-organisms can be readily attributed to the effect of free radicals on the vulnerable purine and pyrimidine bases of the nucleic acids. The dosages required to kill vegetative bacteria lie between 0.5 and 1.0 kGy; bacterial spores require doses in the range 1.0–5.0 kGy. The greater complexity of higher organisms leads them to require much lower doses for lethality: 5–10 Gy for mammals, 10–1000 Gy for insects. More detailed consideration of radiation effects on micro-organisms is properly the concern of microbiologists. The other biological effects of importance are those on the development processes in fruit and vegetables. Depending on the fruit and the exact dosage, small amounts of radiation will accelerate or retard the onset and speed of ripening. Sprouting in stored potatoes can also be inhibited. The actual mechanism of these effects is obscure, but this is hardly surprising when one considers how little is known of the biochemical control of plant development in general, even under normal conditions.

The side effects of irradiation on the chemical components of food are our primary concern. In general the effects on carbohydrates resemble those of heating under alkaline conditions (*see* Chapter 2). Attack by free radicals causes fragmentation to low molecular-weight compounds such as formaldehyde and glyoxal. The glycosidic bonds of oligo- and poly-saccharides are also vulnerable; in model systems amylose solutions have been shown to lose most of their viscosity with radiation doses of 20 kGy. The total amounts of carbohydrate breakdown products obtained are small and do not appear to include any that are not also associated with heating effects.

Much the same situation exists with regard to amino acids and proteins. Losses of amino acids are small, particularly in the dry state, when direct interaction of the γ-ray photons with the target molecules occurs rather than through water-derived products. As with polysaccharides, some depolymerisation occurs in aqueous systems. While this can have a deleterious effect on some physical properties, such as the whippability of egg-white, nutritional value is largely unaffected.

Lipid materials, especially unsaturated fatty acids, are much more vulnerable to irradiation. The susceptibility of lipids to free radical attack was described in Chapter 4, and experiments have shown that irradiation can initiate autoxidation along classical lines with all the expected end-products. Very high levels of fatty acid hydroperoxides are formed when oxygen is abundant. When fatty food materials are irradiated in an anaerobic environment, which is subsequently maintained, the extent of autoxidation and the development of rancidity are greatly reduced.

With the notable exceptions of thiamin, retinol, and tocopherols, losses of vitamins on irradiation of foods are slight. Many vitamins, *e.g.*, ascorbic acid, are quite unstable in simple aqueous solutions but are stabilised by close associations with other food components such as sugars and proteins. Substantial losses of thiamine occur on irradiation because of attack by both $\cdot$OH radicals and e_{aq}^{-} on the C=S and C=O double bonds. Carotenoids as well as retinol suffer from irradiation-induced autoxidation reactions in fatty foods but are more stable in vegetable materials. The same autoxidation reactions will overwhelm the tocopherols.

The overall conclusion to be drawn is that treatments of up to 5–10 kGy will have very little obvious effect on all but fatty foods, except to reduce greatly the rate of microbial spoilage. In general, irradiation of foods when they are frozen minimises deleterious effects on food components without much reduction in the anti-microbial effect. High doses will have very similar effects on the chemical components of food to those of sterilisation by heat. In spite of the reassurance that these observations should give to consumers, there is considerable opposition to the widespread use of irradiation of food preservation. A lot of this opposition is based on the fear, actually unfounded, of induced radioactivity. However, the concern of legislators and consumer organisations is the risk that unscrupulous food processors will be able to pass off as 'fresh' foods to which the popular understanding of this adjective could never apply. Successful legislation will require a reliable test which can show whether or not a foodstuff has been irradiated. Several methods have been developed in recent years but none are universally applicable to all foods.

In spite of the observation that up to 300 mg kg^{-1} radiolytic products may be produced per 10 kGy of radiation the nature of these products varies with the composition of the food. Heat processing can also generate so many of the same compounds that the number that qualify as 'unique radiolytic products' (URPs) is very small. A failure to detect URPs could not be taken as good evidence that food had not been irradiated. A valuable screening technique is examine the microbial flora of a suspect food. Too few bacteria, or more precisely, far more dead bacterial cells (as detected by the 'Direct Epifluorescence Filter Technique', DEFT) than live ones strongly suggests irradiation.

Free radicals generated by irradiation can remain trapped in foods containing dry, rigid matrices, such as bone (including fragments), shells (of shellfish), and some nuts and seeds. Their unpaired electrons can be detected by electron spin resonance (ESR) but the instrumentation is

expensive and the technique highly skilled. Most plant foods carry traces
of silica in soil particles, even after they have been washed, and irradiation
energy is trapped in silica's crystalline matrix. Heating releases this energy
as light, thermoluminescence, which can be measured and related to the
radiation dose.

When foods contain at least some lipid then it is worthwhile seeking
fatty acid breakdown products, as long as they can be distinguished from
those arising from heat treatment or oxidative rancidity (*see* Chapter 4).
The amounts of fragments from long chain unsaturated fatty acids such as
tetradecene ($C_{14}H_{28}$) and hexadecadiene ($C_{16}H_{30}$) have been shown to rise
in relation to the irradiation dose. They are separated from the foodstuff
by vacuum distillation and then identified and quantified by gas chromatography linked to mass spectrometry (GC/MS).

Another class of fatty acid/triglyceride breakdown product that do
qualify as URPs are the alkylcyclobutanones, whose formation is shown in
Figure 9.3. These too can be detected by gas chromatography.

One effect of irradiation on DNA is to cause breaks in the polymer
chain, leading to the formation of small fragments. These fragments can
be detected by a technique known as microelectrophoresis. Although
repeated freezing and thawing also damages DNA the pattern of fragments
produced by irradiation is sufficiently distinctive for this to be a valuable
procedure.

The exhortations from the authorities and the food industry that the
losses of flavour and nutritional value caused by low doses of irradiation

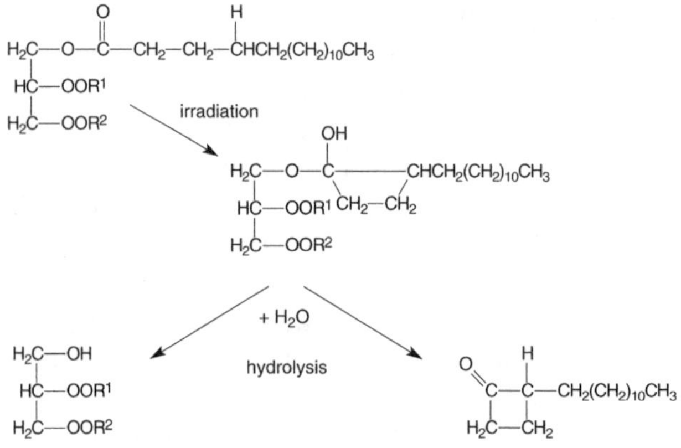

Figure 9.3   *The formation of 2-alkylcyclobutanes from triglycerides during the irradiation
of food. The example shown here is the formation of 2-dodecylcyclobutanone
from a triglyceride containing palmitic acid.*

are trivial are certainly not proving sufficient to quell public unease. Although the use of irradiation is increasing in many parts of the world, *e.g.* irradiated fresh meat is on sale in the USA, in Britain the food industry is very wary of public opinion. It has been said, in bringing irradiated products to the marketplace, that the major food retailers in Britain are in a desperate race 'to be first to be second'. At the present time (2002) herbs and spices are the only irradiated products on sale in Britain. One source of consumer concern is that unscrupulous food suppliers would be able to pass off less than fresh foods by using irradiation to eliminate their high microbial population without any in any way reversing the other signs of excess age, such as nutrient losses and formation of bacterial exotoxins (*see* Chapter 10) or off-flavours.

FURTHER READING

R. Angold, G. Beech and J. Taggart, 'Food Biotechnology', Cambridge University Press, Cambridge, 1989.

J. Stephen and R. A. Pietrowski, 'Bacterial Toxins', Nelson, Walton-on-Thames, 1981.

'Antimicrobials in Foods', ed. P. M. Davidson and A. L. Branen, 2nd edn., Dekker, New York, 1993.

'Natural Antimicrobial Systems and Food Preservation', ed. V. M. Dillon and R. G. Board, CAB International, Wallingford, 1994.

'Foodborne Diseases', ed. D. O. Cliver, Academic Press, London, 1990.

V. M. Wilkinson and G. W. Gould, 'Food Irradiation. A Reference Guide', Butterworth-Heinemann, Oxford, 1996.

E. Lück. and M. Jager, 'Antimicrobial Food Additives. Characteristics, Uses, Effects', 2nd edn., Springer, Berlin, 1997.

'New Methods of Food Preservation', ed. G. W. Gould, Blackie, London, 1995.

J. M. Jay, 'Modern Food Microbiology', 6th edn., Aspen, Gaithersburg, Md., 2000

'Dietary Intake of Food Additives in the UK. Initial Surveillance', Ministry of Agriculture, Fisheries and Food, HMSO, London, 1993.

M. J. Waites, N. L. Morgan, J. S. Rockey and G. Higton, 'Industrial Microbiology: An Introduction', Blackwell Science, Oxford, 2001.

Chapter 10

Undesirables

The previous chapters in this book should have left the reader with no doubt that the foods we eat contain many substances that are neither directly nutritive nor otherwise valuable. While we can suppose that there must be many minor food components to which our bodies are quite indifferent, the rule that substances that are not friends must be enemies will remain useful. This chapter is devoted to a representative selection of the innumerable substances that we find in our food that we would probably be better off without. Obviously there are many others, a good proportion of which feature elsewhere in this book, that do have some adverse effects but have some redeeming nutritional virtues, such as the saturated fatty acids.

So far in this book, substances have been classified in terms of either their chemical structures, in Chapters 2–5, or their functions, in Chapters 6–9. For the undesirable subjects of this chapter neither approach will do. The enormous diversity of both their chemical structures and the nature of their undesirable effects means that we must resort to classification by origin. This gives us the four broad groups listed here and the basis for the overall arrangement of the material in this chapter.

   (i) Endogenous toxins: substances that are normal, natural components of food materials.

  (ii) Microbial toxins: substances that arise through the activity of contaminating moulds or bacteria.

 (iii) Toxic residues: substances that are carried over into food materials from procedures applied (by man) to the living plants or animals that become our food.

 (iv) Toxic contaminants: substances that arise during, or are derived from, food-processing, preservation or cooking operations.

ENDOGENOUS TOXINS OF PLANT FOODS

There is a widespread assumption that if a foodstuff is 'natural' it must be 'safe'. Following from this assumption is the idea that those afflictions of mankind that can be traced back to diet will disappear just as soon as we stop eating 'unnatural' foods and stop processing the 'natural' ones in 'unnatural' ways. For the author there is no clearer demonstration of the naivety of these views than the abundance of naturally occurring toxins in apparently 'natural' plant food materials. In the face of the evidence of these toxins it is hard to maintain the view that any plants were placed on this planet with the primary function of sustaining human life. The physiological similarities between humans and the animals they eat mean that endogenous toxins are encountered more frequently in plants than in animals and more frequently in fish than in mammals. In this section some of the more important plant toxins will be examined, although the selection here cannot be regarded as comprehensive.

One of the best known of all plant toxins is the solanine of potatoes. This steroidal glycoalkaloid (10.1) occurs not only in potatoes (along with much smaller amounts of a variety of related alkaloid derivatives) but also in other members of the Solanaceae family such as the aubergine and the highly poisonous nightshades. Normally potatoes contain 2–15 mg per 100 g fresh weight. When potatoes have been exposed to light and turned green, the level of solanine may reach 100 mg per 100 g, mostly concentrated just under the skin. The growing popularity of commercially prepared potato products where the skin is intended to be eaten is a cause for concern as even in the absence of obvious greening the concentration of total glycoalkaloids in the skin can reach 60 mg per 100 g. Other metabolically active tissues such as the sprouts may contain even higher concentrations. Solanine is not one of the more potent toxins, although over the years several accounts of fatalities caused by it have accumulated in the scientific literature.

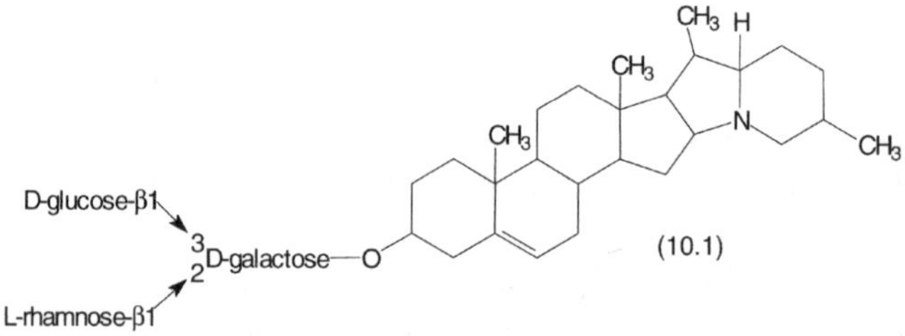

D-glucose-β1
3 D-galactose—O
2
L-rhamnose-β1
(10.1)

Solanine is established as an inhibitor of the enzyme acetyl choline esterase, a key component of the nervous system, and signs of neurological impairment have been recorded after ingestion of the toxin at a level of approximately 2.8 mg kg^{-1}. The relatively high doses required for toxic effects are considered to be the result of its poor absorption from the gastro-intestinal tract, as experimental animals show much greater sensitivity when the toxin is injected directly into the bloodstream.

General public awareness of the dangers of eating greened potatoes has kept incidences of potato poisoning to a low level. This is aided by the observation that potatoes with above about 20 mg per 100 g solanine have a pronounced bitter taste, making them unacceptable anyway. As solanine is fairly insoluble in water and heat stable, none is lost during normal cooking procedures. The widely accepted safety limit for total glyco-alkaloids is 200 mg per kg but lower limits have been suggested. Partly this is due to the wide variations that can be found in samples from the same batch of potatoes and also because of our increasing awareness that some of the milder, but no less unwanted, symptoms of solanine poisoning such as abdominal pain, vomiting *etc.* may well go unreported.

Caffeine (10.2), already mentioned in Chapter 6, is a purine alkaloid. Another important member of this group, otherwise known as the methylxanthines, is theobromine (10.3). It may be argued that these substances, found in tea, coffee, cocoa, and cola beverages, should not be classified as toxins since they are usually regarded as stimulants. In the author's opinion there are two good reasons for treating them as toxins. Firstly, whether their physiological effects are beneficial or not is dependent on the amount consumed; secondly, they could never be regarded as nutrients.

(10.2)   (10.3)

Roasted coffee beans have between 1 and 2% caffeine, but the level in the beverage is highly dependent on the method of preparation and the strength. Values anywhere between 50 and 125 mg per cupful are normal. Black tea leaf (*see* Chapter 6) contains some 3–4% caffeine, giving around 50 mg per cupful, and rather less theobromine, yielding around 2.5 mg per cupful. Cocoa powder contains around 2% theobromine and 0.2% caffeine. The consequent levels in plain chocolate are about one quarter of these figures and in milk chocolate about one-tenth. Nowadays the caffeine content of cola drinks is restricted to a maximum of

$200 \, \text{mg} \, l^{-1}$, although the average value is around $65 \, \text{mg} \, l^{-1}$. In recent years so-called 'energy drinks' containing caffeine have become popular. These often include extracts of the South American plant, guarana (*Paullinia cupana*), which provides additional caffeine to give a final caffeine concentration in the $200–300 \, \text{mg} \, l^{-1}$ range.

Nearly all physiological studies of methylxanthines have concentrated on caffeine, and two of its effects stand out. Caffeine's stimulant action is the result of its stimulation of the synthesis or release of the catecholamine hormones epinephrine and norepinephrine (better known as adrenalin and noradrenalin, respectively) into the bloodstream. Most of the other immediate effects that have been observed on levels of glucose, triglycerides, and cholesterol in the blood of coffee drinkers can be traced back to this hormone release. Caffeine's diuretic action is of course well known, but like so much of the action of these compounds in the human body its details are beyond the scope of this book. Although, even for quite heavy coffee drinkers, these various effects tend to lie within the normal range, it remains questionable whether they can be beneficial in the long term.

There have been too few cases of death directly attributable to caffeine consumption to give useful data on its acute toxicity, but doses from 150 to $200 \, \text{mg} \, \text{kg}^{-1}$ have proved fatal to humans. Nothing like this sort of dosage is going to be achieved even by obsessive coffee drinking. A vast amount of work has been carried out, using both animal experimentation and epidemiological methods, to try and establish whether caffeine consumption has adverse effects on human reproduction. At the present time caffeine has been exonerated from causing either birth defects or low birth weights, but it is still concluded that its ordinary stimulant effects are not beneficial during pregnancy. For many consumers the answer to the real or supposed adverse effects of caffeine consumption is to consume decaffeinated instant coffee.

The caffeine is removed by solvent extraction before the beans are roasted. Water is first added to the beans to bring their moisture content up to about 40%. They are then extracted in a countercurrent system with methylene chloride at temperatures between 50 and 120 °C, conditions fairly selective for caffeine. Residual solvent and moisture are then driven off and the beans roasted. An alternative solvent that is extremely selective for caffeine is liquid carbon dioxide at high pressure, 120–180 times atmospheric. The beans still have to be moist, and the caffeine is removed from the circulating supercritical CO_2 by absorption onto activated charcoal. This process has the advantages that removal of desirable flavour elements is minimised and that there is no question of the possible toxicity of residual solvent.

One other group of alkaloids, the so-called pyrrolizidine alkaloids or

PAs, are more often associated with herbal remedies than foods. They would barely deserve a mention in this book from their very occasional appearance in milk or honey when the cows or bees concerned have fed on unsuitable plants. However, one plant material, comfrey, is often used to make herbal tea, a product on the borderline between food and medicine. Comfrey contains at least twelve different pyrrolizidine alkaloids. Symphytine (10.4) is one of the most abundant of them and illustrates the characteristic chemical structure of these compounds, the fused 5-membered rings with a shared nitrogen atom. A cup of herbal tea made from comfrey root could contain as much as 9 mg total PAs. With so many different compounds occurring in such a diverse range of plants reliable data on the toxicity of PAs are scarce. Symphytine is regarded as possibly carcinogenic and PAs are usually assumed to be the culprits in cases of liver damage associated with the use of herbal preparations.

(10.4)   (10.5)

Substances with adverse effects on human consumers often turn out to be natural insecticides, *i.e.* substances that plants synthesise to ward off insect pests. A good example of this phenomenon are the psoralens of celery. Psoralens are a group of substances (classed as furocoumarins) that widely in the plant kingdom. Their significance as components of edible plants came to notice in the USA in 1986 when numerous reports of severe skin rashes and related conditions in farm workers, packers, shop workers and others were linked to them having handled a new variety of celery, bred for resistance to insect pests. Psoralens, notably 8-methoxy-psoralen (10.5), were found to be present in the green leafy parts of the celery at levels around 6200 p.p.b., compared with 800 p.p.b. in other varieties. The psoralens from the plant are absorbed through the skin and entered the nuclei of cells in the skin. Here they interact with the DNA, finding their way into the space between neighbouring pairs of thymidine residues, a process referred to as *intercalation*. Subsequent irradiation with UV light, as would be experienced by anyone working out of doors, provokes the formation of a link between the psoralen molecule and the pyrimidine ring of the thymidine. Although the human systems for the repair of damaged DNA do recognise psoralen damaged DNA tissue damage still results. Suffice to say this celery variety is no longer grown.

The other food plant that has presented the same problem is the lime. Limes can produce similar UV induced photodermatitis. In this case the usual victims are sunbathers who have whiled away their time on the beach preparing margaritas*, leaving behind traces of fresh lime juice on their skin.

Chocolate is also well known for its ability to bring on migraine headaches in susceptible individuals. This is due to its content of phenylethylamine. Phenylethylamine is one of the so-called vasopressor amines that occur widely in foods, some of the more notable of which are shown in Figure 10.1. They may be normal products of the amino acid metabolism of the plant or, in the case of cheese and wine, the fermenting micro-organism. Normally the body is well equipped to deal with these amines. The activity of the enzyme monoamine oxidase situated in the mitochondria of many tissues converts them into the corresponding aldehydes. It is clear that some individuals are either especially sensitive to these amines or have difficulty in oxidising them. Inhibitors of

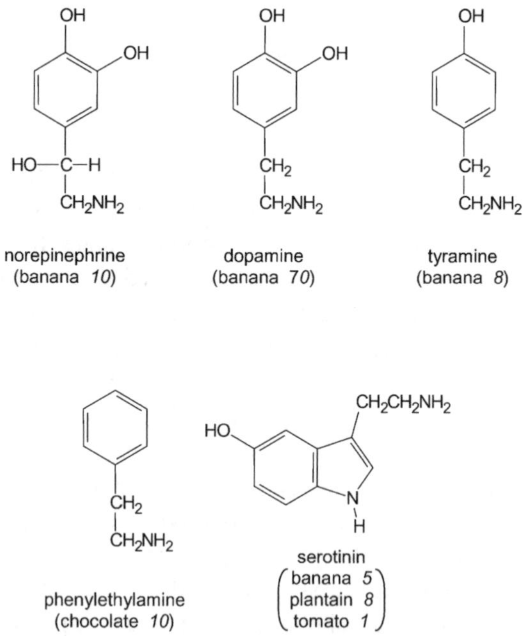

Figure 10.1 *Vasopressor amines. The figures in italics are typical values for the concentrations of the amine in the food material indicated, expressed in mg per 100 g fresh weight. Phenylethylamine is also found in significant amounts in many cheeses and red wines.*

---

* Margarita = a cocktail of lemon or lime juice plus tequila.

monoamine oxidase are important drugs (*e.g.* in the treatment of depression), so that patients taking them are advised to avoid these foods. The usual effect of these amines is constriction of the blood vessels in different parts of the body. Hypertension is the generalised result, but in the brain vasoconstriction causes intense headaches. There is also some support for the suggestion that the prevalence of a certain type of heart disease in West Africa may be caused by the large amounts of plantain, a relative of the banana, in the diet. A typical plantain-based diet could easily give a daily intake of 200 mg serotonin.

Another African staple food, cassava (manioc), poses a quite different toxicological problem, that of cyanogenesis. In Chapter 2 the glycoside amygdalin was briefly described. In fact amygdalin is just one of a group of glycosides that occur in many plant families but particularly the Rosaceae and Leguminoseae (*see* page 302). Although the presence of amygdalin in the stones of some fruit does not pose much of a hazard, the higher levels of α-hydroxyisobutyronitrile and related compounds in cassava do. Fresh cassava may be able to give rise to 50 mg hydrogen cyanide 100 g due to the activity of two enzymes, β-glucosidase and hydroxynitrile lyase (8.46). These both occur in the plant tissue and become active (*cf.* phenolase, Chapter 6) when the tissue is damaged during harvest or preparation for cooking. Although animals do not produce either of these enzymes for themselves, there is evidence that intestinal bacteria are able to release the hydrogen cyanide from cyanogenic glycosides. The need to allow a fermentation period during the preparation of cassava for consumption in order that the endogenous cassava enzymes release the HCN, which, being volatile, will then be lost during cooking, is a well established tradition in West Africa amongst those who depend on this crop.

In spite of these defensive food preparation procedures, it is becoming clear that chronic cyanide poisoning, caused by ingestion of low levels over prolonged periods, is widespread in parts of the world dependent on cassava. Two common diseases, one a degenerative neurological condition (ataxia) and the other a form of blindness, have been attributed to an interaction between cyanide intake and deficiency of the vitamin cobalamin. Another problem is caused by the body's use of the enzyme rhodanase to detoxify small amounts of cyanide by converting it to thiocyanate, CNS^-. Where cyanide intake is high, *e.g.* in those parts of eastern Nigeria where dried, unfermented cassava is popular, high levels of thiocyanate are found in the blood. Thiocyanate is known to interfere with iodine metabolism, resulting in the goitre that is common in eastern Nigeria.

Another important food crop that can have dangerous levels of cyanogenic glycosides is the lima bean, *Phaseolus lunatus*, grown in many parts of the world. The difficulty of preparing these beans in a way that eliminates their toxicity has led to the breeding of bean varieties with reduced levels of glycosides.

Cyanogens are not the only toxins found in legumes: they also have two other types of toxin which are potentially troublesome to those who like their peas or beans raw. The first to be considered are the protease inhibitors. Legume seeds, including peas, beans, soybeans, and peanuts, all contain proteins that will inhibit the proteases of the mammalian digestive tract, notably trypsin and chymotrypsin. The inhibition is the result of a one-to-one binding that blocks the active site of the enzyme. These proteins fall into two groups, those that are fairly specific to trypsin and with molecular weights of around 22 000 and those with molecular weights of between 6000 and 10 000 that inhibit chymotrypsin as well. Studies with experimental animals fed on diets of little besides unheated soya bean meal show reduced efficiency of utilisation (*see* Chapter 5) of proteins. If the diet is persisted with for some time, the pancreas, the organ that synthesises trypsin and chymotrypsin, shows a marked increase in size, presumably some type of compensatory mechanism.

When legumes have been heated, the efficiency with which they are utilised as sources of dietary protein rises, but experiments with soya have shown that prolonged heating does not allow the theoretical efficiency to be reached. This is because the losses of essential amino acids caused by the Maillard reaction start to become significant.

Any examination of the effects of protease inhibitors in legumes is complicated by the presence of the second group of toxic proteins they contain, the lectins. The lectins are characterised by their ability to bind to the surface of certain animal cells, particularly red blood cells, and cause them to clump together; hence their alternative name, haemagglutinins. Because different lectins show a high degree of specificity for particular types of animal cells, *e.g.* red blood cells of different blood groups may be distinguished, they have become popular laboratory tools for animal biochemists and physiologists. Injected directly into the bloodstream, some can be extremely potent poisons with toxic doses in the region of $0.5 \, \mathrm{mg \, kg^{-1}}$.

Lectins are proteins in the high-molecular-weight range, around 10^5. Because of their usefulness to biochemists rather than their dietary importance, some have been studied in great detail, but the results did not, until recently, have much relevance to food science. In recent years many experiments, some successful, have been conducted to determine whether the genes for particular lectins could be inserted into other plant species by recombinant DNA techniques. The idea is that this might be a nonchemical way of enhancing the pest resistance of valuable agricultural crops. Lectins may represent quite a significant fraction of the total protein of legumes, 3% in the case of soybeans. Experimental animals fed diets containing as little as 1% unheated soybeans show impaired growth, but some other legumes produce lethal effects at this level. The differences appear to be caused by the different susceptibilities of different lectins to breakdown by digestive enzymes. The lectin of kidney beans is one of the more resistant. The toxic effect of orally administered lectins is reduced uptake of nutrients from the digestive tract. This is supposed to be the result of damage to the intestinal mucosa. The relevance of lectin toxicity to humans is debatable. From time to time there are bursts of enthusiasm in the media for eating raw vegetables, which, as we have seen, poses a potential hazard. As long as the contribution of uncooked legumes to one's total protein requirement remains small, the danger of lectin-induced illness is likely to be slight. Fortunately for advocates of the culinary value of barely cooked sprouted beans, massive breakdown of lectins occurs as the seeds germinate.

Soya beans are the best known source of another group of biologically active substances that actually quite widely in plants, the phytoestrogens. They were discovered in the 1940s as the cause of fertility problems in Australian sheep that had consumed a diet rich in a species of clover, now recognised as major source of phytoestrogens. Most phytoestrogens are isoflavones. Foods based on soya, such as soya milk, soy sauce*, tofu* and tempeh* are the main sources in human diets. Amounts of total phytoestrogens found in soya beans range between 200 and 3000 mg per kg, mostly genistein or daidzein, (*see* Figure 10.2).

With the exception of infants being fed soya milk as an alternative to human milk or cow's milk based 'formula', human intakes are very low, by a factor of 50 to 100 compared with artificial eostrogen intakes from

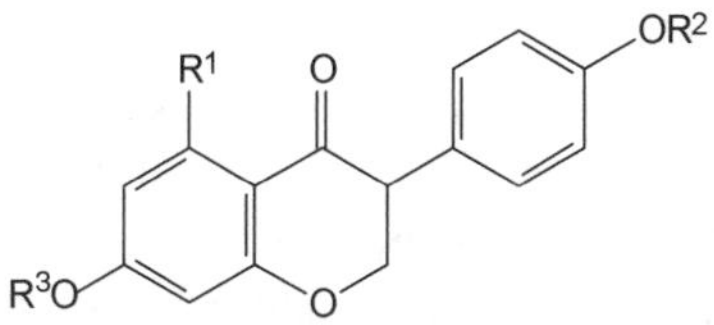

	R^1	R^2	R^3
daidzein	H	H	H
genistein	OH	H	H
glycetein	H	OCH_3	H

Figure 10.2 *The principal isoflavonoid phytoestrogens of soya beans. In the raw beans there is usually a glucose residue attached at position 3. These glycosides are known as daidzein, genistein and glycetein.*

* All products of bacterial or fungal fermentation.

hormone replacement therapy or oral contraceptives. Total phytoestrogen levels in soya based infant formula vary but are normally around 9 mg per litre (when made up). Concern is regularly expressed regarding the possible effects on subsequent reproductive function following high consumption levels from soya milk in infancy (in the region of 4 mg per kg body weight per day) but to date there is insufficient evidence to allay, or substantiate, this concern. There is, however some evidence that the incidence of breast cancer may be lower in populations, such as Asians and vegetarians, whose diet is rich in soya beans and soya products. Studies comparing Japanese and Finnish males show much higher plasma isoflavone concentrations amongst the Japanese.

As an aside to these problems with soya this is an appropriate point to mention the contamination of some soya based products with 3-mono-chloropropane-1,2-diol (3-MCPD, 10.6). 3-MCPD is found from time to time at very low levels ($\sim$10 μg kg^{-1}) in a wide variety of food products. The UK authorities concluded that at this level 3-MCPD is unlikely to present a carcinogenic risk to man, bearing in mind that the 'No Observable Effect Level' (NOEL, *see* page 353) has been established as 1.1 mg per kg body weight per day. However, in some products, notably soy sauce, containing acid hydrolysed soya protein 3-MCPD has been found at higher levels ($>$20 μg kg^{-1}) that are a cause for concern. The 3-MCPD arises from a reaction between the hydrochloric acid used for the hydrolysis and the inevitable residues of soya bean oil fatty acids.

(10.6)   (10.7)

The concluding toxin in this section is myristicin (10.7). This occurs in significant levels in nutmeg and smaller amounts in black pepper, carrots, and celery. There is sufficient myristicin in 10 g of nutmeg powder to induce similar symptoms to a heavy dose of ethanol: initial euphoria, hallucinations, and narcosis. Higher doses induce symptoms that also correspond closely with those of ethanol poisoning, including nausea, delirium, depression, and stupor. Of course the amounts of myristicin contained in the quantities of nutmeg used as a flavouring are way below those required for effects on the brain, but it is worth noting that traditionally the consumption of nutmeg-flavoured foods during pregnancy has been regarded as ill advised.

ENDOGENOUS TOXINS OF ANIMAL FOODS

The reason for the relative shortness of this section has already been indicated. Most animal toxins of significance to human eating habits are found in taxonomic groups distant from man, such as the crustaceans, but one of the most interesting occurs in a number of fish species, notably the infamous puffer (or *fugu*) fish. The toxin tetrodotoxin (10.8) occurs in various organs of the puffer fish including the liver and ovaries (at around 30 mg per 100 g) and to a lesser extent the skin and intestines. The minimum lethal oral dose of tetrodotoxin is estimated to lie between 1.5 and 4.0 mg. The muscles and testes are generally free of the toxin, and it is these that are a popular delicacy in Japan. Unfortunately, great skill is required by the cook to separate the deadly parts of the fish from the merely risky. Stringent licensing of expert cooks has certainly reduced the risk, but fatalities remain a regular occurrence.

(10.8)

The toxin is rapidly absorbed, and the first symptoms may be seen as soon as ten minutes after eating the toxic fish; however, an interval of an hour or so is more usual. The effect of the toxin is to block movement of sodium ions across the membranes of nerve fibres. The result is that nerve impulses are no longer transmitted and the victim suffers a range of distressing nervous symptoms developing into total paralysis and the onset of the respiratory failure that usually leads to death within 6–24 hours. No effective treatment has been discovered, and in the face of continued enthusiasm for eating these fish there is little that can be done except to ensure the best possible training for Japanese chefs.

Of the other toxins associated with marine foods the best known are rather difficult to classify unequivocally as being of animal origin. For example there is a group of toxins that occur under certain circumstances in many otherwise edible types of shellfish such as species of mussels, cockles, clams, and scallops. At certain times of year the coastal waters of many of the hotter parts of the world develop a striking reddish colour due to the massive proliferation of red-pigmented dinoflagellates, a type of plankton. These 'red tides' are referred to in the Old Testament (Exodus 7: 20–21) and may have given the name to the Red Sea. Problems also occur in some cooler regions such as the coasts of Alaska and Scandinavia, and the dinoflagellates are not always red-pigmented. Some species of dinoflagellates contain toxins that quickly find their way up the food chain to shellfish normally caught for human consumption. There are a number of different classes of shellfish toxins, but the best known are those that cause paralysis and other effects on the nervous system that can result in rapid death from doses of only a few milligrams. Saxitoxin (10.9) is one of the best known of the toxins causing PSP – paralytic shellfish poisoning. Most of the others differ only in having minor variations in the groups attached around the rings. A number of different toxins may be encountered in a single outbreak of PSP. The relationship between red tides and the toxicity of shellfish appears to have been well recognised in coastal communities for centuries and as a result the number of outbreaks is not large. However, the number of fatalities that may occur in a single outbreak of paralytic shellfish poisoning is such that the authorities in vulnerable regions such as the Pacific coast of North America maintain routine checks on the toxin levels in shellfish catches. A serious, but fortunately not fatal, outbreak of PSP occurred in 1968 when 78 people were taken ill after eating mussels from sites on the north east coast of England. Since that time shellfish on the north east coast of Britain have been regularly monitored and on one subsequent occasion (ten weeks in the summer of 1990) the levels detected were sufficiently high, *i.e.* greater than 80 g per 100 g of mollusc flesh, for warnings against eating locally caught shellfish to be issued.

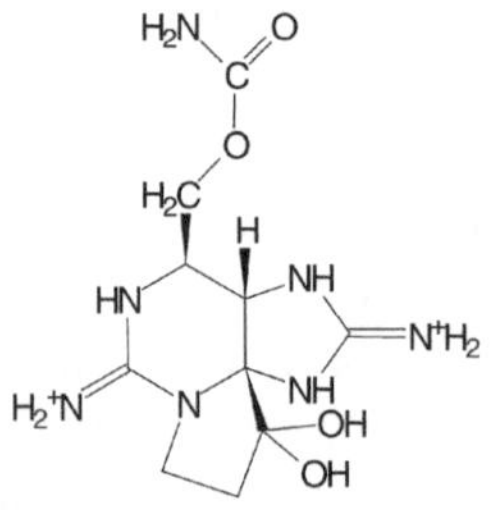

(10.9)

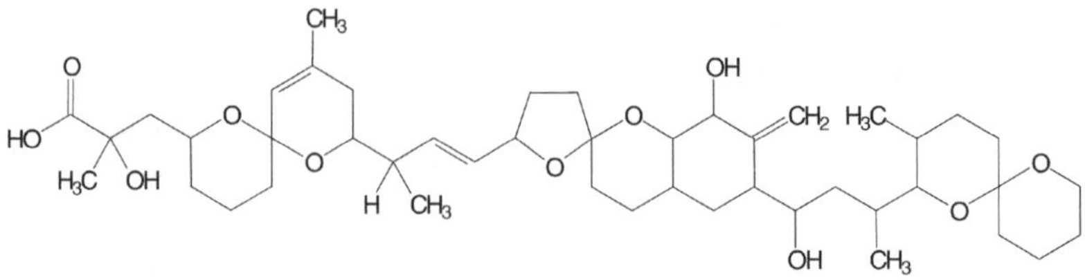

(10.10)

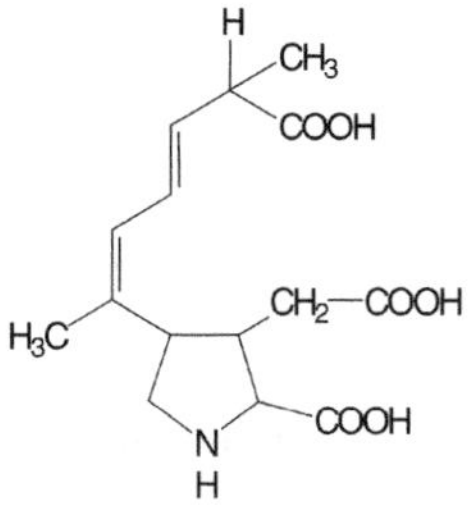

(10.11)

Diarrhetic shellfish poisoning (DSP), involving a rapid onset of diarrhoea, nausea and vomiting that lasts up to 3 days, occurs mostly in Japan but cases do occur world-wide. Although no cases have been recorded in Britain the toxins concerned have been detected in shellfish from British waters. The best known of the DSP toxins is okadaic acid (10.10), again produced by species of dinoflagellates. Cases of amnesic shellfish poisoning (ASP), were first identified in Canada. This can cause severe neurological symptoms that can lead to permanent brain damage. Domoic acid (10.11), the toxin involved, is produced by various diatom species.

A quite different type of poisoning is associated with fish belonging to the family Scombridae. This family includes many commonly eaten fish including mackerel, tuna, and sardines. When freshly caught the fish are not toxic, but if held at temperatures above 10 °C for several hours high levels of histamine (10.12) are sometimes formed from the amino acid histidine that occurs naturally in the muscle. The reaction is catalysed by a decarboxylase that is produced by a number of common bacteria, but that of *Proteus morganii*, an otherwise fairly innocuous marine bacterium, has been particularly incriminated. In spite of the activity of the bacteria in producing levels of histamine of over 100 mg per 100 g, obvious putrefaction of fish may not have occurred. Consumption of the large doses of histamine that might occur in contaminated fish does not necessarily lead to any adverse effects, as histamine is poorly absorbed from the intestine. Nevertheless the collection of symptoms that is seen from time to time, which include severe headaches, palpitations, gastrointestinal upsets, skin flushes, and erythema, correspond closely to the effects of histamine injection into the bloodstream. Also, the symptoms are characteristically relieved by antihistamine drugs. It is assumed that victims of scombrotoxic fish have some minor intestinal lesion or disorder that has let the histamine through.

$$\underset{\substack{N\diagdown\\ \diagup NH}}{}\overset{\substack{CH_2-\underset{|}{\overset{COOH}{\underset{|}{CH}}}\\ NH_2}}{} \longrightarrow \underset{\substack{N\diagdown\\ \diagup NH}}{}\overset{CH_2CH_2NH_2}{} + CO_2$$

(10.12)

Histamine is classified as a biogenic amine, *i.e.* a low molecular weight amine having a recognised biological activity. Other biogenic amines that occur in plant materials, and grouped together by their tendency to raise blood pressure in humans, were described on page 331. Such biogenic amines are not uncommon in foods of animal origin whenever micro-organisms have become involved, as in the spoilage of meat and the maturation of cheese. Histamine, tyramine, cadaverine [$H_2N(CH_2)_5NH_2$, derived from the amino acid lysine] and putrescine [$H_2N(CH_2)_4NH_2$, from glutamine] are all abundant in cheese, typically at levels between 100 and 500 mg kg^{-1}. The levels of amines in meat are generally around one tenth of these figures.

MYCOTOXINS

In the preceding chapter the susceptibility of human and animal foodstuffs to contaminating micro-organisms was explained. While the value to man of the fermentative micro-organisms involved in dairy products *etc.* is generally recognised, most people will still reject food that is obviously 'mouldy'. This rejection may well be made on what appear to be aesthetic grounds, but in fact the growth of a great many species of fungi on foods and raw materials is accompanied by the production of toxins that are often extremely dangerous.

The best known of the diseases caused in this way is *ergotism*. In the Middle Ages this disease reached epidemic proportions in much of Europe, particularly where rye was an important cereal crop. During this century smaller outbreaks have occurred in both Britain and France. It is now established that the cause of the disease is infection of rye and other cereals in the field by the commonly occurring mould *Claviceps purpurae*, or ergot. At one stage during its life cycle the mould produces structures known as *sclerotia*, hard, purplish-black masses of dormant cells. About the same size as a cereal grain, these may be harvested inadvertently along with the grain and, if not recognised and removed, end up being milled into the flour. The sclerotia have been found to contain at least 20 different toxic alkaloids. The most abundant of these are peptide derivatives of lysergic acid, as shown in Figure 10.3.

The biological effects of the ergot alkaloids appear to be the collective

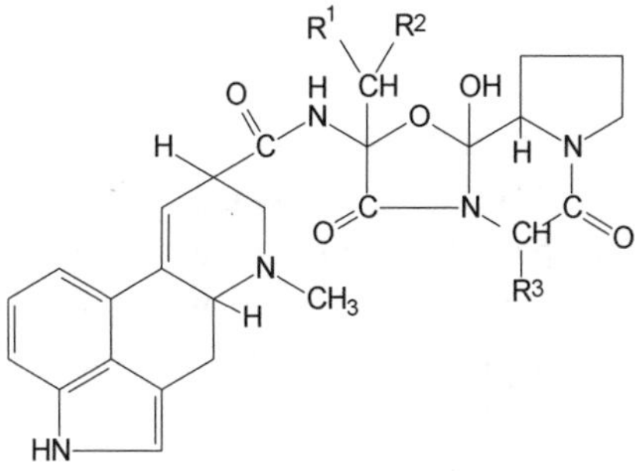

Figure 10.3 *Ergot alkaloids. In ergotamine, one of the most abundant ergot alkaloids, R^1, R^2 and R^3 are H, CH_3 and C_6H_5 respectively. Others have various combinations of H or CH_3 at R^1 and R^2 and R^3 and $C_6H_5CH_2$, $(CH_3)_2CHCH_2$ or C_3H_2 at R^3.*

result of the effects of different alkaloids on separate aspects of human physiology, so that the symptoms will vary from outbreak to outbreak depending on the range of toxins produced by the particular strain of the mould involved. The onset of the disease is usually slow but insidious and generally catastrophic in outcome. Through as yet poorly understood effects on both the nervous and vascular systems the victim initially suffers generalised discomfort and then intense burning pains in the hands and feet. These give way to a total loss of sensation in the limbs that mercifully precedes first a gangrenous withering and blackening and then loss of the limbs. These symptoms are often accompanied by mental derangement and gastrointestinal failure before the victim finally succumbs. It is hardly surprising that in the Middle Ages this awful disease was given the name St Anthony's Fire.

The comparative rarity of the disease today can be ascribed to a number of factors. Modern fungicides reduce the incidence of infected plants to a minimum, and modern grain-drying systems largely prevent post-harvest spread. Nowadays millers inspect all grain shipments for the tell-tale signs of the black sclerotia, and of course modern grain-cleaning methods are effective in separating the heavy grains from the light sclerotia. In spite of the success of these preventive measures, there is no room for complacency. One should beware of 'amateur' millers, especially as they are also likely to favour cereals that were grown without the benefit of fungicide treatments.

To date, some 150 different mould species have been shown to produce toxins when they grow on human or animal foodstuffs. Cereals are their most popular targets. Most of these moulds are members of one of the three genera *Fusarium*, *Penicillium*, or *Aspergillus*. It would obviously be outside the scope of this book to consider even a small selection in the detail given to ergot, but there is one *Aspergillus* species, *A. flavus*, that demands special attention.

In England in 1960 thousands of turkeys were killed in an outbreak of what became known as Turkey X disease. The cause was traced to Brazilian groundnut (*i.e.* peanut) meal infected with *A. flavus*, and it was not long before the group of toxins shown in Figure 10.4 was identified. Although there are some differences in the relative toxicities of the different aflatoxins between different animal species, it is now clear that aflatoxin B1 is one of the most potent liver carcinogens known. A diet containing only 15 p.p.b. will, if fed for a few weeks, induce tumour formation in most experimental animals.

As might be expected, data on the effects of these toxins on humans are scarce, but they are recognised as acute hepatotoxins. They are carcinogenic in many, but not all, animal species. In humans long term exposure is a probable cause of liver necrosis and involvement in liver cancer. There are also links to the occurrence of hepatitis B. A further complication is that the mechanism of aflatoxin induced carcinogenesis quite different in humans from that in rodents. In the tropics numerous cases have now been

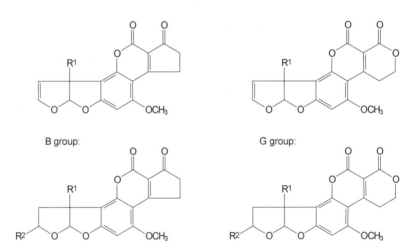

Figure 10.4 *The mycotoxins of* Aspergillus flavus. *The aflatoxins are bifuranocoumarins fused, in the B group, to a cyclopentanone ring or, in the G group, to a lactone ring. Variations in substitution pattern of the bifuranocoumarin system occur in both groups, as shown in the table. Hydroxylation at* R^2 *eliminates toxicity.*

R^1	R^2	B group	G group
H		B_1	G_1
OH		M_1	GM_1
H	H	B_2	G_2
H	OH	B_2a	G_2a

recorded of fatalities due to acute hepatitis and related disorders that were linked to the consumption of mouldy cereals, especially rice, that subsequently were shown to contain between 0.2 and 20 mg kg^{-1} of aflatoxins. Of much greater importance are the epidemiological studies that have now been carried out in many parts of the world, notably Africa and the Far East. These studies have shown a clear positive correlation between the incidence of liver cancer in a community and the level of aflatoxins in its staple foodstuffs. Although few plant foodstuffs appear to be immune from aflatoxin contamination, there is no doubt that peanuts and derived products such as peanut butter can present particular hazards. Levels of aflatoxins in peanut butter produced in some Third World countries may routinely reach 500 μg kg^{-1}. As a result of such widespread contamination it is not uncommon to find the total daily aflatoxin intake in some Third World communities being estimated at some 5 ng kg^{-1} body weight.

Since the climate of western Europe tends not to favour aflatoxin production the regulatory authorities in Britain focus their attention on a small range of imported products, including peanuts and peanut butter, dried figs, animal feeding stuffs, and some other types of nuts which are known to be particularly vulnerable to contamination. The legal limit for aflatoxin contamination in the UK is 4 μg kg^{-1}. The problem that faces the authorities in the detection of contaminated foodstuffs is the wide variation between samples. For example some years ago an analysis of 36 1-kg samples selected at random from a single consignment of imported dried figs showed that the majority were below this limit but three were above 50 μg kg^{-1} and one of these had over 2 mg kg^{-1}!* Wherever contaminated animal feedstuffs are used, aflatoxins can be expected in milk. Aflatoxins B$_1$ and G$_1$ are converted into M$_1$ and GM$_1$ (the 'M' is for milk), respectively, by liver enzymes, which then appear at levels of around 0.05–0.5 g l^{-1} in liquid milk. It is remarkable how much of the aflatoxin in a cow's diet subsequently appears in its milk. The concentration of aflatoxin is generally found to be around 1% of the level in the feed. This may not seem much, but it is sufficient for aflatoxins to be a potentially serious problem for the dairy industry. The concentrations of aflatoxins that come through into meat are less than one-thousandth of feedstuff levels.

The accumulation of detailed knowledge of aflatoxin distribution that the authorities in many parts of the world have achieved has been greatly facilitated by the relative ease with which aflatoxins may be detected and

---

* Quoted in Ministry of Agriculture, Fisheries and Food (1993) Mycotoxins: third report. Food Surveillance report No 36, HMSO.

quantified. Aflatoxins fluoresce strongly in UV light. At around 365 nm the B-types give a blue and the G-types a green fluorescence. Solvent extraction has to be followed by fairly elaborate cleanup procedures before TLC or HPLC can be used for detection at the μg level. Determinations down to 10 p.p.b. (on the original food) are now possible immunochemical techniques such as enzyme-linked immunosorbent assay, ELISA.

In recent years food scientists have begun to recognise the potential hazards caused by a number of other mycotoxins. Ochratoxins, notably ochratoxin A (10.13), were first identified as products of *Aspergillus ochraceus*, which occurs mostly in warm climates. In temperate climates other aspergilli and penicillia, notably *P. verrucosum*, that contaminate cereals and cereal products are more important. Ochratoxins have been implicated in causing kidney damage in pigs but the toxin is destroyed by the rumen bacteria of cattle. A serious form of kidney disease affecting humans and endemic in the parts of the Balkans has also been associated with ochratoxins. Although the total intake of ochratoxins from a typical UK diet is very low, either from cereal products directly or from pigs' kidneys or black pudding made from pigs blood (the two tissues where most toxin accumulates), research is continuing.

(10.13)   (10.14)

Patulin (10.14) is classified as a mycotoxin but it is not regarded as dangerous. Although it has been found to cause mutations in test bacteria grown in the laboratory, animal studies have shown it to be neither carcinogenic nor teratogenic. The possibility that it might have useful antibiotic activity has even been explored. Patulin is particularly associated with apple juice where its presence tends to be regarded as a marker for the use of mouldy fruit rather than as a health hazard in its own right. It is also found in home-made jam that has mould growing on its surface. Patulin is produced by a number of species of *Aspergillus* and *Penicillium* (particularly *P. expansum*) that cause rotting of stored fruit. Most clear apple juice sold in the UK has less than 10 μg kg^{-1} but cloudy juice frequently contains patulin above these levels. The difference is caused by the tendency of patulin to associate with the particulate material in the juice and therefore be removed when the juice is filtered.

The increasing awareness of the health problems that mycotoxins may cause has led to food manufacturers and processors taking a number of preventive steps. The need to discard mouldy cereals and nuts is obvious, but there are many food products that traditionally require ripening by moulds. These include the blue cheeses and many fermented sausages (salamis). The approach now being adopted with these products is to forsake 'natural' contamination from the factory walls, plant, or employees and to use instead pure starter cultures of mould strains that are known not to be toxin producers.

BACTERIAL TOXINS

The growth of moulds on food is often obvious to the naked eye. The growth of bacteria, however, is frequently far from obvious, and the first we know of toxins being produced is when we start to suffer their effects. In any consideration of the diseases caused by food-borne bacteria it is essential to distinguish between infections and intoxications. Infections result when harmful bacteria present in our food are ingested. Once into the gastrointestinal tract, they proliferate and produce toxins that cause the symptoms of the particular disease. The diseases caused by *Vibrio cholerae* and the innumerable species and subspecies of the genus *Salmonella* are among the most important of these food-borne infections.

Our concern in this section is with the intoxications that result from growth of harmful bacteria on food before it is consumed. Although the toxins secreted may well be deadly, it is curious, but hardly any consolation, that the conditions encountered in the intestines of the victim may render the bacteria, but not the toxins they had earlier secreted, quite harmless.

The most feared of the bacterial food intoxications is botulism, caused by the toxin secreted by *Clostridium botulinum*.* The causative organism is commonly found in soil throughout the world, although being an obligate anaerobe it only proliferates in the absence of oxygen. Its vegetative cells are not particularly resistant to heat, unlike the spores. Even the least resistant *C. botulinum* spores can withstand heating at 100 °C for 2–3 minutes. The heating regimes required in food preservation by canning are designed to eliminate all reasonable risk of *C. botulinum* spores surviving. Experience has shown that for safety the entire content of a can must experience at least 3 minutes at 121 °C. Times approaching half an hour are required if the temperature is lowered to 111 °C. To achieve temperatures like these at the centre of a large can may well

* *Botulus* is Latin for sausage.

require prolonged heat treatment for the can as a whole. Acidic foods such as fruit or pickled vegetables present no hazard since their pH is too low for growth and toxin production. Obviously, home bottling of nonacid foods, including vegetables such as beans or carrots, is risky unless salt at high concentration, as a pickling brine of at least 15% w/v NaCl, is also included. Other factors that prevent growth are reduction in water activity (a_w; *see* Chapter 12) to below 0.93 or the presence of nitrite (*see* Chapter 9). If heating is inadequate and other conditions are suitable, some spores, carried into the can with the raw food materials, will eventually germinate, multiply, and secrete toxin.

The species *C. botulinum* is divided into a number of types (A to G) that can be distinguished serologically but more importantly differ in the lethality of their toxins and in their geographical distribution. For example, in the USA to the west of the Rocky Mountains type A predominates whereas to the north in Canada and Alaska type E and elsewhere in North America type B predominate. In Western Europe only type B is found in soil samples. Throughout the world type E is associated with intoxication from seafoods. Types C and D are most often associated with manifestations of botulism in domestic animals rather than man. Ducks are especially vulnerable because of their occasional consumption of rotting vegetation from the anaerobic depths of ponds.

As with the mycotoxins, and for similar reasons, data on the amounts of botulinum toxin that produce adverse effects in humans are scarce. Estimates of the minimum lethal dose are around 1 μg for an adult, *i.e.* approximately 1.4×10^{-2} μg kg^{-1}. These amounts of toxin are small both in absolute terms and also relative to the productivity of the bacteria. Levels of toxin of around 50 μg ml^{-1} have been encountered in an infected can of beans. Laboratory cultures on ideal media may contain concentrations 100 times greater than this. Of the most common toxin types (A, B, and E) B appears to be the most toxic, E the least. The variations in toxicity are principally due to differences in the stability of the toxins in the intestine and rates of absorption into the bloodstream. By direct injection the toxicity is some 10^3–10^6 times higher!

Botulinum toxin consists of a complex of two distinct protein components: a haemagglutinin of molecular weight approximately 5×10^5 and a neurotoxin of molecular weight 1.5×10^5. After absorption the toxin is transported throughout the body by both the blood and lymphatic systems. It reaches the nerve endings where it binds, apparently through the haemagglutinin component, and blocks the transmission of nerve impulses to the muscles. The result is that some 12–36 hours after ingestion of the toxin the first signs of neurological damage appear: dizziness and general weakness, followed over a day or so by development of generalised paralysis and death by respiratory and cardiac failure. When the victim is known to have consumed toxin contaminated food but before extensive symptoms have appeared, it is possible to retrieve the situation by administration of the correct antitoxin for the type of toxin involved. The difficulty of rapid identification of the causative toxin type is such that guesswork informed by geography and clues as to which was the guilty foodstuff may be the best that can be achieved.

The intractability of botulism means that it is a disease to be prevented rather than cured. Although rarely seen in the UK, there are typically 40 or so outbreaks per year in the USA. In the main this difference can be ascribed to the greater enthusiasm for home-based food preservation in rural America rather than to any lack of diligence on the part of the authorities or food manufacturers. Although the spores of *C. botulinum* are resistant to heat, the toxins themselves, in common with most other biologically active proteins, are quickly denatured at high temperatures. Any low-acid, high-water-activity home-bottled or canned food can be boiled for 10 minutes before eating to ensure freedom from this deadly toxin.

In contrast to the severity and rarity of botulism, the consequences of consuming food contaminated by the toxin from *Staphylococcus aureus* are rarely fatal but experienced by almost everyone at one time or another. The staphylococci are a ubiquitous group of micro-organisms that may be detected in air, dust, and natural water. One species, *S. aureus,* is particularly associated with humans and is to be found on the skin and in the mucous membranes of the nose and throat of a high proportion of the population. In the ordinary way we remain quite unaffected by the presence of this microbial guest. However, any human carrier is also an extremely efficient distribution system. Naturally shed flakes of skin from the hands will transfer the organism to anything touched, and the microdroplets from a sneeze will ensure more widespread distribution.

One particular strain (specifically phage type 42E) of *S. aureus* presents special problems. On arrival on a food material that offers the correct environmental characteristics it will start to grow and at the same time secrete an enterotoxin. There are at least five different types of enterotoxin produced by different strains of *S. aureus*. All are proteins with molecular weights of around 32 000. Although they are all very similar in their clinical effects, the different toxins show wide variation in their stability to heating. Thus the least stable, type A, loses 50% of its activity after 20 minutes at 60 °C whereas the most stable, type B, still retains over 50% after 5 minutes at 100 °C. All the enterotoxins show considerable resistance to the proteolytic digestive enzymes of the stomach and small intestine. At the molecular level little is known of the mechanism of action of the staphylococcal enterotoxins. What is known is that 1 μg is sufficient to cause the development of symptoms within an hour or so. The victim suffers from vomiting and sometimes diarrhoea accompanied by a selection of symptoms which may include sweating, fever, hypothermia, headache, and muscular cramps. Only very rarely has the toxin proved fatal; generally the symptoms abate a few hours later.

The physiological requirements of *S. aureus* indicate which foods it is likely to contaminate. Nutritionally it requires most of the amino acids and a few of the B vitamins. Oxygen enhances growth but is not essential. Although growth and toxin production are optimal at around 36 °C, there are indications that toxin is produced at temperatures down to 15 °C and up to 43 °C. Toxin is not produced below pH 5 or at salt concentrations above 10% w/v. These requirements make cooked meats such as sliced ham, prepared custard, and hamburger patties (before grilling) prime targets. Any food not consumed straight after cooking is vulnerable, especially if it is not adequately refrigerated. Studies of food items incriminated in outbreaks of *S. aureus* poisoning have shown numbers of organisms between 10^7 and 10^9 g^{-1}. The discrepancies between the requirements for optimal growth and optimal toxin production and also between the thermal stabilities of the micro-organism and its toxin mean that there is no correlation between cell count and toxin level. Responsibility for the prevention of *S. aureus* poisoning thus appears to lie with food handlers, retailers, and caterers rather than with food processors or manufacturers.

As might be expected, the determination of enterotoxin levels in foodstuffs is not easy. Traditionally, animal assays have been required, often using kittens since they are especially sensitive and show similar symptoms to humans. Serological methods can be applied, but although they are sensitive (down to 0.05 μg ml^{-1}) they take days to complete.

ALLERGENS

Most of the undesirable substances that have been the subject of this chapter so far do not discriminate between their victims, except that the youngest, oldest, or otherwise frail may be more vulnerable. However, a small but not insignificant number of individuals can suffer from food components to which the majority of us are totally indifferent. These are the food allergens. In view of our natural tendency to apply the term 'allergy' to any negative response, especially in the young, to something the rest of us regard valuable* it is as well to remind ourselves that when

---

* 'Hard work' is the classic example.

doctors and scientists talk of allergy they have a very specific phenomenon in mind. Recent surveys have shown that as many as two out of ten people in Britain believe that they are allergic (in this popular sense of an unpleasant reaction of some type or another) to certain foods. However, the same surveys have shown that when proper diagnostic tests are applied to these individuals only one out of ten actually demonstrates a measurable reaction (*i.e.* 2% of the population). Although most of those who believed they suffered from a food intolerance identified food additives as the culprits in fact the estimated incidence of reactions to food additives is much rarer, estimated to affect only 1 in 10 000 people.

Allergy (sometimes referred to as 'hypersensitivity') is defined as an abnormal reaction of the immune system to a foreign (but not infectious) material (the allergen) that leads to reversible or irreversible injury of the body. Immune reactions are normally classified into four types. Reactions involving food components are almost always Type I, involving IgE antibodies. Types II and III also involve antibodies but are not concerned with reactions to food components. Type IV reactions develop slowly and are caused by migrating lymphocytes (white blood cells). The only Type IV reaction of interest to us is coeliac disease, in which wheat gluten causes damage to the lining of the small intestine.

Food allergy is associated with a wide range of symptoms including skin reactions such as eczema and urticaria, and gastrointestinal responses such as vomiting and diarrhoea. In *anaphylaxis*, an exaggerated immune response resulting from re-exposure to an allergen, these symptoms can be very severe and also involve the respiratory system with tissue swelling causing obstruction to the airways. All these allergic reactions follow the release of histamine and other inflammatory mediators provoked by the presence of the allergen. In food allergy this immune response is directed against a specific protein in the diet. It is important to recognise that for an allergic reaction to take place the allergen must penetrate the lining of the digestive tract. A combination of digestive enzymes, the mucous membrane lining, and lymphoid tissue provides the normal barrier that protects us from allergens and invasive micro-organisms that we may consume. Most of the more common types of food allergy are shown only by infants and young children but 'growing out' of the problem is more likely to result from a gradual strengthening of these barriers rather than the body 'getting used' to the allergen.

In general allergens need to have a molecular mass greater than 5000 and below 70 000. Molecules below this size range only provoke a reaction when they have become bound to a larger protein; above this range they are too large to reach the immune system. In most cases the allergenic effect of proteins are not lost when the protein is denatured (*see* page

131). This suggests that the immunological reaction is directed towards relatively short sections of the molecule having a particular amino acid sequence, rather than the tertiary structure or overall shape of a section of the molecule.

Cow's milk allergy (CMA) is well known, affecting up to 7% of infants but very few older children. Any one, or more, of the milk proteins may be the culprit in particular cases. Goat's milk is often suggested as an alternative but in fact the similarity of many of its proteins to their cow's milk counterparts means that victims of cow's milk allergy are likely to react to goat's milk as well. Soy milk, a preparation of proteins extracted from soya beans, does not cause such cross-reactions but infants that have suffered from CMA often go on to become allergic to soy proteins. This is because the damage to the gut lining caused by the CMA can allow excessive interaction between the soy proteins and the immune system.

Coeliac disease is the clinical name given to the allergy to wheat proteins, more specifically the gliadin fraction (*see* page 164) of the gluten proteins. Victims also react to the corresponding proteins in barley and rye and sometimes oats. Coeliac disease occurs throughout the world, wherever wheat is the staple cereal. The disease starts when babies are weaned and wheat based foods begin to appear in their diet. However, the effects of the disease do not normally become obvious until after the child's first birthday. Alongside the appearance in the blood of antibodies to gliadin proteins there is severe damage to the cells lining the jejunum (part of the small intestine). This damage causes severely impaired digestion and absorption of nutrients resulting in a generalized failure to thrive, anaemia, lethargy, gastrointestinal disorders and other problems.

Complete removal from the diet of wheat, (and the other cereals mentioned above) is the only satisfactory treatment. As with the allergens of cow's milk the allergenicity of the gliadins is unaffected by cooking. Fortunately both maize and rice are safe for coeliac sufferers and this makes it possible to provide gluten-free cereal products that can take the nutritional, but only the nutritional, place of the bread, cakes and pastry that the rest of us enjoy. The use of wheat flour in many processed food products which are not obviously 'cereal' in character poses special problems and emphasises the importance of accurate food labelling.

A number of other foodstuffs are well known for their allergenicity, even though the number of sufferers is small. Egg proteins (notably ovomucoid, in the white, *see* page 148) can produce similar symptoms to those produced by milk. The allergenicity of ovomucoid survives cooking. Traces of two other egg proteins, ovalbumin and ovotransferrin, have been found in raw chicken meat but fortunately their allergenicity does not survive cooking.

In recent years particular attention has been devoted to nut allergies, especially to peanuts (*Arachis hypogaea*). Far more cases of peanut allergy occur compared with allergies to 'tree' nuts (*e.g.* almonds, brazils, cashews *etc.*) but not nearly so many of these are consumed, compared with peanuts, and derived products such as peanut butter. One explanation for the increasing incidence of peanut allergies is that peanut (*i.e.* groundnut) oil is nowadays a common ingredient of ointments and creams and if these are applied to the broken skin of infants may provide an entry route for traces of allergenic proteins and thereby iniate an immune response. The edible part of the peanut, the cotyledon, contains around 25% protein, mostly arachin and conarachin. Both of these proteins are aggregates of protein subunits and it appears that certain individual subunits (μ-arachin and conarachin I) are the usual culprits in peanut allergy. Peanut allergens can produce serious, and sudden, symptoms including anaphylaxis in susceptible individuals that may well be life threatening. This allergen is particularly heat resistant and can still be detected in products where only a small amount of peanut flour has been incorporated.

TOXIC AGRICULTURAL RESIDUES

Modern agricultural methods include the use of innumerable products of the modern chemical and pharmaceutical industries. Whether we approve of them or not, we are told that the high yields of agriculture today are dependent on these chemicals and therefore the food scientist must be familiar with them.

(10.15)

The most obviously worrying group of agricultural chemicals are the pesticides. It was found in the 1960s that the insecticide DDT, dichlorodiphenyltrichloroethane (10.15), had penetrated virtually every food chain that was studied. This was the result of the combination of its extreme stability in the environment and the vast scale on which it was used from the time of World War II to the early 1960s. Not only was it very cheap to manufacture but also it was extremely effective in controlling both the insect plagues of mosquitoes carrying malaria and yellow fever as well as numerous agricultural pests. It was the accumulation of DDT in birds of prey and its catastrophic effects on their numbers that caught attention, but it was the realisation that typical samples of human fat and milk were contaminated that led to restrictions in its use. In 1972 in Britain DDT levels of 2.5 p.p.m. were found to be general in human fat. The mean levels of DDT in human milk in Britain have fallen sharply since the 1960s, as shown in Figure 10.5. There are no indications that DDT, at the levels normally encountered in our diet, is a significant contributor to the incidence of cancer.

In general the levels of DDT in cow's milk were around 1% of these figures. The insecticide had reached man travelling in the fat of beef and other meat and also *via* dairy products. Now that the use of DDT is either banned or greatly constrained in Europe, North America and many other parts of the world the levels to be found nowadays in human tissues in these areas are much reduced. In Africa and other tropical regions its cost effectiveness in the control of insect-borne disease means it will remain in use for many years yet. The 1970 report from the National Academy of Sciences (USA) that DDT use had by then prevented 500 million otherwise

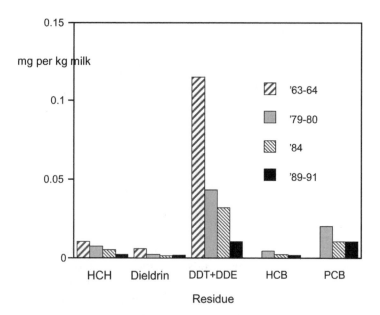

Figure 10.5 *The decline in mean levels of DDT and other undesirables in human milk in the UK since 1963. HCH, hexachlorocyclohexane; DDE, 2,2-bis(4-dichlorophenyl)-1,1-dichloroethylene, the major metabolite of DDT; HCB, hexachlorobenzene; PCB, total polychlorinated biphenyls. PCBs are given further consideration at the end of this chapter. No data for HCB or PCBs were recorded in 1963–64. (Based on data in the Ministry of Agriculture, Fisheries and Food Surveillance Paper No 34, Report on the Working Party on pesticide residues; 1988–90, 1992.)*

inevitable deaths from malaria must be weighed against the effects on our eagles, ospreys *etc.*

From the difficulties with DDT the authorities learnt the necessity for continual monitoring of pesticide levels in all types of human foodstuffs. The disturbing result is that we now know that we are eating food that is almost universally contaminated with a wide range of pesticides and their breakdown products. Many countries, including Britain and the USA, maintain intensive surveillance programmes monitoring the levels of a large number of different pesticides in everyday food items. A summary of the British data for 1999 is shown in Table 10.1. The authorities in different countries have different policies regarding the choice of raw materials and commodities to be tested. In Britain the major contributors to the national diet, such as wheat and potatoes, apples, milk and margarine, where any residues might be expected to make a major contribution to a consumer's total exposure, are monitored more or less continuously. Many fruit and vegetables, where there are hundreds of different types are selected on one of two counts – intensive monitoring of

Table 10.1 *Summarized results of pesticide residue analysis in Britain in 1999. The italicised data are examples picked out from the major sections. Maximum Residue Levels (MRLs) are not set for composite food products or sea food. Data taken the Annual Report of the Working Party on Pesticide Residues: 1999, MAFF*

Commodity	Samples	Total with detectable residues	Total with multiple residues	Number above MRL
Bread	212	18	2	N/A[*]
Milk	214	None	None	None
Potatoes	142	73	21	None
Fruit and veg'	802	388	166	26
Apples	*144*	*68*	*12*	*none*
Celery	*68*	*49*	*18*	*5*
Oranges	*72*	*68*	*67*	*none*
Cereals & cereal products	294	75	21	3
Beer	*72*	*25*	*none*	*n/a*[*]
Grains	*78*	*50*	*21*	*3*
Pasta & infant foods	*144*	*none*	*none*	*none*[*]
Animal products	375	49	12	1[*]
Imported rabbit[†]	*17*	*7*	*4*	*none*
Lamb & pork	*139*	*none*	*none*	*none*
Sea fish	*14*	*2*	*none*	*n/a*
Total	2098	661 (32.5%)	222 (10.6%)	30 (1.4%)

[*] Data includes composite food products. [†] All from one country.

types where pesticide residues frequently present problems, *e.g.* celery and carrots, and a rolling programme of tests every few years on more rarely consumed items. For example persimmons and pomegranates were last scrutinised in 1995, swedes and kohlrabi were scrutinised for the first time in 1999. Not all the substances monitored are necessarily pesticides or fungicides, *e.g.* tecnazene is used to inhibit sprouting in stored potatoes. Depending on the food being examined the total number of different residues actively sought in a single sample ranges from around 8 to more than 90.

Samples are reported as containing an undesirable residue if they exceed the detection limits of the analytical procedure being applied. The sensitivity of modern instrumentation, especially gas chromatography, is such that for most substances these levels are rarely higher than 50 μg per kg (50 p.p.b) and commonly one tenth of this figure. The legal limit for residues are the *Maximum Residue Levels*. In UK legislation these are arrived at by a consideration of what should be achievable by good agricultural practices, rather than from toxicity data alone. Perhaps surprisingly we are assured that this actually results in lower limits for most types of residues.

The degree of pesticide contamination varies from year to year and country to country, mostly due to differences in agricultural practice. For example the number of bread samples in which malathion was detected by the British authorities fell from about 14% in previous years to zero in the period 1988 to 1990, a period of unusually hot dry summers. It has been suggested that the fine weather meant that more home-grown wheat was suitable for bread-making, thereby reducing our use of wheat imported from countries that commonly treat wheat with malathion prior to shipping.

A special problem for the UK is that so much of our food, especially fruit and vegetables, is imported from overseas. Some of the problems of sampling imported produce were mentioned earlier in this chapter. Data of this type is informative about particular foodstuffs but since the sampling does not set out to be representative of entire diets says nothing about the pesticide intake of a typical consumer. Surveys to supply this information are obviously going to be massive undertakings and data from the most recent studies in the USA and UK (on a smaller scale in terms of the number of residues sought) are shown Table 10.2. These are often referred to as a 'Total Diet Survey' or 'Shopping Basket Survey'. Interpretation of this type of data is difficult as an individual's actual intake may differ considerably from the mean, being affected by factors such as age, gender, economic and ethnic factors, as well as personal preferences, but they do give a valuable picture of the overall situation.

Table 10.2 *Pesticide residues in the total food intake. The Acceptable daily Intake (ADI) figures are those assigned by the FAO/WHO expressed per kg of body weight. (The nutritionist's 'typical adult male' weighs 70 kg.) Intake covers all dietary sources based on data published by the FDA in the USA or MAFF in the UK. This 1987 US data discriminates between different sections of the population; by way of illustration data shown here is for 14–16 year-old males. UK data refer to the average of the whole population in either 1984–5 or 1989–90. Maximum Residue Levels (MRLs) are those current in UK legislation, where levels have been set. Different food stuffs have different MRLs with cereals at the top end of most of the ranges shown here*

Pesticide*	ADI	US intake	UK intake	UK MRL
	μg per kg body weight per day			$mg\,kg^{-1}$
Dieldrin	0.1	0.0045	<0.001	0.006–0.2
Carbaryl	10	0.0173	nd[†]	0.5
Chlordane	0.5	0.0018	nd	0.002–0.05
Chlorpyrifos-methyl	10	0.0066	0.01	0.01–10
DDT	20	0.0192	0.0017	0.04–1
Diazinon	2	0.0123	nd	0.02–0.7
Endosulfan	8	0.0206	nd	0.1
Fenitrothion	3	0.00051	nd	0.05–0.5
Heptachlor	0.5	0.0018	nd	0.01–0.2
Malathion	20	0.119	<0.0017	8
Pirimiphos-methyl	10	0.0016	0.05	10
Quintozine	7	0.00112	0.000029	–
Tecnazene	10	0.0001	0.055	–

* Useful details of the chemical structure and other aspects of these pesticides are given in the book by Ware quoted in Further Reading.
[†] nd = no data, either 'not determined' or 'determined but not published'.

Acceptable Daily Intake (ADI) figures have been established internationally (by the combined efforts of the Food and Agriculture Organisation (FAO) and the World Health Organization (WHO)) and represent the maximum daily exposure that results in no appreciable risk. They are arrived at by dividing by 100 the highest level of exposure that has not resulted in any detectable toxicity in the most sensitive animal species tested, the so-called No Observable Effect Level or NOEL. Such tests normally involve continuous exposure over the full natural lifetime of the laboratory animal used, not simply on a single test exposure. NOELs are not readily established for carcinogens as current understanding of carcinogenicity implies that even one molecule of a carcinogen will have some probability of causing cancer. In these cases risks are calculated on the basis of the number of additional cancers that a lifetime's exposure to a pesticide at particular level can be expected to cause. In the USA the 'negligible risk' standard being applied in the evaluation of particular pesticides is one excess cancer per million of the population. This should be set against the overall incidence of cancer of approximately one in four of the population, 250 000 per million. For suspected carcinogens the term ADI is being replaced by TDI, *i.e.* Tolerable Daily Intake, in recognition of the theoretical absence of a No Observable Effect Level.

One obvious inference to be drawn from data like these is that pesticides are now so universally distributed through our environment that for the foreseeable future they are going to be unavoidable components of our diet. Although little work appears to have been done on the subject it seems likely that trying to restrict one's diet to items that are claimed to have been produced 'without the use of chemicals' will have only a marginal effect on one's personal pesticide intake. This does not rule out the influence that may be exerted on suppliers through the workings of the 'market'. There is nowadays a much greater willingness to accept that the fungicides used on peanut crops are far less hazardous than the aflatoxins those fungicides prevent. One must remember that the levels that we normally encounter in our diet are exceedingly low; generally some two orders of magnitude below the maximum ADI levels established by studies with experimental animals. That we are able to measure such low levels at all is a credit to the skills of the analytical chemists. It must also be pointed out that the recorded cases of human poisoning by pesticides have tended to result from relatively massive intakes following accidental cross-contamination of food containers or factory accidents. These observations should not be taken to imply complacency. What is required is constant vigilance and a willingness to accept that sometimes the dangers of using certain pesticides may be outweighed by the benefits.

Another view of this balance is offered by a consideration of the use of hormones in animal husbandry. It is well established that castration of male animals leads to a higher proportion of fat in the carcass in relation to lean, *i.e.* muscle, tissue and to less efficient conversion of feed into meat. Early research showed that the use of male sex hormones, androgens, as additives in animal diets was expensive and relatively ineffective and that there were serious risks in overdosage. In the 1950s it was discovered that the female hormones, the oestrogens, had the same effect as the androgens in increasing the output of growth hormone. It rapidly became the norm to supply one of the synthetic oestrogens, hexoestrol (10.16) or diethylstilbestrol (DES) (10.17), to meat animals, particularly cattle and poultry. The hormone was either added to the animals' feed or implanted as a pellet under the skin.

(10.16)   (10.17)

Concern about the use of these hormones in food arose in the 1970s when an unusual number of cases of vaginal cancer in young women began to appear. It was realised that all of the mothers of these young women had been treated with DES for various disorders of their pregnancy. Subsequently, prenatal exposure to these hormones has been shown to cause a wide range of disorders of reproductive organs of both sexes. In view of the uproar these findings generated, the banning of DES was inevitable. However, what was overlooked was that the clinical doses of DES had ranged from 1.5 to 150 mg day^{-1}. Even allowing for poor transmission across the placental barrier, the actual levels found in meat were way below that which had caused problems. In the mid 1970s only a very small proportion of beef livers from treated cattle were found to have DES levels above 0.5 μg kg^{-1}. Levels in muscle tissue were estimated to be about one-tenth of this.

Some other hormones were regarded as less hazardous and continued to be permitted in Britain until recently. At the present time hexoestrol is permitted for use in poultry. A pellet containing the hormone is implanted under the skin at the top of the neck of young cockerels to simulate the effects of castration and produce a capon.

Modern agricultural practices can present much more insidious problems than hormones. In Chapter 9 the problems posed by the nitrosamines in cured meats were discussed. However, the formation of nitrosamines in the human digestive tract from nitrate in our food is no less important. Of a typical daily intake from food of around 60 mg vegetables, including potatoes, contribute over 75%. As shown in Table 10.3 some vegetables contain spectacular amounts of nitrate but wide variations are found between different regions, reflecting local soil types, the levels in the water supply and fertiliser use. Solid food normally accounts for more than 80% of total nitrate intake. However, where the nitrate level in drinking water is high (above 100 mg l^{-1}) this takes over as the major source and total intake can be increased by 2–3 times. An individual's average intake of nitrate from drinking water in Britain is around 13 mg daily but the figure for South East England is nearly three times this and that for North Wales only one twentieth the average. The UK limit for nitrate in drinking water is 80 mg l^{-1}, averaged over a period of three months.

Table 10.3 *Typical nitrate levels in fresh vegetables. The nitrate content of individual samples of these vegetables may vary between one third and three times these values. Cooking in boiling water can reduce these figures by up to 75%*

	Nitrate, as NaNO₃ (mg per 100g)
Beetroot	121
Celery	230
Spinach	163
Lettuce	105
Potatoes	16
Tomatoes	2

In Chapter 9 (*see* page 308) the carcinogenicity of nitrosamines was considered. It is now recognised that the reaction of nitrite and secondary amines to form nitrosamines can, and does, occur *in vivo* and is probably an important cause of gastric cancer. Nitrate entering the gastro-alimentary tract can be converted into nitrite:

$$NO_3^- + 2[H] \rightarrow NO_2^- + H_2O,$$

at two points. Nitrate, first absorbed from food and drink by the small intestine, is secreted in the saliva and exposed to the activity of the nitrate reducing bacteria that are part of the mouth's natural flora. In the stomach the acidity is normally sufficient to prevent this continuing. However, reduced hydrochloric acid secretion, *achlorhydria*, does lead to the stomach being at a sufficiently high pH for bacterial nitrate reduction to occur. Achlorhydria is, like stomach cancer, associated with poor levels of nutrition over a long period and old age. The incidence of stomach cancer amongst the elderly is significantly higher in regions of Britain where high levels of nitrate in the water coincide with economic deprivation. The frequent absence of an association between nitrate intake and stomach cancer in some parts of the world may well be the result of good nutrition.

The final class of agricultural residue to be mentioned is the antibiotics. Antibiotics are used in agriculture in two quite different roles. The obvious use is in the treatment of bacterial diseases in farm animals. Good agricultural practice demands that milk from a cow treated for mastitis will be discarded, and there are regulations governing the maximum permitted levels of antibiotics such as penicillin in milk. The occasional lapses will have the potential for causing allergic reactions in sensitive individuals, but the major problem will be for manufacturers of dairy products like yoghurt and cheese. These require lactobacilli to generate their lactic acid content, and even very low levels of antibiotics will impair the growth of this unusually sensitive group of bacteria.

Less obvious is the practice of including antibiotics in animal feedstuffs to act as growth promoters. A number of different antibiotics have been used for this latter purpose over the past 30 years. For this use antibiotics are added to the feed at $5-50$ g ton^{-1} depending on the antibiotic and the animal species. Although the mechanism by which they cause growth rate increases of as much as 50% and feed conversion some 15% better is obscure, the benefits are clear. The carry-over of antibiotic residues from the animal's diet into either meat or dairy products from this use is so small as to be insignificant, but that does not mean the practice is without risk.

The difficulty is that the animal receiving an antibiotic for a prolonged period becomes an ideal breeding ground for strains of bacteria that are resistant to the antibiotic. This is an especial problem with the enteric bacteria such as *Escherichia coli* and the salmonellae. It is the ability of these bacteria to exchange genetic material, including the so-called R-factors that carry antibiotic resistance, that creates havoc. Resistance developed within a harmless, or even beneficial, bacterial strain may be transferred to a dangerous pathogen with serious long-term results for animal or human health. Some years ago it was realised that the appearance of strains of disease causing bacteria that were resistant to a wide range of different antibiotics could be blamed, at least in part, on antibiotic use in agriculture. In consequence the use of antibiotics for growth promotion is now heavily restricted in many parts of the world. Within the EC there is a list of seven antibiotics that may be used, but they are all ones that have little or no application in the treatment of human disease. Current UK regulations list the maximum levels of residues of a long list of antibiotics, sulfonamides, and other medicinal products that are permitted in meat and meat products.

TOXIC METAL RESIDUES

The consideration of metal residues requires departure from our classification of undesirables by origin. Toxic metals may reach our food from a number of sources. The more important of these are:

(i) the soil in which human or animal feedstuffs are grown,

(ii) sewage sludge, fertilisers, and other chemicals applied to agricultural land,
(iii) the water used in food processing or cooking,
(iv) contaminating dirt, *e.g.* soil on unwashed vegetables, and
(v) the equipment, containers, and utensils used for food processing, storage, or cooking.

The actual contribution of these different sources to the burden of undesirable metals in our diet will become more apparent as we consider a selection of different metals. Another problem is the distinction between desirable metals, required for health much as vitamins are, and toxic ones. As will be seen, the distinction is frequently one of amount rather than nature, and where a metal is normally regarded as a beneficial nutrient at the levels normally encountered in food it will not be considered here but in the next chapter.

Lead

Lead is undoubtedly the metal that springs to mind first when the question of metal contamination of food is raised. The ancient Greeks were well aware of the health hazards it posed to miners and metal workers, and, ever since, the need for legislation to protect both lead workers and the general public has been recognised, although not necessarily provided. On a world-wide basis at least 5 million tons are used annually, so that it is hardly surprising that considerable amounts find their unwanted way into our lives. Although organic lead compounds such as the tetraethyl-lead used in motor fuels are nowadays probably the most worrying source of lead in our environment, we must restrict our consideration here to lead that reaches us *via* our food and drink. Repeated examinations of crops from fields close to busy roads have failed to demonstrate the extra contamination that might be expected from vehicle exhaust fumes. Similarly the milk from cows grazing close to busy roads shows no increase in lead content when compared with that from apparently less contaminated environments.

The levels of lead in untreated water supplies vary widely depending on the lead content of the rock the water has encountered; most supplies are around 5 p.p.b. The WHO recommended maximum for drinking water is 50 p.p.b. The use of lead piping and lead-lined tanks in domestic water supplies can lead to much higher levels if the water supply is particularly soft, such as the acidic moorland water still used in many parts of upland Britain. The problem of acid water dissolving lead from pipework is exacerbated if the water that has sat in the household pipework overnight is not discarded. This water may well contain over 100 p.p.b. of lead. In the past the use of lead pipes and tanks in breweries and cider factories gave rise to frequent incidences of lead poisoning and was blamed in 1767 for what was then described as the 'endemic colic of Devonshire'. Nowadays lead contamination of beverages is largely restricted to illicitly produced distilled spirits such as the 'moonshine' whisky and 'bathtub' gin of the USA and the 'poteen' of Ireland. Lead levels in excess of 1 mg l^{-1} are still commonly encountered in American 'moonshine'. Home-made stills are often made with lead piping. Car radiators with metal joints sealed with lead-rich solder are sometimes used as condensers.

Another source of lead in beverages is pottery glaze. If earthenware pottery is glazed at temperatures below 120 °C the lead compounds in the glaze will not be rendered sufficiently insoluble. There is then a serious risk that lead salts will be leached into acidic materials stored in such pottery vessels. There are numerous cases in the literature of fatal or near-fatal poisonings resulting from the storage of pickles, fruit juices, wine, cider, and vinegar in the products of amateur potters. In one case of fatal lead poisoning apple juice was reported to have reached 130 mg l^{-1} after 3 days' storage in an inadequately glazed earthenware jar.

Canning is an obvious potential source of lead contamination of food. Steel coated with tin (tinplate) was the original canning material, first used about 160 years ago, and still is. Although lead-based solder was used to seal the joins in the cans, the relative positions of tin and lead in the electrochemical series ensured that the tin was dissolved first. Unless solder was splashed into the can during manufacture, only a very small surface area would normally be exposed to leaching by the can's contents. An excellent illustration of this effect is the analysis of a can of veal, a century after it was prepared for use on the Arctic expedition in 1837. The veal contained $3 \text{ mg lead kg}^{-1}$, $71 \text{ mg iron kg}^{-1}$, and $783 \text{ mg tin kg}^{-1}$.

Nowadays much of the tinplate used for canning is lacquered. The lacquers used are thermosetting resins which are polymerised onto the surface of the tinplate at high temperature before the can is fabricated. Unfortunately lacquering is not as successful as one might expect. Blocking the reaction between the tin and the can's contents can apparently lead to an enhanced loss of lead from the solder. The lead content of canned foods is also affected by the time and temperature of storage as well as the acid content and pH of the food. It is thus very difficult to be precise about the lead content of canned food, but most analyses yield values between 100 and 1000 g kg^{-1}.

Another source of lead in the diet, though, it must be admitted, in that of only a small section of the population, is the lead shot in game birds, such as pheasants and wild ducks, and in rabbits and hares. There is always the possibility of inadvertently swallowing the odd pellet lodging unnoticed in a roast pheasant breast, but in metallic form lead is so poorly absorbed for this not to constitute a real hazard. The problem arises from the dispersion of the metal in the mildly acidic muscle fluids. This can give rise to levels approaching $10 \, \text{mg kg}^{-1}$ in game meat and products such as game pâté. Unless the consumption of game were to be banned completely (not a likely prospect when one considers the lifestyles of so many of our legislators), this is one source of dietary lead some of us are going to have to live with. However, it may be reasonable to add a toxicological argument to the purely economic one for not encouraging young children to eat such expensive food. A more valid objection to game consumption is that in its pursuit a vast amount of lead is fired off into the environment that never actually hits anything edible.

Most unprocessed food and foods whose processing has not added to the lead content have lead levels in the range $50-500 \, \mu\text{g kg}^{-1}$. In the UK most foods are subject to regulations restricting their lead content to $1.0 \, \text{mg kg}^{-1}$. Food products specifically sold for infants must be below $0.2 \, \text{mg kg}^{-1}$ but in recognition of their regrettably high, but inevitable, normal content both shellfish and gamebirds have their limit set at $10 \, \text{mg kg}^{-1}$. The total lead intake from food and beverages has been estimated for adults in various industrialised countries to be in the region of $250-300 \, \mu\text{g day}^{-1}$. The very low water solubility of most lead compounds is a major factor in its poor absorption from the gastrointestinal tract; indeed 90% of intake is immediately excreted in the faeces. Young children may absorb a much higher percentage.

The inevitable result of all this lead in our diet is that even before birth the human body contains a measurable amount of lead. The lead content of the body builds up as we get older, and typical adults will contain between 100 and 400 mg, all but about 10% bound tightly, but not irreversibly, in the bones. The remaining lead is held in the soft tissues of the body and the blood, in equilibrium. Lead is lost from the body in the urine (accounting for about three-quarters of that absorbed), sweat, hair, and nails. It is widely believed that serum lead levels below 40 g per 100 ml have no deleterious effects on health. From 40 to 80 μg per 100 ml there are biochemical indications that the body is failing to cope with an excessive lead burden. Levels higher than 80 μg per 100 ml are associated with chronic lead poisoning. Clinical studies indicate that an adult male needs to assimilate over 100 μg daily before his serum level rises above 40 μg per 100 ml. This corresponds to a dietary intake of about 1 mg daily – data from the UK 1994 Total Diet Study indicate an average intake of 24 μg day^{-1}.

Ordinary consumption in the diet will not lead to acute lead poisoning.

Furthermore the obvious symptoms of chronic lead poisoning such as anaemia are rarely associated with intake from food or drink. More worrying are the effects on young children of serum levels somewhat lower than the so-called safe level quoted above. There are indications that various neuropsychological indicators, including IQ test performance and aspects of social and learning skills, show definite negative correlations with serum lead level when all other variables have been allowed for. It is difficult at the moment to assess the contribution that dietary lead is making to the damage that some children may be suffering. This appears to be a particular problem for children reared in an 'inner-city' environment, and it may well be that tetraethyl-lead in exhaust fumes is the prime culprit rather than food.

Mercury

Mercury presents food scientists with problems similar to those of lead but on a rather different scale. Like lead, it occurs in three different forms: the free metal (Hg^0), inorganic mercuric (Hg^{2+}) salts, and alkyl mercury compounds. From time to time liquid mercury has found its way into food, from broken thermometers for example, but in this form mercury cannot be regarded as particularly toxic. There are reports of people drinking 500 g and suffering no more than occasional diarrhoea! Mercuric salts are more hazardous (the lethal dose of Hg^{2+} is about 1 g), though still very rare as food contaminants. It is much harder to dismiss the problem posed by the organic compounds of mercury.

Alkyl mercury compounds reach our food from two sources. On occasion there have been serious outbreaks of mercury poisoning caused by consumption of cereal grains intended for planting and dressed with antifungal mercurial compounds. The practice of also dyeing treated seeds lurid colours has not prevented their misuse, and many countries now ban the use of mercurial seed dressings. The most important source of alkyl mercury compounds is industrial pollution of coastal waters and consequently of fish and other seafoods. The problem of marine pollution by mercury compounds first came to world attention in the mid-1950s when numerous cases of mercury poisoning in Japan were traced to the contamination of Minimata Bay by methylmercury, CH_3Hg^+. In spite of the identification of the nature of poisonings, the pollution continued until the 1970s, by which time there had been over 700 cases of poisoning, including 46 deaths. Fish and shellfish in the bay had mercury levels of up to 29 mg kg^{-1}, levels that pushed the local average daily intake to over 300 μg per head. A more normal daily intake would be unlikely to be above 10 μg. Data from the 1994 Total Diet Study indicate an average intake in Britain of 4 μg day^{-1}.

Subsequently even higher levels have been found in fish from some polluted Swedish and North American lakes, but local fishing bans prevented further tragedies.

Dangerous levels of methylmercury will accumulate even if the pollutant is an inorganic mercury salt or the free metal. Various anaerobic methane-producing bacteria that occur in muddy sediments will carry out the reactions (10.18):

$$Hg^{2+} \xrightarrow{+ 2e} Hg^0 \xrightarrow{+ 2\,[CH_3]} (CH_3)_2Hg \xrightarrow{+ H^+} CH_3Hg^+ + CH_4 \quad (10.18)$$

as part of their methane-synthesising system. The methyl groups are donated by methylcobalamin. There is evidence that similar mercury methylating systems occur in many types of anaerobic bacteria.

Pollution is not always to blame for high mercury levels in seafood. Some deep-sea fish such as tuna appear to accumulate mercury, as CH_3Hg^+, to around 0.5 mg kg^{-1}. Fish from the most polluted of Britain's coastal waters have similar levels. The widespread occurrence of mercury in fish leads inevitably to its accumulation in the tissues of marine mammals. These animals are apparently able to detoxify the mercury by converting it into an as yet uncharacterised complex involving mercury and selenium. Formation of a similar complex in man may afford a degree of protection for those exposed to mercury over long periods.

It is generally recognised that clinical signs of mercury poisoning are seen in adults when the intake of organic mercury is above 300 μg day^{-1}, and therefore a safe limit for adults is assumed to be approximately one-tenth of this. There is every reason to suppose that children are more sensitive. The symptoms of mercury poisoning are variable, but all of them point to damage to the central nervous system. In the case of the Minimata poisonings many mothers who themselves showed no clinical signs of poisoning nevertheless gave birth to children who went on to suffer cerebral-palsy-like symptoms.

Arsenic

Arsenic is another element that occurs at low, probably innocuous, levels in most of our food in consequence of its widespread distribution in nature. Ordinary drinking water in most parts of the world has around 0.5 p.p.b., but some thermal spring and spa waters may have as much as 1 mg l^{-1}. The amounts in food rarely exceed 1 mg kg^{-1} except in seafoods.

The tendency of fish and crustaceans to accumulate toxic metals is very difficult to explain, but as we have seen in the case of mercury it can cause problems. In British coastal waters prawns have been caught that had 170 mg kg^{-1}! Data from the UK 1994 Total Diet Study indicate an average intake of 63 μg day^{-1}. Of this fish accounted for nearly 75%. An important factor in the consideration of elevated levels of arsenic in fish is that most of the arsenic to be found in fish is apparently in the form of arsenobetaine, $(CH_3)_3As^+CH_2COO^-$, an organic derivative that is not metabolized by man.

Most cases of arsenic poisoning *via* food have occurred following large-scale accidental contamination. A classic incident occurred in 1900 when 6000 cases of arsenic poisoning, including 70 deaths, occurred amongst beer drinkers in the Manchester region. The cause was so-called 'brewing sugar', a glucose-rich syrup used to supplement expensive malted cereals that had been made from starch by hydrolysis with crude sulfuric acid contaminated with arsenic. In Japan, in 1955, 12000 infants were poisoned by dried milk in a bottle-feeding 'formula'; at least 120 died. This was caused by the use of sodium phosphate contaminated with arsenic trioxide as stabiliser in the milk powder. It is far from clear at what level arsenic in food starts to become hazardous. At the moment British legislation sets a limit of 1 p.p.m. for most foods except seafoods, for which no limit is specified.

Cadmium

If it were not for one outbreak in Japan which came to a head in the late 1960s, cadmium would hardly get a mention as a serious metal contaminant of food. The amount in different foodstuffs varies widely but for most diets averages out at around 50 p.p.b. Data from the UK 1994 Total Diet Study indicate an average intake of 14 μg day^{-1}.

The WHO suggests that 50 μg daily is a maximum tolerable intake for an adult, so that ordinary diets do not present any problems. The Japanese outbreak occurred over a period of twelve years during which rice paddies were irrigated with water contaminated with cadmium from a mining operation. The contamination of the rice resulted in daily intakes of cadmium of 100–1000 μg. The cadmium caused an extremely painful demineralisation of the skeleton, especially amongst post-menopausal women. The disease, known as *itai-itai* from the Japanese equivalent of 'ouch-ouch', was exacerbated by dietary calcium deficiency.

Two other sources of cadmium should be noted. One is the use of cadmium-plated components in food-processing machinery. While specialist manufacturers of this machinery can be expected to avoid the use of

Box 10.1   The Total Intake of Undesirable Metals

On page 352 the concept of the 'Total Diet Survey' was introduced in connection with the total exposure of an individual to pesticide residues in the diet. In 1997 the British Ministry of Agriculture, Fisheries and Food conducted such a survey of the intake of a wide range metals to assess their possible risk to health. The contribution of a vast range of different food items to the nutrient content of the British diet had been surveyed in 1986/87 (J. Gregory, K. Foster, H. Tyler and M. Wiseman, M., 'The Dietary and Nutritional Survey of British Adults.' The Stationery Office, London, 1990.) Representative samples from each of the 20 food groupings that were used in this survey were then analysed for each metal. This analytical data was then combined with the 1986/87 data on the intake of different foods to establish the average total intake by an adult for each metal. The results for metals discussed in this chapter are shown in Table 10.4.

With the obvious exception of arsenic the margins between the mean intake and the tolerable intakes are reassuringly large. This also remains true when the calculated intakes for those at the top end of the intake range (in statistical terms the 97.5[th] percentile), which are not included here, are considered.

It is not easy in routine analysis to distinguish between the dangerous inorganic forms of arsenic and the relatively benign organic forms such as arsenobetaine. The intake data shown here refers to the total arsenic content of the diet whereas the tolerable intake figure refers only to the inorganic forms. Since the bulk of dietary arsenic is in the organic form the data given here are in no way as alarming as they might first appear.

Table 10.4 *Undesirable metals in the total diet. Tolerable Daily Intake figures are based on those recommended by the FAO and WHO*

	Mean Intake	Tolerable daily intake for a 70 kg person
	(mg per day)	
Aluminium	3.2	70
Arsenic	0.063	0.14
Cadmium	0.014	0.07
Lead	0.024	0.24
Mercury	0.004	0.050
Tin	1.9	140

cadmium plating, there will always be risks if non-food machinery is adapted to use in contact with food. Another source of cadmium that should be guarded against is zinc plating or galvanising. Zinc usually contains some cadmium. If galvanised buckets *etc.* are used to store acid foods, then some zinc, itself harmless, will dissolve and carry traces of cadmium with it into solution.

Tin and Aluminium

The occurrence of tin in canned foods has already been referred to in connection with lead. Most studies of tin toxicity have concluded that even quite high intakes are harmless; the minimum toxic dose is around 400 mg for an adult.

The other important packaging metal is aluminium. There is no substantial evidence that ingested aluminium is harmful in any way. We consume it in considerable quantities from a number of sources including toothpaste, baking powder, and alkaline indigestion remedies. A number of aluminium compounds are used as food additives. For example sodium aluminium phosphate is a slow acting leavening agent when used in combination with sodium bicarbonate, and sodium and calcium aluminium silicates are used as anti-caking agents. The general level in fresh unprocessed food averages below $10 \, \text{mg kg}^{-1}$ (many foods have less than $1 \, \text{mg kg}^{-1}$) but when levels higher than this are encountered in some processed foods this is almost invariably due to the use of aluminium containing additives such as those just mentioned. Tea is often quoted as an exception but in spite of levels around $500 \, \text{mg kg}^{-1}$ in the leaves the aluminium remains undissolved so that tea infusions have only around $3 \, \text{mg l}^{-1}$. There is no more aluminium in beer from aluminium casks and cans than is found in beer from wooden casks or glass bottles.

Concern was expressed some years ago about the levels of aluminium in infant formulae based on soya protein, compared with cow's milk based formulae, averages of $0.98 \, \text{mg l}^{-1}$ compared with $0.11 \, \text{mg l}^{-1}$. Reformulation of soya milk formulae succeeded in halving this figure but the natural aluminium content of soya beans limits what can be achieved. The amount of aluminium leached from aluminium cooking utensils is actually quite small. Eliminating the use of aluminium cooking utensils and foil can reduce the total daily intake by about 2 mg.

The amounts of aluminium consumed as the hydroxide in indigestion remedies and buffered aspirin preparations can be massively higher than the intake from food, up to 1 gram per day. The only question mark over aluminium's safety record is the recent suggestion that it accumulates in the diseased brain tissue found in Alzheimer's disease. Although several epidemiological studies have hinted at an association between the incidence of Alzheimer's disease and the concentration of aluminium in local drinking water the small contribution of water to total intake makes these results very hard to interpret. We need to know much more about the absorption of the metal from the gut before useful conclusions can be drawn. At the present time there is every reason to believe that this accumulation is a secondary and unimportant effect of this distressing condition rather than a cause.

TOXINS GENERATED DURING HEAT TREATMENT OF FOOD

A substantial proportion of the food we eat is subjected to heating before consumption. The benefits of this will by now have been seen to be considerable. As well as the physical effects of tenderisation *etc.*, cooking provides a measure of protection against food-borne micro-organisms and many of their toxins. Unfortunately we cannot automatically assume that the application of heat to our food is itself an entirely blameless procedure. The polyaromatic (or polynuclear) hydrocarbons (PAHs) were mentioned in the previous chapter as being components of smoke. It is now becoming clear that these carcinogens can be found in other types of food. When carbohydrates or, more significantly, fatty acids are heated above 500 °C, they give rise to unsaturated fragments such as:

$$CH_2{=}CH_2 \qquad CH_2{=}CH{-}CH{=}CH_2 \qquad CH_2{=}C(CH_3)CH{=}CH_2$$

and their radicals. These condense to form progressively larger polyaromatic systems until compounds such as 3,4-benzpyrene (9.6) are formed. While smoked fish, bacon, and sausages typically have levels of 3,4-benzpyrene below 5 p.p.b., barbecued meats, hamburgers, chops, steaks, *etc.* frequently have rather higher levels, up to around 100 p.p.b. These elevated levels appear to be caused by fat dripping out of the meat onto the burning charcoal of the barbecue and decomposing in the familiar puff of smoke and flames. Designs for barbecues and grills that prevent dripping fat from reaching the fire are obviously called for. 'Clean' fuels such as charcoal are obviously to be preferred over wood and coal.

There is little doubt as to the carcinogenicity of the PAHs, at least in laboratory animals. There is real doubt, however, as to whether the levels actually ingested from barbecued and grilled meats are sufficient to cause tumours in humans. Much higher levels of them occur in the tar of tobacco smoke and diesel exhaust fumes, and tackling these will do much more for human health than outlawing the barbecue.

In recent years many new classes of potential carcinogens have been identified in cooked protein-rich foods such as meat, fish, soya products, and cereals. These substances are identified as potential carcinogens on the basis of their ability to promote mutations in special strains of *Salmonella typhimurium* in the 'Ames test'. At the present time there is little or no information on the carcinogenicity of these substances in animal systems. Some types appear to be derived from the direct pyrolytic breakdown of amino acids, particularly tryptophan, and others from a variety of amino acids caught up in the ramifications of the Maillard reaction. It is really too early to start speculation as to the involvement of these substances in the elevated incidence of cancer of the gastrointestinal tract in countries where a lot of meat is eaten. The possibility that the very same substances may be important contributors to the attractive flavours of, for example, roast meat will not help resolution of this issue.

PACKAGING RESIDUES

The plastics that are used to package our food can be the source of two types of food contaminant. The plastics themselves are polymers of extremely high molecular weight so that the question of traces of the actual polymer occurring in food products does not really arise. However, polymer materials also include low molecular weight substances that are potentially problematical. One group of these are the monomers, *i.e.* the molecular building blocks from which the polymer was synthesised. It is inevitable that traces of these will remain in the polymer, from which they may migrate into food they contact. The tendency of vinyl chloride (often referred to as vinyl chloride monomer), $H_2C=CHCl$, to leach out of PVC, poly(vinyl chloride), is a potential problem if food is stored for a long period in a PVC container. Although there have been occasional incidents when much higher concentrations have been detected UK legislation requires that PVC used for packaging does not lead to concentrations of vinyl chloride above $10~\mu g\,kg^{-1}$ in food. Cooking oils *etc.* are much more likely to become contaminated with vinyl chloride than non-fatty materials, such as mineral water.

Of greater concern are the plasticisers. These are the substances incorporated into plastics that prevent them from becoming too brittle. In recent years particular attention has focused on 'cling films' used for wrapping food. These usually consist of PVC although some are made from polyethylene (polythene) or VDC (vinylidene chloride). They are around 10 mm thick and not only have the clinging property that makes wrapping irregular food items and temporarily covering glass vessels easy but also the right degree of permeability to oxygen and water vapour. The plasticiser most commonly used in PVC cling film, and the one that has received most attention from the authorities, is di-2-ethylhexyl adipate (10.19), otherwise known as dioctyl adipate, DEHA or DOA. Since each type of plastic packaging tends to contain different plasticisers there are far too many to accommodate a full account here and we will concentrate on DEHA.

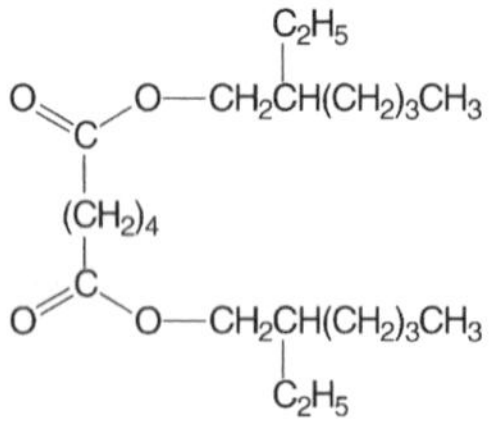

(10.19) 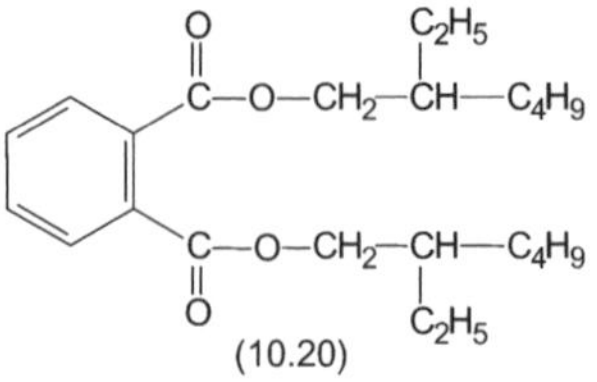   (10.20)

Extensive surveys conducted in the late 1980s showed that significant levels of DEHA occurred in a wide range of foods consumed in Britain. Taking into account typical consumption of these foods the maximum daily intake of DEHA was calculated to be 8.2 mg per person in 1988.* This figure was half that recorded in 1986. The fall was largely due to changes in the formulation of the films but also the manufacturers ensured that the labelling of the films made clear that they were unsuitable for wrapping food when it is heated in microwave ovens. Continuing advances in the formulation of plastics are expected to bring this figure down substantially. Although these figures still seem very high the minimum levels needed to induce cancer in studies with rodents are at least three thousand times higher.

Phthalates, in particular di(2-diethylhexyl) phthalate (DEHP) (10.20), are also widely used as plasticisers in food packaging polymers that we encounter in our daily lives. Although they do not appear to be carcinogenic the possibility is becoming recognised that they may mimic female human sex hormones, and be implicated in the fall in male fertility and the increase in the incidence of other disorders of the male reproductive system that has occurred over recent decades, in both humans and other animal species. Regardless of its ultimate resolution this issue is a valuable, and overdue, reminder that toxicologists need to maintain their vigilance beyond our current preoccupation with carcinogenicity.

The complexity of the packaging residue issue will be clear to anyone

---

* Ministry of Agriculture, Fisheries and Food. Plasticisers: Continuing Surveillance. Food Surveillance Paper No. 30. HMSO, 1990.

who reads the MAFF report mentioned here. A major cause of the complexity is the enormous variety of different polymers in use nowadays and their tendency to turn up unexpectedly. An innocent looking glass jar will probably have a lid that includes a plastic liner to ensure that it seals properly. All that can reasonably be expected from the readers of this book is an awareness that the problem exists; a real understanding of the situation demands special expertise in an area of applied chemistry rather foreign to the food scientist.

ENVIRONMENTAL POLLUTANTS

Some environmental pollutants demand the attention of food chemists even though their origins have little or nothing to do with food. The dioxins, the closely related dibenzofurans (often referred to in this context as 'furans'), and the so-called 'dioxin-like' polychlorinated biphenyls (PCBs) are the most important. Their structures are shown in Figure 10.6. In view of the widely differing toxicities of the numerous variants, analytical data on the dioxins, furans and PCBs are reported as 'TEQs', *i.e.* '2,3,7,8-TCDD Toxic Equivalents'. By analogy with the 'retinol equivalents' used for vitamin A measurements (*see* page 290) 1 g of TEQs consists of a quantity of dioxins *etc.* whose total toxicity is equivalent to that of 1 g of 2,3,7,8-TCDD.

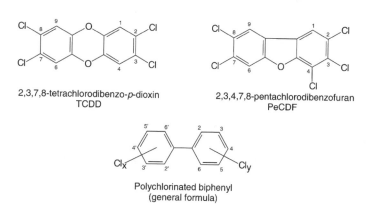

2,3,7,8-tetrachlorodibenzo-*p*-dioxin
TCDD

2,3,4,7,8-pentachlorodibenzofuran
PeCDF

Polychlorinated biphenyl
(general formula)

Figure 10.6 *Dioxins, dibenzofurans and polychlorinated biphenyls. The dioxin shown is only one of 210 'congeners' differing in their pattern of chlorine substitution. There is a similar abundance of dibenzofuran congeners. Polychlorinated biphenyls are not normally referred to specifically in terms of the numbers or location of their substituent chlorine atoms. When commercial PCBs were sold to the electrical industry they were identified merely by the average number of carbon atoms and chlorine atoms per molecule in a batch.*

PCBs were manufactured for use as the dielectric fluid in transformers and capacitors. In the light of their toxicological problems and persistence in the environment their manufacture had ceased by 1980 and their use is now banned. Existing electrical equipment that contains them is gradually being replaced and their PCB content destroyed. Dioxins and dibenzofurans arise from numerous sources. By-products of the industrial production of chlorinated chemicals account for about a third of the total. Bleaching of wood pulp with chlorine can also generate them but the major source is the combustion of materials that contain chlorinated compounds. This occurs in waste incinerators, metal refining, domestic and industrial coal burning, and vehicle exhausts. Replacing the dumping of sewage sludge at sea with incineration or spreading it on the land may constitute an important future source. There are suggestions that wholly natural events such as forest fires also contribute dioxins to the environment. With the exception of workers in the chemical industry, particularly the victims of accidents such as the one at Seveso in Italy, the human exposure to dioxins is almost entirely through food. However, their ubiquitous character, and their persistence in the environment, makes it impossible to identify particular routes by which they reach our food. Once into the food chain their solubility in fats, and extreme insolubility in water, leads them to accumulate in the fatty tissues of animals, fish, and humans. As Figure 10.7 shows levels of these compounds in milk Britain have fallen steadily over recent years. The difference between cow's milk and human milk is a simple reflection of the difference between the diets of cows and humans.

Total Diet Studies of dioxins and PCBs, similar to those type referred to earlier in this chapter with reference to pesticide residues, have been carried out by the UK authorities on three occasions. The results, shown in Table 10.5 that the decline in levels in milk is reflected in the total exposure from the diet as a whole.

The current Tolerable Daily Intake (TDI) figure is 10 pg TEQ (dioxins plus furans) per kg body weight per day. No comparable figure for the PCBs has been established. In 1998 consultations were initiated by the WHO to bring in a TDI range for dioxins plus PCBs of 1–4 pg TEQ per kg body weight per day. The figures quoted in Table 10.5 may be regarded as being uncomfortably close to these TDI figures (in sharp contrast to the situation with pesticide residues and food additives mentioned elsewhere in this book). Of course lowering a TDI figure does not make an undesirable substance any more hazardous than it was before, it merely provides even greater incentives on the authorities world-wide to maintain the policies which have resulted in the sharp declines of the last two decades. Reducing the level of dioxins in the diet is regarded as a priority

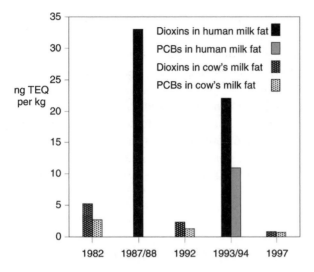

Figure 10.7 *The decline in the levels of dioxins and PCBs in milk fat in Britain. Levels of dioxins plus dibenzofurans (Dioxins) and dioxin-like polychlorinated biphenyls (PCBs) are expressed as ng of Toxic Equivalents (TEQ) as discussed in the text.*
(Based on data in Food Surveillance Information Sheets Number 105, publ. by the Food Safety Directorate of MAFF, 1997 and Number 4/00, publ. by the Food Standards Agency, 2000.)

Table 10.5 *Total Diet Studies of dioxins and PCBs in the diet of adults and children in Britain. Estimated levels of dioxins plus dibenzofurans (dioxins) and dioxin-like polychlorinated biphenyls (PCBs) are expressed as pg of Toxic Equivalents (TEQ) as discussed in the text.*
(Based on data in Food Surveillance Information Sheet Number 4/00, publ. by the Food Standards Agency, 2000)

	Children (aged 1.5–4.5 years)		*Adults*	
	Dioxins	*PCBs*	*Dioxins*	*PCBs*
	pg TEQ per kg body weight per day			
1982	17	33	7.2	13
1992	5.6	9.2	2.5	4.3
1997	4.0	6.9	1.8	3.1

for governments in Europe and elsewhere but it will be achieved through controls on environmental pollution rather than by the actions of farmers or food processors. In mitigation it must be pointed out that degree of carcinogenicity of these substances in humans is still unclear. The difficulty of measuring the low levels of individual dioxins and PCBs in foods also means that the dietary exposure figures tend be to be over-estimates; where a particular congener is not detectable it is assumed, for the purpose of these calculations, to be present at a level corresponding to the limit of detection.

The high levels in human milk result in elevated levels for breast fed infants. However, the TDI figures relate to exposure over a full natural lifespan and the period of breast feeding normally occupies less than 1% of that. Furthermore, the nutritional advantages of human milk over bottle feeding based on cow's milk are still regarded by the authorities as very much more important.

FURTHER READING

'Toxicological Aspects of Food', ed. K. Miller, Elsevier Applied Science, London, 1987.

J. M. Concon, 'Food Toxicology', Dekker, New York, 1988.

'Natural Toxicants in Food', ed. D. H. Watson, Sheffield Academic, Sheffield, 1998.

'Natural Toxins: Animal, Plant and Microbial', ed. J. B. Harris, Clarendon, Oxford, 1986.

G. W. Ware, 'The Pesticide Book', Freeman, New York, 1978.

C. Reilly, 'Metal Contamination of Food', 2nd edn., Elsevier, London, 1991.

'Food Contaminants, Sources and Surveillance', ed. C. Creaser and R. Purchase, The Royal Society of Chemistry, Cambridge, 1991.

D. Watson, 'Safety of Chemicals in Food', Ellis Horwood, Chichester, 1993.

'Nitrates and Nitrites in Food and Water', ed. M. J. Hill, Ellis Horwood, 1991.

'Chemicals in the Human Food Chain', ed. K. C. Winter, J. N. Seiber, and C. F. Nuckton, Van Nostrand Reinhold, New York, 1990.

'Food Chemical Risk Analysis', ed. D. R. Tennant, Chapman & Hall, London, 1997.

'Environmental Contaminants in Food', eds. C. F. Moffat and K. J. Whittle, Sheffield Academic Press, Sheffield, 1999.

Chapter 11

Minerals

In the previous chapter we saw that there are many inorganic elements whose presence in our diet is undesirable. The food chemist has special responsibilities to ensure that these elements are not consumed in harmful amounts and therefore needs to understand how they reach our diet, from the environment, raw materials, processing operations, *etc.* When we turn to nutritionally desirable elements, the food chemist's task is somewhat simpler and can be examined under a number of headings:

(i) The analysis of foodstuffs to determine how much of the various inorganic nutrients they contain in relation to our nutritional needs. For each mineral element there are many different laboratory techniques that could be applied, depending on the analytical objectives and the resources available.*

(ii) Studies of the form in which certain inorganic elements occur in foodstuffs where it is believed that this may affect our ability to absorb them from the gastrointestinal tract.

(iii) Studies of the actual behaviour of inorganic elements in food in those cases where this may affect other aspects of food quality.

The list of bulk and trace elements required in the diet is a long one but, as Figure 11.1 shows, is largely restricted to elements of low atomic number. As has been remarked before, the fact that our diet consists almost exclusively of materials that were once living organisms, coupled with the broad similarity of the biochemistry of all forms of animal and plant life, means that we can expect any reasonably mixed diet to supply our mineral needs in about the right proportions.

---

* A treatment of this aspect of food chemistry that was actually useful would be outside the general scope of this book. Readers are recommended to refer to some of the more specialised texts listed in Appendix II.

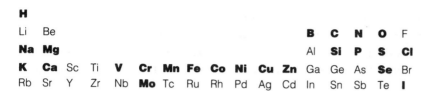

Figure 11.1 *The elements essential for animal life. The elements are listed as an extract of the Periodic Table; those almost certainly not required are shown in lighter type.*

THE BULK MINERALS

Sodium

The alkali metal sodium has a well established role in animal physiology and is abundant in both human tissues and human diets. It occurs at levels between 50 and 100 mg per 100 g in most foods of animal origin including milk, meat, and fish, although eggs have around 150 mg per 100 g. In their raw state vegetables and cereals have very much less sodium, between 1 and 10 mg per 100 g, but the near-universal practice of adding sodium chloride during cooking or at the table brings their sodium content, as eaten, up to similar levels to those in animal foods.

It is difficult to state with any certainty a minimum sodium intake necessary for good health. Special low-sodium diets prescribed for patients suffering from heart disease and some other conditions bring the daily intake of salt down from a typical figure of about 9 g to around 2 g, but this requires considerable distortion of normal eating habits. A strict vegetarian diet without added salt could approach such a figure but at the cost of very low palatability. One should not overlook the fact that our taste buds are able to distinguish sodium chloride from other salts (*see* Chapter 7) and that there is a sound physiological basis for all the other responses of our sense of taste.

The importance of sodium chloride's use as a preservative was discussed in Chapter 9, and in foods where it has been used in this role, *e.g.* bacon, kippers, or butter, levels ten times those just quoted may be encountered. There is considerable pressure from the medical profession for us to reduce the salt content of our diet. There is no doubt of the importance of reduced-sodium diets in the management of a number of physiological disorders, notably hypertension, but the actual effects of reduced sodium intake can be quite small. A reduction of sodium intake by 6 g per day (a substantial proportion of a typical intake) can reduce systolic blood pressure by 2–15 mmHg. Reductions at the top end of this range are associated with older people with high blood pressure. It cannot be stated with any confidence that a high-salt diet actually causes hypertension in someone who otherwise would not suffer from it. Potassium chloride, often suggested as an alternative for culinary and table use, simply does not bear comparison with the real thing, especially in important products such as bread.

Potassium

Like sodium, the alkali metal potassium also has an important part to play in animal physiology and occurs abundantly in both human tissues and human diets, at more uniform levels throughout most foodstuffs of animal and plant origin. Almost all foods, except on the one hand the oils and fats, with virtually no mineral content at all, and on the other hand the seeds and nuts, with 0.5–1.0% potassium, lie within the range 100–350 mg per 100 g. The character of one's diet therefore has little effect on total potassium intake. The only food in which the potassium level is of particular interest is jam. This has nothing to do with potassium's nutritional role but does provide some insight into that most elusive quality parameter of jam, its fruit content. The essential similarity of the sugars, the organic acids, pectins, pigments, and many of the more easily determined flavour compounds makes analytical confirmation of the jam-maker's assertions about his ingredients virtually impossible. It is routine to include acids and pectins from sources other than the fruit named on the label. The level of potassium in a jam will provide a rough measure of the total content of fruit (and vegetable) tissues, assuming no potassium metabisulfite has been used as a preservative. The weakness of this entire area of food analysis will be apparent when one considers that counting and identifying the pips is still a routine procedure in the quality assessment of many types of jam.

Magnesium

Like potassium, magnesium is widely distributed in foodstuffs of all types (at 10–40 mg per 100 g), although foods derived from plant seeds such as wholemeal flour, nuts, and legumes have over 100 mg per 100 g. It is impossible to envisage a diet otherwise nutritionally adequate that could lead to a deficiency of either potassium or magnesium. Beer brewed in many parts of Britain (notably Yorkshire) owes some of its bitterness, and some of its cathartic effect, to the naturally high levels of magnesium sulfate in the water used.

Calcium

When we turn to calcium, we find a much more complex situation. The calcium content of different foodstuffs covers a very wide range. At the bottom end we find some fruit and vegetables such as apples, peas, and potatoes with less than 10 mg per 100 g. Other vegetables have very much more, *e.g.* broccoli and spinach with around 100 and 600 mg per 100 g respectively. Meat and fish generally have low calcium contents, around 10 mg per 100 g. Cereals are naturally quite low in calcium; for example wholemeal flour has around 35 mg per 100 g, but white flours are fortified with additional calcium carbonate. Current UK regulations require that white flour for retail sale and most direct food manufacturing applications* shall contain between 235 and 390 mg of calcium carbonate per 100 g. Self-raising flour contains calcium carbonate anyway as the source of carbon dioxide.

There is evidence from some parts of Britain that even in today's enlightened and affluent world many children's diets would contain dangerously low levels of calcium without this fortification. The question of adult calcium requirements is a complex one largely beyond the scope of this book. A mild calcium deficiency cannot be diagnosed, as the body tends to maintain serum levels at the expense of skeletal reserves. Similarly, the level of calcium in milk during lactation is maintained by depletion of reserves. The uptake of calcium from the intestine and its movement between the blood, skeleton, and other tissues are under hormonal control. When deficiencies in calcium levels in the body do occur, resulting in rickets and other disorders, they are most frequently found to be caused by the impaired alimentary absorption that vitamin D deficiency causes.

Dairy products are always regarded as our primary source of calcium. The calcium content of cow's milk varies quite a lot, but the average is about 120 mg per 100 g. The calcium of milk is concentrated in hard cheeses such as Cheddar to levels around 800 mg per 100 g, but in soft cheeses levels around 450 mg per 100 g are more typical. The biological reason for the presence of large quantities of calcium in milk is obvious: the newborn mammal has a skeleton to build. The skeleton's primary raw material is calcium phosphate, $Ca_3(PO_4)_2$, which presents a problem. One of the features of calcium phosphate that commends it for skeleton construction is its insolubility at neutral pH values. However, the phosphate must also be delivered by milk, and it is not easy to see how both calcium and phosphate ions could share the same delivery system without

---

* Flours for making communion wafers and matzos are exempted.

insoluble crystals forming. Nature's answer lies in the subtle interactions of these two incompatible ions with the caseins discussed in Chapter 5. An average batch of cow's milk contains:

0.065 mol dm^{-3} inorganic phosphate (0.62 g per 100 cm^3, as PO$_4$$^{3-}$),
0.024 mol dm^{-3} organic phosphate (0.23 g per 100 cm^3, as PO$_4$$^{3-}$),
          essentially the phosphoserine of α- and β-caseins,
0.009 mol dm^{-3} citrate, and
0.030 mol dm^{-3} calcium (0.12 g per 100 cm^3 as Ca^{2+})

At fresh milk's normal pH of 6.6 the inorganic phosphate is almost entirely present as HPO$_4$$^{2-}$ and H$_2$PO$_4$$^-$, in the ratio 1 : 3. The virtual absence of PO$_4$$^{3-}$, the significant proportion of the HPO$_4$$^{2-}$ and H$_2$PO$_4$$^-$ being protein bound, and the intervention of the citrate (largely citrate^{3-} at pH 6.6) rule out the formation of the highly insoluble Ca$_3$(PO$_4$)$_2$. Instead we have the loose network of anions and cations that holds the submicelles of the casein together.

The interaction of calcium with phytic acid (3.10) and polysaccharides such as pectins and alginates was discussed in Chapter 3.

Phosphorus

It seems inevitable that a consideration of phosphorus will always follow close behind that of calcium. As both inorganic phosphate and organic esters, phosphorus is a component of all living organisms. Of course in animals there is a concentration in the skeleton, but it should not be forgotten that the intermediary metabolism of sugars is actually the metabolism of their phosphate esters. Sugar phosphates are also essential components of the nucleotide coenzymes and the nucleic acids.

The metabolic role of phosphate is reflected in its distribution in foodstuffs. For example, lean meat (muscle) has around 180 mg of phosphorus per 100 g whereas the corresponding figure for liver is 370 mg. Of plant foodstuffs those derived from seeds, such as legumes, cereals, and nuts, also have high levels of phosphorus, from 100 to 400 mg per 100 g. Other vegetables, such as potatoes or spinach, and fruit, such as apples, have levels of phosphorus below 100 mg per 100 g. The phosphorus in the wheat grain is concentrated in the bran fraction so that white flour has only about one-third of the level found in wholemeal flour, *i.e.* 130 compared with 340 mg per 100 g. Most of the phosphorus in the bran is in the form of phytic acid, whose role is simply that of compact phosphate reserve for the germinating seed. Nuts are similar in that 70–85% of their phosphorus content is also in the form of phytic acid. The human digestive system is normally equipped with low levels of the enzyme phytase, which liberates the phosphate from the phytic acid, but there is evidence to suggest that more of the enzyme is synthesised by people whose diet has a high phytic acid content over a long period. The high phosphorus content of milk has already been mentioned, and so the presence in many types of cheese of around 500 mg of phosphorus per 100 g will be expected. Eggs are another good source of phosphorus, with around 220 mg per 100 g. The wide-spread distribution of phosphorus in foodstuffs means that phosphorus deficiency is not encountered in other-wise satisfactory diets.

Not all the phosphorus in our food is of entirely natural origin. Polyphosphates are popular additives in many meat products, especially ready-sliced ham, luncheon meat, and prepacked frozen poultry. Polyphos-phates, such as tetrasodium diphosphate (11.1) are manufactured by heating orthophosphates:

(11.1)

Those most commonly used have two or three phosphorus atoms per molecule, but polymeric forms with more than twenty phosphorus atoms have some applications. Those with two phosphorus atoms are known either as pyrophosphates or diphosphates. Polyphosphates enhance the water-binding pro-perties of muscle proteins. Apart from the obvious advantage to the processor of improving the yield of meat products from a given weight of raw meat, the greater retention of muscle water and its solutes during cooking will enhance the juiciness, and flavour, of products such as burgers, ham, and bacon. In spite of the apparent addition of water the use of polyphosphates entails, and the superficial wetness of many pre-packaged ham type products, their final water content can still end up lower than that of the uncured meat. This results in the curious statement occasionally seen on labels of cured meat packages that the product contains 110% meat! This is the unavoidable consequence of the meat content being calculated on the basis of the Kjeldahl nitrogen protein determination and the factors used to convert the protein content into meat content being based on the protein content of fresh, *i.e.* uncured meat. An incidental advantage of polyphos-phates is that they appear to delay the onset of rancidity in some products. As they are powerful chelating agents, it is assumed that this effect is exerted through their removal of iron and copper, which would otherwise promote autoxidation of meat lipids (*see* Chapter 4). They are normally used at levels of about 0.1–0.3%. At these levels they pose no health problems since they appear to be hydrolysed to orthophosphates by the pyrophosphatases present in most animal and plant tissues.

THE TRACE MINERALS

Iron

Of all the metals iron is probably the one which the layman is most aware of as a nutrient and also the one which is perceived as potentially in short supply in the diet. One reason for this is that most people are familiar with the need for iron in the blood even though a rather smaller number of people will know what it is doing there, in the myoglobin of muscle and in respiratory enzyme systems generally. Iron is the most important transition metal in the animal body, where it occurs almost entirely in elaborate co-ordination compounds based on the porphyrin nucleus, notably the haem pigments (*see* Figure 5.6), which carry oxygen. Alternation between its two oxidation states, Fe^{II} (ferrous) and Fe^{III} (ferric), is an essential feature of its contribution to the oxidation/reduction reactions involving the other iron porphyrin proteins active in respiration, the cytochromes. A deficiency of iron is always manifested as anaemia, *i.e.* abnormally low blood haemoglobin level.

Iron is generally abundant in most foodstuffs, of plant as well as animal origin. Lean meat contains between 2 and 4 mg per 100 g, mostly as myoglobin, so that the relative redness of different cuts is a fair guide to the relative abundance of the metal. Chicken, lamb and ox liver has rather more, around 9 mg per 100 g; pig liver around twice as much! The iron in liver is not present as myoglobin or haemoglobin but is bound by specialist iron binding storage proteins, notably ferritin. The ferritin molecule has a core that can contain up to around 4000 iron atoms in a crystalline structure resembling the mineral ferrihydrate, which has the overall formula $[FeO(OH)]_8[FeO(H_2PO_4)]$. Leafy green vegetables, legumes, nuts, and whole cereal grains all have between 2 and 4 mg per 100 g. Most fruit, potatoes, and white fish such as cod have between 0.3 and 1.2 mg per 100 g. Although spinach is one of the vegetables at the upper end of the range for iron content ($\sim$4 mg per 100 g) sadly there is no scientific basis for Popeye's dietary enthusiasm. Much more remarkable is the figure of 26 mg per 100 g reported for the humble cockle, but whether this observation is remotely relevant to anyone's regular diet is highly dubious. The iron content of cow's milk is remarkably low, 0.1–0.4 mg per 100 g, but it is to some extent concentrated in dairy products such as butter and cheese, to around 0.16 and 0.4 mg per 100 g, respectively. Eggs have around 2 mg per 100 g.

It is tempting to deduce from all these data that iron is abundant in the diet and that iron deficiency should be rare. Certainly most people's diets contain between 10 and 14 mg per day, and the absence of any specific mechanism for iron secretion from the body would appear to support this assumption. Losses during menstruation can reach 15–20 mg of iron per month, but other losses are much smaller. A total of up to 1 mg daily may be lost *via* the urine, sweat, and epithelial cells shed from the skin and the lining of intestines. The difficulty is that iron is poorly absorbed from the intestine, and the degree of absorption is highly dependent on the nature of the iron compounds in the diet. It has been estimated that world-wide approximately 80% of women of child-bearing age are iron deficient. Only about 10% is absorbed from typical European diets. The behaviour of ferric and ferrous ions in aqueous solution is at the root of the problem.

In the acidic environment of the stomach ferrous and ferric ions occur as the soluble, hydrated ions $Fe(H_2O)_6^{2+}$ and $Fe(H_2O)^{3+}$, but as the pH rises to 8 in the duodenum much less soluble hydroxides are formed, $Fe(OH)_2$ and $Fe(OH)_3$. Ferrous hydroxide is soluble up to 10 mmol dm^{-3}, but ferric hydroxide is essentially insoluble, the maximum concentration in solution being 10^{-18} mol dm^{-3} *i.e.* 1 μg dm^{-3}. Clearly there will be little prospect of absorbing useful quantities of ferric iron at pH 8. Studies with radioactive isotopes of iron, ^{55}Fe and ^{59}Fe, have shown that the presence of ascorbic acid in the duodenum greatly enhances the extent of iron uptake. At least in part this is due to its action in reducing Fe^{III} to Fe^{II}. The acid conditions in the stomach will also help to ensure that any insoluble ferric iron in the food is solubilised. The other role that ascorbic acid has is shared with many other substances, notably sugars, citric acid, and amino acids. All these are effective chelating agents that form soluble complexes with iron, keeping it available for absorption. Homemade beer, with a significant residual sugar level and brewed in iron vessels, has been blamed for the excessive iron intakes which are one cause of the cirrhosis of the liver that is common amongst the urban Bantu population of South Africa. Whenever iron vessels are extensively used for food preparation, there is a likelihood that iron overloading will occur.

Although there is no suggestion that toxicity has been the result, it is worth noting that some canned fruits and vegetables may have high levels of dissolved iron if poorly tinned cans are used. An instance where canned blackcurrants had 130 mg per 100 g of iron has been recorded. The high nitrate level in the tropical paw paw (papaya) solubilises so much iron that this fruit is not regarded as suitable for preservation by this means. It is commonly observed that discolorations in canned foods only appear if the contents are stored in the can after it has been opened. When air is admitted to the can, any ferrous ions dissolved from the material of the can will be oxidised to ferric ions. These, but not their ferrous counterparts, form a black coordination compound with a flavanoid commonly found in fruit and vegetables, rutin (11.2), a glycoside of the flavonol quercitin. Rutinose is L-rhamnose-$\alpha 1 \rightarrow 6$-D-glucose.

(11.2)

Some natural drinking water contains quite high levels of iron; if water is stored in iron tanks or passes through cast-iron pipes, objectionable concentrations of iron may be reached. Iron gives a bitter or astringent taste to drinking water and beverages.

The nature of the substances complexing the iron in more typical European diets has a considerable influence over the proportion that is absorbed. When haem iron, *e.g.* the myoglobin of meat or fish skeletal muscle tissue, is the dietary source, 15–25% is absorbed. No more than 8% of the iron in plant foods, including cereals, legumes, and leafy vegetables, is found to be absorbed. The figure for spinach has been reported to be around 2.5%. The iron in plant foods is mostly present as insoluble complexes of Fe^{III} with phytic acid, oxalates, phosphates, and carbonates. One reassuring aspect of iron absorption from the intestine is that its efficiency rises when the body's iron reserves are low, and *vice versa*. In view of the absence of any mechanism for the excretion of excess iron, such regulation of intake would appear essential even though it may not always be wholly effective.

The low level of iron in both human and cow's milk has already been mentioned. The iron in milk is carried by a specialised protein, lactoferrin. This is a very similar protein to transferrin, the non-haem iron protein that transports iron in the bloodstream. The most curious aspect of this protein is that, as it occurs in milk, its full iron-binding capacity is not taken up, even when there is no suggestion of iron deficiency in the lactating animal. For their first few months of life, humans, and many other mammals, have to rely on the reserves built up in the foetus before birth. The most likely explanation for this phenomenon is that having low levels of iron in the milk has other advantages. Lactoferrin undersaturated with iron maintains an exceedingly low free-iron concentration in milk and, in consequence, the upper reaches of the gastrointestinal tract. This is believed to provide the new-born mammal, or possibly the mammary gland, of the mother with protection against undesirable bacteria that have a strict requirement for iron. Amongst these are members of the Enterobacteriaceae, including pathogenic strains of *Escherichia coli* that can cause severe diarrhoea in infants. Other more welcome bacteria such as the lactobacilli do not have the same demands for iron. It is worth noting that the concentration of lactoferrin in human milk is approximately 100 times that in cow's milk, even though the total iron content in human milk is only a little greater than that in cow's milk.

Copper

Copper is a mineral that nowadays poses few problems. It is widely distributed, as a component of various enzymes, in foodstuffs of all kinds, at levels between 0.1 and 0.5 mg per 100 g. Milk is notably low in copper, at around 0.2 mg per 100 g, and mammalian liver is exceptionally high, at around 8 mg per 100 g. The daily intake from normal adult diets is between 1 and 3 mg, which roughly corresponds to the intake level recommended by most authorities. As with iron, it appears that nature expects newborn infants to get by on the reserves built up before birth until they are weaned. The occasional cases of copper deficiency, manifested as a form of anaemia, that do occur in infants are always associated with other physiological or nutritional disorders.

Copper is not as toxic as one might imagine. Large doses of copper sulfate are emetic, and a dose of 100 g is reported to have resulted in liver and kidney damage. Long-term intake of moderately excessive amounts of copper can arise from regular drinking of water from the hot tap in houses with copper plumbing, but the only likely outcome is gastrointestinal upset. This fairly optimistic attitude to the toxicity of copper salts has not always been the rule. Elizabeth Raffald, an 18th century forerunner of Mrs Beeton, warned against the practice of using copper salts when pickling vegetables:

'... for nothing is more common than to green pickles in a brass pan for the sake of having them a good green, when at the same time they will green as well by heating the liquor without the help of brass, or verdegrease of any kind, for it is a poison to a great degree ...'[*]

It was not until 1855 that public pressure began to force the commercial

[*] E. Raffald, 'The Experienced English Housekeeper', 8th edn., Baldwin, London, 1782. Published in facsimile by E. and W. Books, London, 1970.

pickle manufacturers to abandon the use of copper, but for a time sales actually went down until consumers got used to the more natural brown colour. It is ironical that nowadays caramel has to be added to enhance the brown colour that was once so undesirable.

Zinc

The distribution of zinc in our foodstuffs has much in common with copper. It is an essential component of the active sites of many enzymes, and it is therefore not surprising to find it at high levels in animal tissues, such as lean meat and liver, (around 4 mg per 100 g.) Wholemeal flour, dried peas, and nuts have similar levels but white flour has only one third of this. Other fruit and vegetables have around 0.5 mg per 100 g or less. Of the dairy products only eggs, with 1.5 mg per 100 g, can be regarded as a good source of zinc. Cow's milk has around 0.35 mg per 100 ml, but human milk has less than a third of this. As with so many other minerals, the new-born infant has to rely on reserves laid down in the latter stages of foetal development. Typical diets have been calculated to supply between 7 and 17 mg per day, of which about 20% is absorbed.

The potential for zinc absorption to be reduced by the presence of large quantities of fibre in the diet does not seem to be realised in the case of ordinary European diets. Although the occurrence of zinc deficiency and its symptoms are well recognised in veterinary practice, there is little to indicate that it is a significant problem in human diets. There is some evidence to suggest that zinc deficiency is common amongst some poor Middle Eastern peasant communities, a major part of whose diet consists of unleavened bread made from wholemeal flour. When the otherwise poor diet is supplemented with zinc, a number of symptoms, including delayed puberty and reduced stature, tend to disappear.

Selenium

Selenium was regarded as toxic decades before it was recognised, in the 1950s, as an essential nutrient. The toxic effects of selenium were identified in farm animals in a number of parts of the world where there are unusually high levels of the metal in the soil. There are a few species of pasture plants that are able to accumulate extraordinary amounts, over $1 \, g \, kg^{-1}$ (dry weight), when growing on selenium-rich soil. Wide variations are found in the selenium content of human diets. At one extreme is New Zealand, where much of the agricultural land is deficient and the average per capita daily intake has been calculated to be 25 μg. The corresponding figure for the UK, from the 1995 Total Diet Survey is 34 μg, also low by international standards and apparently low when compared with the Reference Nutrient Intake (*see* Appendix I) figures of 60 and 75 μg day^{-1} for adult females and males respectively. However, the Total Diet Survey figure is the average for the entire population and makes no allowance for the contribution to the average of younger members of the population. In different parts of the North American continent diets range from 60 to 220 μg. Even within the British Isles a wide range of selenium contents can be found in the same commodity produced in different regions. In the UK the selenium content of cereals lies in the range 0.03–0.23 mg kg^{-1}, meat and fish 0.09–0.38 mg kg^{-1}, fruit and vegetables and also milk ~0.01 mg kg^{-1}. Nuts are generally very rich in selenium; some samples of Brazil-nuts have been found with 53 mg kg^{-1}. Just one of these nuts daily would supply the intake of selenium recommended by some authorities. Care must always be exercised in the pursuit of a healthy selenium intake, as it is believed that prolonged intake of over 3 mg daily is sufficient to invoke symptoms of selenium poisoning. Proposals for maximum permitted limits for selenium in foods (1.0 mg kg^{-1} for most foods) have been made in Australia.

The biological function of selenium was mentioned in Chapter 8 as part of the discussion of the antioxidant role of vitamin E. Evidence is beginning to accumulate that selenium intakes at the top of the normal range are beneficial. For example, the incidence of certain cancers in the USA is being tentatively linked to selenium levels in the local diet. At the time of writing studies are being carried out in the USA to explore the possibility that selenium supplementation may afford protection against prostate cancer. However, the relatively narrow safety margin between beneficial and harmful intake rates will make wide-scale dietary supplementation extremely difficult and self-medication decidedly hazardous.

Iodine

Iodine is another nutrient whose concentration in the diet can vary dramatically from one locality to another. However, unlike selenium, the variation is sufficient to result in iodine deficiency being widespread throughout the world. Some years ago the World Health Organisation estimated that at least 200 million people are affected. Iodine has only one role in the body, as a component of the pair of thyroid hormones thyroxine (11.3) and triiodothyronine (which differs from thyroxine only in lacking the arrowed iodine atom). Through the release of these two hormones the thyroid gland in the neck regulates metabolic activity and promotes growth and development. Inadequate levels of these hormones are associated with a lowered metabolic rate and what might be regarded as corresponding physical symptoms, listlessness, constipation, a slow pulse, and a tendency to 'feel cold'. Fortunately the simple goitre that is common in iodine-deficient parts of the world does not lead to a marked hormone deficiency, only a massive, and unsightly, over-development of the thyroid gland in the neck. The thyroid gland enlarges in an attempt to synthesise more thyroxine and triiodothyronine. More severe symptoms occur if iodine intake is very low during critical stages of growth. The mental and physical abnormalities described collectively as cretinism are found in the children of mothers who suffered from severe iodine deficiency immediately before or during their pregnancy. Children who have very low intakes of iodine during their growing years suffer from a related disorder known as myxoedema.

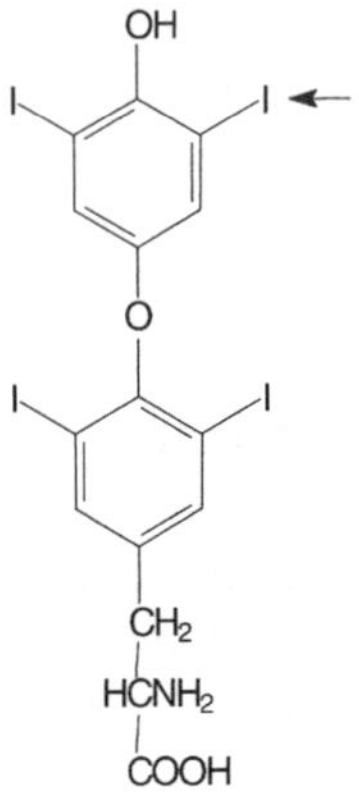

(11.3)

The incidence of goitre is closely related to the levels of iodine in the local soil and water supply and is most prevalent in mountainous regions including the Alps, Pyrenees, Himalayas, and parts of the Andes and the Rocky Mountains. In Britain, where goitre has been known as 'Derbyshire neck', it is particularly associated with parts of the Pennines. Amongst flatter regions of the world that are also potentially vulnerable to endemic goitre are the Great Lakes region of North America and the Thames Valley in Britain.

The wide variations in the iodine content of foodstuffs (where it always occurs as iodide) mean that lists of typical values are fairly meaningless. For the record the 1995 UK Total Diet Study found that the average intake was in the range 151–209 μg per day. What can be said is that most foods are poor sources of iodine. Fruit and vegetables, meat, freshwater fish, cereals, and dairy products usually contain between 20 and 50 μg kg^{-1}. Seafood is the only consistently useful source, with some popular fish species being recorded as containing several thousand μg kg^{-1}. Goitre is almost unknown in Japan, where seaweed, with 4–5 mg of iodine per

100 g, is regularly consumed. Some studies from parts of the world where regions free from goitre are close to others where it is endemic have shown clearly the influence of local food supplies. For example, a study of two such regions in Greece showed differences in iodine content across a wide range of foodstuffs ranging from 1.66 times for cow's milk to 7.05 times for eggs. Superimposed on these factors was the proximity to the sea, and its products, of the low goitre region.

A number of different routes have been utilised to supply iodine to populations of deficient areas. The only consistently successful one is to add it to the salt sold for domestic use. A good example of what can be achieved is provided by data from Michigan in the USA. Before iodised salt (at 0.02% w/w) was made available in 1924, nearly 40% of Michigan schoolchildren showed evidence of goitre. By 1928 this figure had fallen to 9% and in 1951 to 1.4%. Sadly, by 1970 the incidence was back up to 6% because of complacency on the part of the public and less intensive health education. Whether uniodised salt should ever be on sale at all in such regions is of course questionable.

In Britain it is generally concluded that the average diet contains sufficient iodine. One reason for this is the popularity of sea-fish, and there is no evidence that fish fingers lose iodine during manufacture. The increasing degree of industrialisation of our food supply also means that we are much less dependent on the produce of our immediate neighbour-hoods than in years past. Nevertheless, endemic goitre remains a problem in parts of Derbyshire, although not on the scale of the past.

Other Trace Minerals

There are a number of other essential trace elements that have not been discussed here: boron, silicon, vanadium, chromium, manganese, cobalt, nickel, and molybdenum. Our need for these has been inferred from their occurrence in various enzymes isolated from animal sources rather than the identification of human diseases caused by deficiencies in the human diet. For the foreseeable future food chemists are going to have much more important matters to concern them than the supply of these elements in human diets.

FURTHER READING

R. W. Hay, 'Bio-inorganic Chemistry', Ellis Horwood, Chichester, 1984.

E. J. Underwood, 'Trace Elements in Human and Animal Nutrition', Academic Press, London, 1977.

M. S. Chaney, M. L. Ross, and J. C. Witschi, 'Nutrition', 9th edn., Houghton Miflin, Boston, 1979.

'Nutrient Availability: Chemical and Biological Aspects', ed. D. A. T. Southgate, I. T. Johnson and G. R. Fenwick, The Royal Society of Chemistry, Cambridge, 1989.

L. R. McDowell, 'Minerals in Animal and Human Nutrition', Academic Press, San Diego, CA, 1992.

Chapter 12

Water

The final chapter of this book is devoted to the most abundant, and most frequently overlooked, of all food's components, water. Although we are quick to recognise our biological need for drinking water, we tend to overlook the presence of vast quantities of water in our solid foods. Table 12.1 lists the water content of a range of foodstuffs (and a few drinks), and it will not be difficult to assess from these data the size of the contribution that solid foods can make to one's total water intake.

Table 12.1 *Typical water content of food and drinks.*

Food	Water % (w/w)	Food	Water % (w/w)
Lettuce, tomatoes	95	Dried fruit (currants, sultanas *etc.*)	18
Cabbage, broccoli	92	Butter, margarine	16
Carrots, potatoes	90	Low fat spread	50
Citrus fruit	87	Wheat flour	12
Apples, cherries	85	Dried pasta	12
Raw poultry	72	Milk powder	4
Egg white (yolk)	88 (51)		
Raw lean meat	60		Water
Cheese	37	*Drink*	% (v/v)
White bread	35		
Salami	30	Beer	90
Jam (& other preserves)	28	Fruit juices	87
Honey	20	Milk	87
		Scotch whisky	60

389

WATER STRUCTURE

The chemical structure of water, H_2O, seems at first sight exceedingly simple, but all is not as it seems. The first point is that water cannot be automatically treated as a single pure compound. Present in all natural water are innumerable variants caused by the natural occurrence of isotopes such as ^{17}O, ^{18}O, ^{2}H, and ^{3}H. Fortunately for food chemists, their quantitative and qualitative contributions are small enough to be ignored. Furthermore the ionisation of water, to hydroxyl (OH^-) and hydrogen ions (as H_3O^+), while obviously important to the behaviour of other ionising molecules, still does not move the composition of water very far from 100% H_2O, except at extremes of pH irrelevant to food materials.

The problems arise from the structure and behaviour of H_2O itself. Figure 12.1 shows the dimensions of the water molecule. If one visualises the oxygen atom being at the centre of a tetrahedron, the σ-bonds point towards two of the corners and the orbitals of non-bonding electron pairs of the oxygen atom point towards the other two corners.

By comparison with similar molecules involving neighbouring atoms in the Periodic Table (*e.g.* H_2S), one would expect water to melt at ~-90 °C and boil at ~-80 °C. As these are obviously not the figures we observe, we must conclude that in the liquid and solid states water molecules associate together in some way so that their effective molecular weight appears, in some aspects of their behaviour, to be much higher. The large forces of association between water molecules can be traced to the polarised nature of the O–H bond (40% partial ionic character) and the V shape of the molecule. This facilitates hydrogen bonding between the oxygen atom, carrying a small negative charge, and the correspondingly positively charged hydrogen atoms of two other water molecules. It is this

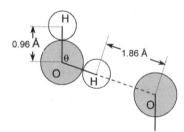

Figure 12.1   *The dimensions of the water molecule and its hydrogen bonds. The angle θ formed by the H–O–H σ-bonds depends on the state of the water: In the gaseous state θ = 104° 52'. In the liquid state θ = 105° 0°. In the solid state (ice) θ = 109° 47'. In a perfect tetrahedron θ = 109° 28'. The other preferred angles are: O–H---O ~180°, (i.e a straight line), H---O---H and H---O–H ~109°.*

ability of water molecules to form links with four others that allows its formation of three-dimensional association structures. In ice an extended three-dimensional lattice occurs as shown in Figure 12.2.

The structure of ice is by no means static. Even at the lowest temperatures regularly utilised in food preservation ($-20\,°C$) the hydrogen atoms oscillate as shown above and a considerable fraction of water molecules will not be fully integrated into the lattice structure. There is plenty of evidence to show that water molecules can diffuse slowly through ice. The OH^- and H_3O^+ ions will also join with entrapped solute molecules in disrupting the lattice. As the temperature is raised, the

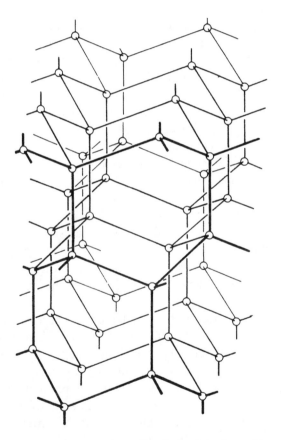

Figure 12.2 *The lattice structure of ice. For clarity only the positions the oxygen atoms are shown, at the junctions of the tetrahedrally arranged bonds. Layers of oxygen atoms are arranged into an array of chair structures resembling those of pyranose sugars. The links between the layers create rings in the boat conformation. There is a continual shift of the hydrogen atoms between their two possible positions bound covalently to neighbouring oxygen atoms:*
O–H---O ↔ O---H–O

proportion of water's hydroxyl groups not involved in hydrogen bonding rises slowly. When liquid at 0 °C, water still has over 90% of its hydroxyl groups involved in hydrogen bonding at any one moment. This figure only falls a little below 80% as the normal boiling point is reached. There is considerable evidence to suggest that the hydroxyl groups not involved in hydrogen bonding are not evenly, or randomly, distributed throughout the bulk of the liquid. Instead water appears to consist of irregular clusters or aggregates of molecules arranged in a related lattice structure. The surfaces of these clusters are marked by the orientation defects that result from numbers of non-hydrogen-bonded hydroxyl groups. At around room temperature the average number of molecules in a cluster is estimated to be between 200 and 300.

The structure of the liquid water clusters has attracted considerable attention in recent years. Although attractive the idea that the clusters have the same lattice structure as ice seems unlikely. One proposal (from the author's colleague Professor Martin Chaplin, *see* M. F. Chaplin, *Biophys. Chem.* 1999, **83**, 211) is that the clusters contain, as in ice, arrangements of water molecules in rings (*see* Figure 12.2). However, where the ice lattice consists only of hexamers, in chair and boat conformations, the liquid water cluster also contains substantial proportion of more or less planar pentamers of water molecules. (The conformations of pyranose and furanose monosaccharides, *see* Chapter 2, provide a useful model.) The H-O-H, 108°, angle in a five membered ring is even closer to the supposed ideal (in the gaseous state) of 104° 52' than that in ice, 109° 47'. In their idealized, virtually spherical, state Chaplin's clusters have 280 water molecules in a total of 70 pentamers and 116 hexamers (80 chairs and 36 boats). Figure 12.3 shows 14 water molecules linked together forming, in this case, two planar pentamers and three hexamers in the boat conformation.

It should never be overlooked that the lattice structure of liquid water is an essentially dynamic one. The lifetime of an individual hydrogen bond must be measured in picoseconds as hydrogen atoms oscillate from one oxygen atom to another. Studies of the 'dielectric relaxation time' suggest that a molecule of liquid water at 0 °C changes its orientation with respect to its neighbours about 10^{11} times per second. In spite of these apparently exceedingly brief periods of continuous stability, the lifetimes of individual clusters are much longer, up to 10^{-8} seconds. This is the result of the co-operativity between nearby hydrogen bonds in maintaining the lattice structure. Although the bonds holding one molecule in place may be broken, the structure as a whole will still be maintained by the bonding between surrounding molecules.

Although the detailed lattice structure of water clusters lattice is

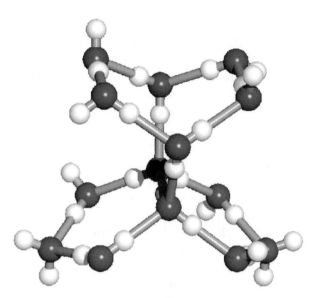

Figure 12.3 *Hydrogen bonded water molecules arranged into two pentamers and three hexamers in a boat conformation. Oxygen and hydrogen atoms are represented by black and white spheres respectively. No attempt has been made to discriminate between covalent and hydrogen bonds and the location of the hydrogen atoms closer to one or the other of the two oxygen atoms is arbitrary.*

(Based on a computer generated drawing kindly provided by Professor Chaplin.)

unresolved there is general agreement that in the liquid state there are always two types of lattice present. One, typified by that described above is referred to as *expanded*, being very open and offering the possibility of non-hydrogen water molecules penetrating into its pores. The other is a *collapsed* structure in which the rigid geometry imposed by the hydrogen bonds is to some extent lost as they weaken at higher temperatures and other forces draw the water molecules closer together. The hexamer and pentamer rings become distorted with values for θ closer to 90°.

As the temperature rises the proportion of water in the collapsed state rises so that the co-ordination number, *i.e.* the average number of nearest neighbours which a single molecule has, rises from 4.0 in ice to 4.4 in water at 1.5 °C and 4.9 at 83 °C. A water molecule having more than the four nearest neighbours expected in the formal lattice structure would still not be expected to be hydrogen bonded to more than four others. As might be expected, temperature also affects the average distance between neighbouring water molecules. In ice at 0 °C and in water at 1.5 °C and 83 °C the O$\leftrightarrow$O distances are, respectively, 2.76 Å, 2.9 Å, and 3.05 Å.

There are a number of consequences of this behaviour that affect the properties of food materials, but one of the most obvious is the change in density. The difference in density between ice at 0 °C, 0.9168 g cm^{-3}, and liquid water at the same temperature, 0.9998 g cm^{-3}, is readily accounted for by the more compact packing shown by the rise in the co-ordination number. Up to 3.98 °C the increase in co-ordination number is the dominating influence on the density, but above this the nearest neighbour distance takes over. These data indicate that water will increase its volume by 9% as it freezes. This density change is of course of profound significance to the evolution of life in aquatic environments. However, food scientists have to be content with its more mundane influences, as on, for example, the design of moulds for ice lollies. The expansion on freezing is also partially responsible for the destructive effect that freezing and thawing can have on the structure of soft fruit such as strawberries.

The inter-relations between water in its liquid, solid, and gaseous states are summarised in the phase diagram shown in Figure 12.4. The diagram shows that, if the pressure is low enough, ice will sublime into the vapour state. This is the basis of the important method of food dehydration known as freeze drying. Delicate food extracts, for example that of coffee in the manufacture of the instant product, are first frozen and then subjected to a high vacuum. The latent heat of sublimation for ice is 2813 J g^{-1}. This

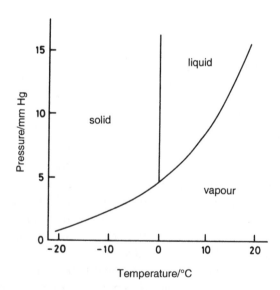

Figure 12.4 *The phase diagram for water. At least five different ice crystal structures have been found to occur at particular pressure/temperature combinations but since the pressures required for these are all very high they are irrelevant to food studies and have been ignored here.*

high figure means that when the vacuum is first drawn on the frozen material the temperature can drop to such an extent that the water is no longer removed at a satisfactory rate. In the usual commercial process, known as 'accelerated freeze drying', this is counteracted by applying radiant heat to the frozen material until almost all the water is removed. Of course the temperature is maintained well below 0 °C throughout the process.

INTERACTIONS OF WATER WITH FOOD COMPONENTS

The interactions of water with other food components are at the root of many problems with the properties of food systems. The interactions at molecular level will be considered first, with particular attention to carbohydrates, lipids, and proteins.

Some features of the interaction of sugars with water were mentioned in Chapter 2. There it was pointed out that in aqueous solutions some water molecules are quite tightly bound to sugar molecules. The value of 3.7 for the number of water molecules bound by glucose is accounted for in the structural diagram shown in Figure 12.5. The distance between the pairs of hydroxyl oxygen atoms on the same face of the pyranose ring of

Figure 12.5 *The insertion of a β-D-glucopyranose molecule into a water lattice. For clarity only the oxygen and carbon atoms are shown; open circles – water oxygen, grey circles – glucose oxygen, black circles – glucose carbon. Comparison with Figure 12.2 will show that at, at least in an ice type lattice, the insertion of the glucose molecule has not disrupted the relationships of water in the layers above and below. The possibility of water molecules binding to the ring and –CH₂OH oxygen atoms of the glucose molecule remains an open question.*

β-D-glucose (*i.e.* those on carbon-1 and -3 and on carbon-2 and -4) is 4.86 Å. This is remarkably close to the distance of 4.9 Å that separates the oxygen atoms in the same plane of the water lattice. Not only are the dimensions therefore correct but also, for equatorially disposed hydroxyl groups, the geometry is correct. The figure of 3.7 is close to the average number of equatorial hydroxyl groups of D-glucose in aqueous solution (4 on the more abundant β-anomer and 3 on the α-anomer).

It has been suggested that the tendency of glucose to retard ice crystal formation in frozen dessert products is a result of this close fit with liquid water but not with ice. In ice at ordinary pressures neighbouring oxygen atoms are 4.5 Å apart. The 'anti-freeze' properties of other substances such as glycerol ($CH_2OH.CHOH.CH_2OH$) and ethylene glycol [$(CH_2OH)_2$], which have α-glycol oxygen atoms capable of similar orientation, may have a similar basis.

In solution isolated glucose molecules will clearly have a strong stabilising effect on the clusters of water molecules around them, and it is tempting to extend this concept to the properties of polysaccharide-based systems. The discussions of gel structure in Chapter 3 should have left the reader in no doubt as to the underlying strength of the polysaccharide/water interaction. For example, a good-quality agar for microbiological use will give a firm gel at a polysaccharide concentration of 1.5% w/v. The average molecular weight of the monosaccharide units of agarose (the gel-forming and major component of agar – *see* Figure 3.10) is about 150. This implies that each monosaccharide unit has been responsible for the effective immobilisation of at least 550 water molecules! The lifetime of an individual water cluster is far too short for one to envisage the stability of the gel being the result of water clusters becoming ensnared in the polymer network. Capillary attraction is a far more plausible, with the gel having some of the character of a wet sponge. As was suggested in Chapter 3 (see page 54), it will not be difficult to modify our perceptions of this model of gel structure to account also for the behaviour of the gums as well as the less soluble fibre polysaccharides.

It will be no surprise that the interaction of water with non-polar molecules is a more aloof affair. It seems axiomatic that water and oil do not mix, but what is not so obvious is why. What is the advantage, in free-energy terms, for two small oil droplets suspended in water to coalesce and become one larger one?

The answer lies in the collective strength of the hydrogen bonds of the water. If a single, small, non-polar molecule is 'dissolved' in water, it will have to lie in a cavity in a mass of associating water molecules. To form this cavity, numerous hydrogen bonds will need to be broken, requiring a large amount of free energy. The free energy of the system is then partially reduced (by compensating entropy changes) as the water molecules reorganise around the cavity to maximise the number of hydrogen bonds between themselves. There is evidence that this reorganisation can involve water structures based on strained five-membered rings. At room temperature the transfer of a small hydrocarbon molecule from an aqueous solution to a non-polar solvent is accompanied by a change in free energy, ΔG°, of around -10 to $-15\,\text{kJ}\,\text{mol}^{-1}$. The much larger molecules of typical food lipids can be expected to require correspondingly larger free-energy changes.

When an isolated non-polar solute molecule is considered, it will be clear that the larger the surface area of the solute molecule the less soluble it will be in water; in other words, larger non-polar molecules will be more hydrophobic. However, when two non-polar molecules come to share the same cavity in the water, the total disruption of hydrogen bonding will be less than when they were apart and the free energy of the system will be lower. This principle readily extends to the coalescence of the oil droplets suspended in water.

The interaction of proteins with water has received a great deal of attention from biochemists for many years. However, this is mostly because of the importance of water to the structure and behaviour of proteins (*in vitro* as well as *in vivo*) rather than the influence of proteins on the behaviour of water. A protein molecule presents three kinds of surface group to its aqueous environment, depending on the nature of the amino acid side chain: polar charged groups such as those of glutamic acid ($-CH_2CH_2COO^-$) or lysine [$-(CH_2)_4NH_3^+$], polar neutral groups such as those of serine ($-CH_2OH$) or glutamic acid at low pH ($-CH_2CH_2COOH$) and non-polar groups such as those of valine [$-CH(CH_3)_2$] or methionine ($CH_2CH_2SCH_3$). For the reasons discussed above, the non-polar side chains are usually found buried in the interior of the protein and will not concern us further. The pH value will obviously be a major influence on the properties of ionisable side chains.

Studies using sophisticated physical methods have probed the water-binding properties of amino acid side chains and obtained results comparable with those referred to earlier concerning sugars. When ionised, the acidic side chains, aspartate and glutamate, bind about six molecules of water; on protonation this figure drops to two, a similar figure to that of the polar but unionised side chains of asparagine, serine, *etc.* The positively charged side chains of lysine and histidine each bind about four molecules of water. These figures refer to the situation when the polypeptide chain is surrounded by an abundance of water. The high values for the charged side chains probably result from the tendency of the charged side chain to orientate multiple layers of the polarised water molecules. These effects of pH on the water-binding capacity of amino acids are reflected in the behaviour of entire proteins, a phenomenon first referred to in Chapter 5 in connection with the loss of water from fresh meat. A homogenised muscle preparation may double its water-binding capacity if its pH is lowered from 5.0 (approximately the mean isoelectric point of the muscle proteins) to 3.5.

As a general rule only a small proportion of the polar groups of a native (*i.e.* undenatured) protein will be unable to bind water, because of being buried in the largely hydrophobic interior of the molecule. Thermal denaturation is assumed to lead to the unfolding of the polypeptide chain, which can cause some extra water binding as the last few polar groups are exposed. The newly exposed hydrophobic regions of the polypeptide chain also show a limited water-binding capacity, partly due to the hydrophilic nature of the peptide bond regions of the backbone. Thus many proteins show a modest increase in water binding on denaturation, from around 30% to around 45% w/w. However, when denaturation leads to aggregation, as unfolded polypeptide chains become entangled around each other, there is sometimes a tendency for protein/water interactions to be usurped by protein/protein interactions that lead to an actual loss of water-binding capacity. The release of bound water that accompanies the roasting of meat provides a good example of this phenomenon.

The behaviour of protein solutions is markedly affected by the presence of low-molecular-weight ions, notably the anions and cations of inorganic salts. Up to concentrations of about $1 \, mol \, dm^{-3}$ many salts, such as sodium or potassium chloride, enhance the solubility of many proteins. Indeed the old-fashioned division of proteins into albumins and globulins was based on the former's ability to dissolve in pure water whereas the latter would only dissolve in a dilute salt solution. The mechanism of this 'salting-in' effect is straightforward: charged groups on the surface of the protein attract and 'bind' anions and cations more strongly than they do water. However, these ions will still bring with them an ordered cluster of their own solvating water molecules, which will then function to maintain the protein molecule in solution.

The solubility-enhancing effect of increasing salt concentrations does not continue indefinitely. At concentrations above $1-2 \, mol \, dm^{-3}$ many salts cause proteins to precipitate out of solution, a phenomenon known as 'salting-out'. For biochemists this is the basis of the time-honoured procedure for the fractionation of crude protein mixtures known as 'ammonium sulfate precipitation'. The use of ammonium sulfate stems from its high solubility (1 litre of water will dissolve 760 grams!) rather than from any special aspect of its interaction with proteins. The explanation is a simple one. As these salts reach high concentrations, the proportion of the total water that they bind starts to become significant. The ions of the salt compete with the protein for the water necessary for the protein to remain in solution, and as the salt concentration is raised the different proteins precipitate out.

Salting-out occurs only at salt concentrations that have little relevance to food systems, but salting-in is a very important phenomenon in food processing. When pork is cured by the Wiltshire method (see Chapter 5), sodium and chloride ions are injected, as brine, into the muscle tissues. These bind to the muscle proteins, particularly the abundant actin and myosin of the myofibrils, and enhance their water-binding capacity. This gives the ham its attractive moistness, as well as quite incidentally increasing its weight. Many modern ham-type products, prepared for 'off-the-bone' slicing by machine for pre-packaged sale, are subjected to the additional process of 'tumbling', in which large pieces of ham are literally massaged for periods of up to half an hour. This has the effect of bringing dissolved myofibrillar proteins out of the confines of the cell structure so that they are able to give the moulded or pressed finished product a firmer texture, based on gels formed by these proteins.

In products such as luncheon meat and sausages, based on homogenised or macerated muscle tissue, the solubilisation of proteins by inorganic salts takes on special importance. Whilst in these products water binding is not without significance, the binding of fat is of paramount importance. The ability of the hydrophobic regions of proteins to bind fat molecules makes proteins excellent emulsifying agents. Getting myofibrillar proteins into solution, and into action as emulsifiers, is the task of added so-called 'emulsifying salts'. Besides the ubiquitous sodium chloride we find sodium and potassium phosphates and polyphosphates, citrates, and tartrates in this role. The phosphates and polyphosphates were considered in greater detail in the previous chapter. We also find emulsifying salts used in processed cheese, where they release casein molecules from their micelles to assist in binding the added fat.

INTERACTIONS OF WATER WITH FOOD MATERIALS

When we come to consider water's interactions with foods as a whole we are forced to leave molecular niceties behind and look instead at the behaviour of water in foods as a whole. Food scientists need to describe this behaviour because it is clear that so many important properties of food, but most particularly its susceptibility to microbial growth, are related to the water present. What is also clear, from data such as those in Table 12.1, is that the water content of a foodstuff is not in itself the major influence on stability. What is important is the availability of the water to micro-organisms, not its abundance. The concept of 'water activity' is nowadays universally adopted by food scientists and technologists to quantify availability.

Water activity, a_W, is defined as follows:

$$a_W = p/p_0$$

where p is the partial pressure of water vapour above the sample (solid or liquid) and p_0 is the partial pressure of water vapour above pure water at the same, specified, temperature. At equilibrium (which in most cases is reached only very slowly) there is a fairly obvious relationship between the a_W of a food material and the equilibrium relative humidity (ERH) of air enclosed above it:

$$a_W = ERH/100$$

This relationship between ERH and a_W enables us to predict which foods will gain or lose water, when exposed to air of a particular humidity.

While the concept of water activity is regarded by some as lacking the full rigour of classical physical chemistry, the inescapable fact is that it does provide a satisfactory basis for the description of the water relations of foodstuffs.

The data in Tables 12.2 and 12.3 give an illustration of the application of the water activity concept. One important feature of these data is the poor correlation between water content and water activity that will be seen when Tables 12.2 and 12.1 are compared. The data in Table 12.3 show the importance of the role of water activity in controlling microbial growth. The halophilic (*i.e.* salt-loving) bacteria are those normally associated with marine environments and the fish that we get from them. The xerophilic moulds and osmophilic yeasts are specially adapted to environments of

Table 12.2 *The water activities of various food-stuffs.*

Foodstuff	Typical water activity a_W
Fresh meat	0.98
Cheese	0.97
Preserves	0.88
Salami	0.83
Dried fruit	0.76
Honey	0.75
Dried pasta	0.5

Table 12.3 *The minimum water activities for the growth of micro-organisms.*

Micro-organism	Water activity a_W
Normal bacteria	0.91
Normal yeasts	0.88
Normal moulds	0.80
Halophilic bacteria	0.75
Xerophilic moulds	0.65
Osmophilic yeasts	0.60

low water activity. Their cells contain extra-high concentrations of particular solutes so that they do not lose water (by osmosis) to their surroundings. In the bacteria these 'protective solutes' are usually the amino acids proline or glutamic acid and in the fungi polyols such as glycerol or ribitol. These fungi and yeasts can pose serious problems for the storage of some foods that are normally regarded as stable, notably dried fruit. Fortunately they have not, so far, been associated with mycotoxin production, only more obvious physical deterioration.

Microbial growth is not the only important phenomenon affected by water activity. The rates of both enzymic and nonenzymic reactions are also markedly affected by water activity. However, to gain a clearer understanding of the mechanisms of these influences, we need to look more closely at the question of water binding.

WATER BINDING

The term 'bound water' is commonly used in discussions of the properties of food materials, but it rarely means exactly the same thing to two different scientists. Linked to this term is another, 'water-holding capacity', which refers with similar vagueness to the maximum amount of bound water a material may contain. Fennema has described a sequence of four categories for the water found in a solid food material which covers the range of definitions that might be applied to bound water. The percentage figures given with each category are the relative contributions of the different categories to the total water in a typical high moisture food, *i.e.* one with a water content of about 90%.

(i) Vicinal water ($0.5 \pm 0.4\%$): this is the water that is held at specific hydrophilic sites on molecules (*i.e.* other than water itself) in the material, by hydrogen bonding or by the attraction of charged

groups for the water dipole. Vicinal water may give an uninterrupted monolayer surrounding a hydrophilic molecule. Vicinal
water is regarded as unfreezable down to at least $-40\,°C$ and
incapable of acting as a solvent.

(ii) Multilayer water $(3 \pm 2\%)$: this is the water that forms several
additional layers around the hydrophilic groups of the material.
Water-solute hydrogen bonds do occur, but water-water bonds
predominate. Most multilayer water does not freeze at $-40\,°C$,
and the remainder still has a very low freezing point. Multilayer
water does have limited solvent capacity.

(iii) Entrapped water (up to ~96%): this water has properties similar
to that in dilute salt solutions, with only a slight reduction in its
freezing point and virtually normal solvent capability. However, it
is prevented from flowing freely by the matrix of gel structures or
in tissues where it is held by capillary attraction.

(iv) Free water (up to ~96%): this is essentially similar to pure water
in all its properties. While it will remain held within the structure
of a food material as a consequence of capillary action or trapping
by layers of fatty material, it does not require much pressure to
squeeze it out.

Starting with an absolutely dry food material, one can visualise the
gradual addition of water filling each of these categories in turn. The
'layer' of vicinal water is considered to be complete at an a_W value of
approximately 0.25. The boundary between multilayer water and either
entrapped or free water lies at an a_W value of approximately 0.8. This
sequence of events is normally expressed in the form of a sorption
isotherm, a plot of water content against water activity. A number of these
are shown in Figure 12.6.

It is tempting to present speculative explanations for the differences
between the shapes of these curves in terms of the molecular properties of
the different materials. However, the reader's guesses are unlikely to be
any less reasonable than the author's, so none will be offered. The value of
plots like these lies in their practical application rather than theoretical
interpretation. Whenever two or more ingredients are confined together in
airtight space, water will move between them and their biological,
chemical, and physical properties are liable to change in ways unexpected
unless the sorption isotherms have been carefully studied. Dehydrated
'instant' meals containing pieces of meat, vegetables, *etc.* together inside
the same compartment are obvious problem areas. To make matters even
more complicated, the sorption isotherm is just that, an isotherm, recorded
at some particular temperature. Low-temperature storage of a product

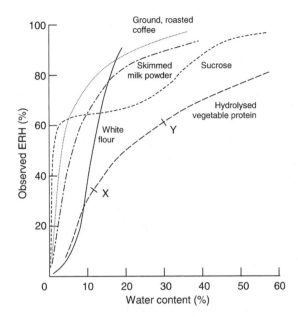

Figure 12.6 *Sorption isotherms. These are representative curves for the food materials shown, at or near to room temperature. The points marked X and Y on the hydrolysed vegetable protein isotherm mark the approximate transitions, first from a free flowing powder to an increasingly sticky mass (at X) and then to a viscous 'solution' (at Y).*

could change all the relationships so carefully established at the temperature of the production line.

Another complication is that with many substances the isotherm obtained as a material is dried (the desorption isotherm) differs from the one obtained as the material gains water (the resorption isotherm). As a general rule the a_W at a given water content will be higher as a material gains water than as it loses water. There are numerous explanations of why this should be, including the time dependence of the process, capillary collapse, and swelling phenomena, and it seems probable that eventually all will be shown to be involved, to some degree or other.

The growth of micro-organisms is by no means the only 'process' that is affected by the water activity of a foodstuff: a number of important enzyme-catalysed and chemical reactions have also been shown to be affected. For example, the rate of enzymic hydrolysis of lipids in foods such as meat becomes negligible (*i.e.* only about 1% of the high a_W rate) at a_W values below 0.4. Chemical reactions can show different effects. The rate of the Maillard reaction (*see* Chapter 2) rises sharply as the water activity falls to ~0.8 and only falls off again below ~0.3. The usual explanation for this is that initially enhancement of the reaction rate occurs as the removal of water effectively concentrates the reactants. The subsequent fall is the result of the loss of reactant mobility in the residual multilayer water. The loss of reactant mobility appears to be the dominant influence on reaction rates at low water activities. One interesting exception to the general trend is the autoxidation of unsaturated lipids in food materials (*see* Chapter 4). Although this process does show a steady decline in rate down to $a_W \sim 0.4$, it then rises again, to about 10% of its maximum rate, at a_W values approaching zero. It is assumed that the removal of vicinal water from hydrophilic sites in the material actually leads to the exposure of vulnerable lipid molecules. Knowledge of the sorption isotherm of a material thus enables food processors to identify the water content of the material at which the effects of degradative reactions can be minimised, usually the point at which multilayer water has been removed but the monolayer of vicinal water is intact.

WATER DETERMINATION

It will now be clear that there are necessarily two quite distinct approaches to the analysis of the water in food materials. The first is to determine the amount of water the material actually contains. Quite often the primary motive of this is to establish what proportion of the food material is not taken up by more valuable nutrients. The second approach is to determine the activity of the water in the material. As we have seen, this will give us useful information about the properties and behaviour of the material. Of course in many cases it is sufficient to know the water content of a material to be able to predict much about its behaviour. For example we knew how far to lower the water content of cereal grains in order to prevent mould growth or sprouting long before the concept of water activity was introduced.

There are three basic methods for the determination of total water content of a food material: gravimetric, volumetric, and chemical. Gravimetric methods simply depend on the difference in weight of the sample following an appropriate drying treatment. The actual drying method must be chosen with care. High temperatures will obviously hasten water loss but at the risk of causing the decomposition of some food components such as sugars; other volatiles such as short-chain fatty acids may also be removed. Before drying it is often necessary to reduce the particle size of the material. Cereal grains need to be milled and wet materials such as jam need to be dispersed onto a particulate matrix such as sand. Without these steps being taken, the drying period can be inordinately long. Whatever drying method is adopted, it is essential that it is continued 'to constant weight', *i.e.* until the sample shows no further weight loss when drying is continued. For many purposes, especially where accounting for the last 1% of the moisture content is not critical, an ordinary laboratory fan oven is sufficient; for precise work a vacuum oven is required. Using temperatures around 100 °C and pressures below 50 mm of mercury, the drying time for cereals can be brought within 16 hours. It must never be forgotten that, having been dried at elevated temperatures, the sample must be cooled in a dry environment such as a desiccator before it is weighed.

Volumetric methods depend on distilling off and condensing any water present when the sample is subjected to reflux distillation with a water-immiscible solvent such as toluene. A special still such as the modified Dean and Stark apparatus shown in Figure 12.7 is used.

The only chemical method for water determination that has found regular use is the Karl Fischer titration. This can be applied to most

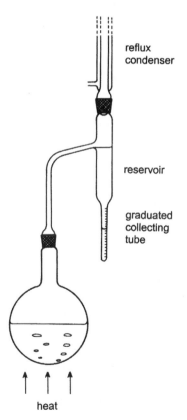

reflux condenser

reservoir

graduated collecting tube

heat

Figure 12.7 *A modified Dean and Stark apparatus. As the distillation proceeds the water collects in the graduated tube while the toluene, being less dense, overflows from the reservoir back into the main flask containing the sample.*

foodstuffs except structurally heterogeneous materials such as fruit, vegetables, and meat, which would be impossible to disperse in the solvent. It is especially effective with heat-sensitive materials, those rich in volatiles, and also sugary materials in which the water is particularly tightly bound. The method depends on the reaction between sulfur dioxide and iodine in the presence of water, pyridine, and methanol:

$$SO_2 + I_2 + 2H_2O \rightarrow H_2SO_4 + 2HI$$

Although it is not shown in this equation, the pyridine appears to form loose compounds with the sulfur dioxide and the iodine, which are the actual reactants. All the reagents for this technique need to be specially prepared to be totally free of water, and the titration apparatus is designed so that water vapour from the atmosphere is excluded. The food sample is dispersed or dissolved in a combined methanol : sulfur dioxide : pyridine reagent and then titrated with standardised iodine in methanol solution. The end-point (the appearance of excess iodine) can be detected visually, but more often an electromeric method is used.

Although the range of materials to which they can be applied is limited, there are two instrumental methods of water determination that deserve a mention. One of these is near-infrared (NIR) spectrophotometry. Water has a number of absorbance bands in the near-infrared of which the one at 1.93 μm is most often used. Comparison of the absorbance at this wavelength (or at 1.94 μm where interference is minimal) with that at 2.08 μm (outside the major absorbance bands) provides a sensitive and very rapid measure of the water content of cereal grains, flour, and similar materials. One special virtue of this technique is that, by recording the absorbance at other wavelengths at the same time, one can also obtain data for the fat and protein content at the same time. Low-resolution nuclear magnetic resonance (NMR) can also be used for water determination. Although the instrumentation is expensive and the theory of the method intimidating, NMR is proving increasingly popular, especially in the cereal industry, as it is both rapid (one minute may suffice for an entire determination) and non-destructive. Even whole grains can be examined.

The determination of the water activity of food materials is not easy. All methods depend on the relationship between water activity and equilibrium relative humidity (ERH) and measure, by one means or another, the tendency of the test material to gain or lose water when exposed to air at a fixed temperature and relative humidity. The usual technique is to place the sample in one small open vessel alongside another containing a solution of known a_W. Both vessels are placed together in a sealed container and kept at constant temperature for a few days. At intervals the

Table 12.4 *The water activity of saturated salt solutions as 25 °C.*

Salt	a_W	Salt	a_W
Potassium dichromate	0.980	Potassium chloride	0.843
Potassium nitrate	0.925	Ammonium sulfate	0.790
Barium chloride	0.902	Sodium chloride	0.753
Sodium benzoate	0.880	Ammonium nitrate	0.618

container is opened so that the weights of the two vessels can be compared to determine whether the sample is gaining or losing moisture. If a range of different standard solutions are available, it is possible to arrive at the a_W value of the sample by interpolation. Table 12.4 shows the a_W values of a number of useful salt solutions. Other values may be obtained by the use of standard solutions of concentrated sulfuric acid or glycerol, but of course the water activity of these changes as they absorb water from the sample under test.

The various instruments that are marketed for the determination of a_W use a variation of this approach. The sample is allowed to equilibrate with a small known volume of air whose relative humidity can be monitored by a special sensor built into the sample container. One common type of sensor contains a tiny crystal of lithium chloride. This rapidly gains or loses water of crystallisation as the relative humidity changes and shows corresponding changes in its electrical conductivity. Although it may take several hours for even a small sample to reach equilibrium modern instruments record several readings over the first few minutes of equilibriation and then use computerised mathematical modelling to extrapolate, with considerable accuracy, to the final result that would be obtained at equilibrium.

FURTHER READING

F. Franks, 'Water, a Matrix of Life.', The Royal Society of Chemistry Paperbacks, Royal Society of Chemistry, London, 2000.

'Water Relations of Foods', ed. R. B. Duckworth, Academic Press, London, 1975.

'Physical Properties of Foods I', ed. R. Jowett, F. Escher, *et al.*, Applied Science, Barking, 1983.

'Properties of Water in Relation to Food Quality and Stability', ed. L. B. Rockland and G. F. Stewart, Academic Press, London, 1981.

'Water and Food Quality', ed. T. M. Hardman, Elsevier Applied Science, London, 1989.

Appendix I

Nutritional Requirements and Dietary Sources

This book does not set out to be a textbook of nutrition. Extensive consideration of the amounts of different food components, as nutrients, that are required in a healthy diet is beyond its scope. Nevertheless it is obviously valuable to be able to place the amounts, stability, and availability of different nutrients as discussed in this book in a context of what is considered by nutritionists to be desirable. Such data have hitherto been presented as Recommended Daily Amounts (RDAs). Unfortunately RDA values were frequently misused, being interpreted as absolute values which could be applied in decisions on the adequacy of an individual's diet. In 1991 the UK Government's Committee on Medical Aspects of Food (COMA) introduced Dietary Reference Values (DRVs). For any given nutrient (*e.g.* a particular vitamin), and defined section of the population (*e.g.* males aged 11–14 years), there are normally three DRVs.

(i) **The Estimated Average Requirement (EAR)** – just that. Statistically approximately half the population will actually require less than the EAR and the remainder more.[*]

(ii) **The Reference Nutrient Intake (RNI)** – the nutrient intake that is enough, or more than enough, for almost all the population. This is not dissimilar from the old RDA and is arbitrarily set at 2 Standard Deviations above the EAR. This statistical jargon assumes that the spread of actual requirements between individual members of the population follows regular statistical rules, *i.e.* is 'normally distributed' and means that 'almost all' in the preceding sentence actually means 97.5%.

---

[*] Statistically literate readers will recognise that the Average (or Mean) is not the same as the Median, which really would divide the population into two exactly equal halves.

409

(iii) **The Lower Reference Nutrient Intake (LRNI)** – the nutrient
intake that is enough, or more than enough, for only a 'small
proportion' of the population. This is arbitrarily set at 2 Standard
Deviations below the EAR so that the 'small proportion' actually
means 2.5%.

The DRVs that have been established for most vitamins and minerals
are presented in Table AI.1. An important feature of the DRVs for three of
the vitamins is that they are quoted in terms of the intake of other
nutrients. This is a reflection of the close involvement of these vitamins in
particular aspects of metabolism; thiamin and niacin with energy metabolism, pyridoxine with protein metabolism. A full set of DRVs have not
been established for some vitamins and minerals. This is largely because

Table AI.1 *Dietary Reference Values for the daily nutrient intake of minerals and
vitamins. The values shown here are merely to illustrate the principles
and refer to the requirements of males aged 11–14.*

	LRNI	EAR	RNI	Units
Thiamin (per 1000 kcal)	230	300	400	μg
Or assuming the EAR for energy of 2220 kcal day^{-1} is met	–	–	900	μg
Riboflavin	0.8	1.0	1.2	
Pyridoxine (per gm protein)	11	13	15	μg
Or assuming the EAR for high quality protein of 33.8 day^{-1} is met	–	–	51	μg
Niacin (per 1000 kcal)	4.4	5.5	6.6	mg
Or assuming the EAR for energy of 2220 kcal day^{-1} is met	–	–	15	mg
Folic acid	100	150	200	μg
Cobalamin	0.8	1.0	1.2	μg
Ascorbic acid	9	22	35	mg
Vitamin A (retinol equivalents)	250	400	600	μg
Calcium	0.45	0.75	1	g
Phosphorus (intake should match calcium on a mole : mole basis)	350	580	770	g
Magnesium	180	230	280	mg
Sodium (no EAR set)	0.46	–	1.6	g
Potassium (no EAR set)	1.6	–	3.1	g
Zinc	5.3	7.0	9.0	mg
Iron	6.1	8.7	11.3	mg
Iodine (no EAR set)	65	–	130	μg
Copper (no LRNI or EAR set)	–	–	800	μg
Selenium (no EAR set)	25	–	45	μg

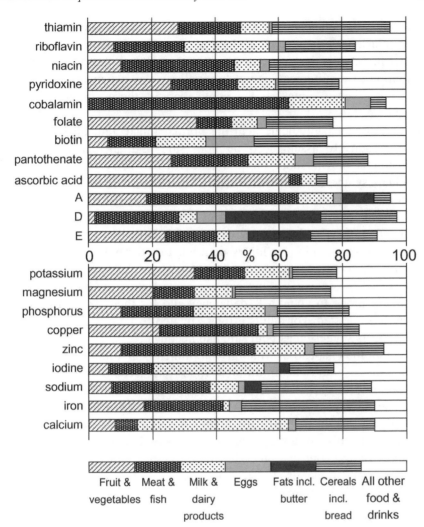

Figure AI.1 *The percentage contribution of different foods to the vitamin and mineral content of the British diet in 1986/7. Based on data in Of the total vitamin content of the diet: (i) The Dietary and Nutritional Survey of British Adults, J. Gregory, K. Fostyer, H. Tyler, and M. Wiseman, (MAFF) HMSO, London, 1990. (ii) The Dietary and Nutritional Survey of British Adults – Further Analysis, MAFF, HMSO, London, 1994. All data are the average of a representative (apart from the omission of pregnant women) sample of 2197 UK men and women 16–64.*

Of the total vitamin content of the diet: (a) breakfast cereals provided 11%, (b) (for males only) beer provided 19%, (c) (for males only) beer provided 16%, (d) coffee provided 10%, (e) baked beans provided 10%, (f) potatoes provided 16%, (g) fruit juice provided 18%, (h) liver provided 43%, (i) oily fish provided 21%. Nutrients that are contributed by the fats incorporated into cereal based products such as cakes and pastry etc. are included with cereals. Similarly nutrients contributed by the pastry component of a pie are included with the principal component of the pie, e.g. meat, rather than with cereals.

the absence of deficiency symptoms in the UK (and elsewhere) leads to the conclusion that even poor diets are supplying sufficient amounts. Another problem is that for some nutrients (particularly iron) the statistical distribution of requirements is not 'normal' (in the statistical sense of the word) but 'skewed', making the application of these simple statistical concepts invalid.

In the case of pantothenic acid the normal intake rate of 3–7 mg per day is reasonably assumed to be both safe and adequate. A similar conclusion is drawn about biotin intakes of 10–200 μg per day. Most individuals synthesise their own vitamin D, as described in Chapter 8, but for individuals confined indoors the RNI is set at 10 μg per day. The special needs of infants are reflected in RNI values of 8.5 μg per day for the first 6 months and 7 μg per day for the following 18 months. The requirement for vitamin E is highly dependent on the polyunsaturated fatty acid (PUFA) content of the diet but the indications are that at any reasonable PUFA intake 3–4 mg per day is adequate. New born infants are particularly prone to vitamin K deficiency and for them an RNI of 10 μg per day is suggested. For adults a figure related to body weight (1 μg per day per kg) is suggested as both safe and adequate.

Although the chapters on vitamins and minerals give information on the amounts of these nutrients in different foods one needs information on the contribution of different foods to the average diet to appreciate such data fully. The survey of British diets carried out by MAFF in 1986/7 has provided just this. It includes data on the effects of region, social class and many other factors on diet and health. Figure AI.1 summarises data from the survey on the contribution of broad classes of foodstuffs to the supply of vitamins and minerals. Readers are strongly recommended to consult the original data for more detailed information.

FURTHER READING

Department of Health, 'Dietary Reference Values for Food Energy and Nutrients for the United Kingdom', HMSO, London, 1991.

'Dietary Reference Intakes for Thiamine, Riboflavin, Niacin, Vitamin B$_6$, Folate, Vitamin B$_{12}$, Pantothenic Acid, Biotin and Choline', Institute of Medicine Standing Committee on the Scientific Evaluation of Dietary Reference Intakes, National Academy of Sciences, Washington D.C., 1998.

'Dietary Reference Intakes. Calcium, Phosphorus, Magnesium, Vitamin D, and Fluoride.', Institute of Medicine Standing Committee on the Scientific

Evaluation of Dietary Reference Intakes, National Academy of Sciences, Washington D.C., 1997.

'Dietary Reference Intakes for Vitamin C, Vitamin E, Selenium, and Carotenoids', Institute of Medicine Food and Nutrition Board, National Academy of Sciences, Washington D.C., 2000.

Appendix II

General Texts for Further Reading

A selection of useful texts that deal with topics that span several of the chapters of this book are listed here.

Food Analysis

D. Pearson, R. S. Kirk, and R. Sawyer, 'Pearson's Composition and Analysis of Foods', 9th edn., Longman, London, 1991.

Y. Pomeranz and C. E. Meloan, 'Food Analysis: Theory and Practice', AVI, Westport, 1988.

'Analyzing Food for Nutrition Labelling and Hazardous Contaminants', ed. I. J. Jeon and W. G. Ikins, Dekker, New York, 1995.

B. Faust, 'Modern Chemical Techniques' Royal Society of Chemistry, Cambridge, 1992.

C. S. James and S. Ceirwyn, 'Analytical Chemistry of Foods', Blackie, London, 1995

Nutrition

'Nutritional Evaluation of Food Processing', 3rd edn., ed. E. Karmas and R. S. Harris, Van Nostrand, New York, 1988.

B. Holland, A. A. Welch, I. D. Unwin, D. H. Buss, A. A. Paul, and D. A. T. Southgate, 'McCance and Widdowson's The Composition of Foods', The Royal Society of Chemistry, Cambridge, 1991.
    (see also the subsequent supplements: Vegetable dishes – 1992; Immigrant foods, Vegetables, herbs and spices – 1991; Amino acids –

1980; Fruit and nuts – 1992; Fish and fish products – 1993; Miscellaneous foods – 1994; Meat, poultry and game – 1995; Meat products and dishes – 1996; Fatty acids – 1998).*

'Human Nutrition and Dietetics', 10th edn., ed. J. S. Jarrow, W. P. T. James and A. Ralph, Churchill Livingstone, Edinburgh, 2000.

'Diet and Heart Disease' ed. M. Ashwell, British Nutrition Foundation, London, 1993.

Food Enzymes

'Enzymes in Food Processing', 3rd edn., ed. T. Nagodawithana and G. Reed, Academic Press, San Diego, 1994.

'Enzymes in Food Processing', ed. G. A. Tucker and F. J. Woods, Blackie, Glasgow, 1991.

D. W. S. Wong, 'Food Enzymes – Structure and Mechanism', Chapman and Hall, New York, 1995.

'Enzymes in Food Technology', ed. R. J. Whitehurst and B. A. Law, Sheffield Academic Press, Sheffield, 2002.

Legislation

D. J. Jukes, 'Food Legislation of the UK', 4th edn., Butterworth, London, 1997.

Culinary Matters

H. Charley, 'Food Science', 2nd edn., Wiley, New York, 1982.

H. McGee, 'On Food and Cooking', Unwin Hyman, London, 1984.

Historical Perspectives

J. Burnett, 'Plenty and Want. A Social History of Diet in England from 1815 to the Present Day', 3rd edn., Routledge, London, 1989.

M. Toussaint-Samat, 'History of Food', Blackwell, Cambridge (USA), 1992.

*Readers should be aware that a new edition of 'McCance and Widdowson' (*The Composition of Foods*, 6th Edition, published by the Royal Society of Chemistry, Cambridge, and the Food Standards Agency, London, 2002) was published while this book was 'in press'. Therefore some of the nutritional data quoted here may now not be the most up to date available.

Subject Index

Page references to structural formulae are italicised.
With the exception of sodium chloride salts are indexed by the name of the anion rather than the cation. Thus ammonium persulfate and sodium nitrate are indexed as 'persulfates' and 'nitrates' respectively.
Prefixes to chemical names (e.g. 1-, α-, cis-, p-) are ignored in the ordering of the index.